Statistical Inference
An Integrated Approach
Second Edition

CHAPMAN & HALL/CRC
Texts in Statistical Science Series

Series Editors

Francesca Dominici, *Harvard School of Public Health, USA*
Julian J. Faraway, *University of Bath, UK*
Martin Tanner, *Northwestern University, USA*
Jim Zidek, *University of British Columbia, Canada*

Texts in Statistical Science

Statistical Inference
An Integrated Approach
Second Edition

Helio S. Migon
Dani Gamerman
Francisco Louzada

CRC Press
Taylor & Francis Group
Boca Raton London New York

CRC Press is an imprint of the
Taylor & Francis Group an **informa** business

A CHAPMAN & HALL BOOK

CRC Press
Taylor & Francis Group
6000 Broken Sound Parkway NW, Suite 300
Boca Raton, FL 33487-2742

© 2015 by Taylor & Francis Group, LLC
CRC Press is an imprint of Taylor & Francis Group, an Informa business

Printed on acid-free paper
Version Date: 20140520

International Standard Book Number-13: 978-1-4398-7880-4 (Hardback)

Library of Congress Cataloging-in-Publication Data

Migon, Helio dos Santos.
   Statistical inference : an integrated approach / Helio S. Migon, Dani Gamerman,
Francisco Louzada-Neto. -- Second edition.
      pages cm
   "A CRC title."
   Includes bibliographical references and index.
   ISBN 978-1-4398-7880-4 (hardcover : alk. paper) 1. Mathematical statistics. I.
   Gamerman, Dani. II. Louzada-Neto, Francisco. III. Title.

QA276.M442 2014
519.5--dc232014019200   2014019200

Visit the Taylor & Francis Web site at
http://www.taylorandfrancis.com

and the CRC Press Web site at
http://www.crcpress.com

To Mirna, our sons and grandchildren (H.S.M.)
To Bernardo (i.m.) and Violeta (D.G.)
To Gleici, Lucas, Caio, Samuel and Francisco (F.L.)

Contents

Preface to the Second Edition

It has been more than a decade since the 1st edition was released. Over this period the need for a more balanced account of the main schools of thought to statistical inference became clear. The main purpose of the book has always been a balanced and integrated presentation of these approaches to the subject. But the first edition was written only by the first two authors of this edition; both were assumedly Bayesian and mentioned that in the Preface for that edition.

Experience gathered in the meantime indicated to us some ingredients to get back to our original goal. Certainly a more detailed and comprehensive presentation of the pros and cons of each approach had to be pursued. The inclusion of a co-author who was more involved with the frequentist school was also recommended to achieve a more balanced presentation. We were fortunate to add both ingredients and present the current version of our work after these additions.

We have tried to eliminate (or reduce, more realistically) the typos present in the 1st edition. Additionally, we have included more explanation, more illustration and more exercises for the concepts already present here. More importantly, new material has been included. The main novelties are associated with a new section on empirical Bayes and penalized likelihoods and their impact on the regression models in the last chapter. Interestingly, these are topics that lie closer to the border between the two schools of thought. The content of this new material has also been connected to many other parts of the book, throughout its chapters. Additionally, we have expanded material in many sections, specially on hypothesis testing, method of moments and bias correction.

This has been an interesting journey for the authors that is far from over. Statistics is alive and growing as a discipline with an ever increasing impact of the growth in computational power. We have no illusion of having reached a complete and final text on the subject, but hope to have reached a stage where readers can grasp the main task ahead of them in their search for an integrated understanding of Statistical Inference.

There are a number of people that we would like to thank and show our gratitude. We all thank our common friend Luis Raul Pericchi for the suggestion of the 3rd author name and many other suggestions. Thanks are also due to our lifelong friend Basilio Pereira for much and relentless advice with references and perspective. There is no way to avoid acknowledging the fundamental help of Rob Calver. He has been keen on making this project viable since the release of the 1st edition. There were many many moments that he was the only one believing it. His persistence and enthusiasm was crucial to have us reached this stage. He is complemented by a number of competent colleagues at CRC who have helped us in many different ways. We also thank colleagues and students who have helped shape the work in various ways, and our families for their encouragement and support. To all of them, our sincerest gratitude. We hold none of them responsible and take full account for the views expressed in this book.

H.S.M., D.G. and F.L.
Rio de Janeiro and São Carlos, May 2014

Preface to the First Edition

This book originated from the lecture notes of a course in Statistical Inference taught at the M.Sc. programs in Statistics at UFRJ and IMPA (once). These have been used since 1987. During this period, various modifications have been introduced until we arrived at this version, judged as minimally presentable.

The motivation to prepare this book came from two different sources. The first and more obvious one for us was the lack of texts in Portuguese, dealing with statistical inference to the desired depth. This motivation led us to prepare the first draft of this book in the Portuguese language in 1993. The second, and perhaps the most attractive as a personal challenge, was the perspective adopted in this text. Although there are various good books in the literature dealing with this subject, in none of them could we find an integrated presentation of the two main schools of statistical thought: the frequentist (or classical) and the Bayesian. This second motivation led to the preparation of this English version. This version has substantial changes with respect to the Portuguese version of 1993. The most notable one is the inclusion of a whole new chapter dealing with approximation and computationally intensive methods.

Generally, statistical books follow their author's point of view, presenting at most, and in separate sections, related results from the alternative approaches. In this book, our proposal was to show, wherever possible, the parallels existing between the results given by both methodologies. *Comparative Statistical Inference* by V. D. Barnett (1973) is the book that is closest to this proposal. It does not, however, present many of the basic inference results that should be included in a text proposing a wide study of the subject. Also we wanted to be as comprehensive as possible for our aim of writing a textbook in statistical inference.

This book is organized as follows. The first chapter is an introduction, presenting the readers with the way we find most appropriate to think of Statistics: discussing the concept of information. Chapter 2 presents some basic concepts of statistics such as sufficiency, exponential family, Fisher information, permutability and likelihood functions. Another basic concept specific

to Bayesian inference is prior distribution; that is separately dealt with in
Chapter 3.

Certain aspects of inference are individually presented in Chapters 4, 6, and
7. Chapter 4 deals with parameter estimation where, intentionally, point and
interval estimation are presented as responses to the summarization question,
and not as two unrelated procedures. The important results for the Normal
distribution are presented and also serve an illustrative purpose. Chapter 6 is
about hypotheses testing problems under the frequentist approach and also
under the various possible forms of the Bayesian paradigm.

In between them, lies Chapter 5 where all approximation and computationally based results are gathered. The reader will find there at least a short
description of the main tools used to approximately solve the relevant statistical problem for situations where an explicit analytic solution is not available.
For this reason, asymptotic theory is also included in this chapter.

Chapter 7 covers prediction from both the frequentist and Bayesian points
of view, and includes the linear Bayes method. Finally in Chapter 8, an introduction to Normal linear models is made. Initially the frequentist approach
is presented, followed by the Bayesian one. Based upon the latter approach,
generalizations are presented leading to the hierarchical and dynamic models.

We would like to alert the readers from the onset that the our preferred
point of view is the Bayesian one. So, a natural emphasis is given to this
approach. We tried, however, to develop a critical analysis and to present the
most important results of both approaches commenting on the positive and
negative aspects of both. As it has already been said, the level of this book
is adequate for an M.Sc. course in Statistics, although we do not rule out the
possibility of its use in an advanced undergraduate course aiming to compare
the two approaches.

This book can also be useful for the more mathematically trained profes-
sionals from related areas of Science such as Economics, Mathematics, Engi-
neering, Operations Research and Epidemiology. The basic requirements are
knowledge of calculus and probability, although basic notions of linear algebra
are also used. As this book is intended as a basic text in statistical inference,
various exercises are included at the end of each chapter. We have also included sketched solutions to some of the exercises and a list of distributions
at the end of the book, for easy reference.

There are many possible uses of this book as a textbook. The first and
most obvious one is to present all the material in the order it appears in
the book and without skipping sections. This may be a heavy workload for a
one-semester course. In this case we suggest dropping Chapter 8 for a later

course. A second option for exclusion in a first course is Chapter 5, although we strongly recommend it for anybody interested in the modern approach to statistics, geared towards applications. The book can also be used as a text for a course that is more strongly oriented towards one of the schools of thought. For a Bayesian route, follow Chapters 1, 2, 3, Sections 4.1, 4.4.1 and 4.5, Chapter 5, Sections 6.3, 6.4, 6.5, 7.1, 7.3.1, 7.4, 8.1, 8.3, 8.4 and 8.5. For a classical route, follow Chapter 1, Sections 2.1, 2.2, 2.5, 2.6, 4.2, 4.3, 4.4.2 and 4.5, Chapter 5, Sections 6.1, 6.2, 6.4, 6.5, 7.2, 7.3.2, 7.4, 8.1 and 8.2.

This book would not have been possible without the cooperation of various people. An initial and very important impulse was the typing of the original lecture notes in TeX by Ricardo Sandes Ehlers. Further help was provided by Ana Beatriz Soares Monteiro, Carolina Gomes, Eliane Amiune Camargo, Monica Magnanini and Otávio Santos Figueiredo. Besides these, many of our former students helped with suggestions and criticism. Careful proofreading of this manuscript was made by our past M.Sc. students and present colleagues Alexandra Mello Schmidt, Hedibert Freitas Lopes and Marco Antonio Rosa Ferreira. Many useful suggestions and comments were provided at this later stage by Steve Brooks, Eduardo Gutierrez-Peña and Gabriel Huerta. We also had the stimulus of several colleagues; in particular, we would like to mention Basílio de B. Pereira. I would also like to thank Nicki Dennis for her support and encouragement throughout all the stages of preparation of this book and for making us feel at home with Arnold. Our families also played the important roles of support and understanding, especially in the weekends and late nights spent trying to meet deadlines! To all of them, our gratitude.

Finally, the subject of the book is not new and we are not claiming any originality here. We would like to think that we are presenting the subject in a way that is not favored in many textbooks and that will help the readers to have an integrated view of the subject. In our path to achieve this goal, we have had the influence of many researchers and books. We have tried to acknowledge this influence by referring to these books whenever we felt it provided a worthwhile reading description of a topic. Therefore, for every major subject presented in our book we tried to relate it to books that treated the subject in a more complete or more interesting way. In line with a textbook character, we opted to favor books rather than research papers as references. We would like to think of our book as a basis for discovery and will feel our task is acomplished whenever readers understand the subject through the book alone, its references or a combination of both.

H.S.M. & D.G.

Rio de Janeiro, December 1998

Chapter 1

Introduction

Before beginning the study of Statistics, it is relevant to characterize the scope of the area and the main issues involved in this study. We avoid directly defining the subject, which is a hard and polemical task. Some of the components involved in this area of science will be presented in the hope that at the end the reader will have a clear notion of the broadness of the subject under consideration. The fundamental problem towards which the study of Statistics is addressed is that where randomness is present. The statistical methodology to deal with the resulting uncertainty is based on the elaboration of probabilistic models in order to summarize the relevant information available.

There are many concepts used in the last sentence that deserve a clarified explanation of their meaning in order to ensure a unified understanding. A model is a formal collection of coherent rules describing, in a simplified way, some real world problem. The language used to describe precisely a model in which uncertainty is present is probability. The meaning of summarization, in this context, refers to the ability to describe a problem in the most concise way (under the assumption that there are many possible forms to describe a problem). The art involved in model building is the desire to balance the need to include as many aspects of reality as possible while keeping it not very complex. A related concept, which is useful to have in mind, is that of parsimony. This means that the model must have an appropriate level of complexity. A very simple model can be misleading since it is eventually missing relevant aspects of the reality. On the other hand, if the model is highly complex it will be hard to understand and extract meaningful information from it. From the previous discussion it is not difficult to guess that information is the main input for Statistics. However, this is a hard concept to define.

Among the objectives of our study, we can identify the two main ones as being to understand reality (estimation and hypothesis testing) and to make a decision (prediction). A strong limitation of many presentations of statistical inference is being mainly concentrated in estimation and testing. In this context, it only deals with quantities that can never be observed in the present or in the future. Nevertheless, our view of Statistics is that it must be concerned with observable quantities. In this way, one is able to verify the model adequacy in an irrefutable way. For the sake of completeness, however, we will present the main results available from all of the topics above.

1.1 Information

As we have already said, the notion of information is present in all the studies developed in Statistics. As far as uncertainty is one of the main ingredients in our models, we need to gather as much information as possible in order to reduce our initial uncertainty. A fundamental question which we are concerned with is about the type of information that is relevant and must be retained in the analysis. A possible reply to this question is that all available information is useful and must be taken in consideration. Another answer is to avoid arbitrariness and take in consideration only objective observation coming from a sampling process. For this all the subjective information must be discarded.

These two points of view roughly form the bases for two different forms of statistical analysis: the Bayesian (or subjectivist) and the classical (or frequentist) approaches, respectively. As we will see in the next section the divergence among these two approaches is much stronger, beginning with the interpretation of the concept of probability. This is always the starting point of a statistical model.

An example to illustrate and clarify these points follows.

Example 1.1 *Consider the situation described in Berger (1985) concerning the following experiments:*

1. *A fine musician, specialized in classical plays, tell us that he is able to distinguish if Hayden or Mozart composed some classical song. Small excerpts of the compositions of both authors are selected at random and the experiment consists of playing them for identification by the musician. The musician makes 10 correct guesses in exactly 10 trials.*

2. *A drunk man says that he can correctly guess in a coin toss what face of the coin will fall down. Again, after 10 trials the drunk man correctly guesses the outcomes of the 10 throws.*

3. *An old English lady is well known for her ability to distinguish whether a cup of tea is prepared by first pouring the milk or the tea. Then cups filled with tea and milk, well mixed and in a random order, are presented to her. She correctly identifies all ten cups.*

It is not difficult to see that the three experiments provide the same information, and therefore, any test to verify the authenticity of the person's statement would result positive for all of them, with the same confidence.

This does not make any sense! We have more reason to believe in the authenticity of the statement of the musician than of the old lady and, certainly, much more than of the drunk man. There is no doubt that the experimental outcome increases the veracity of the statements made. But we cannot reasonably say that we have the same confidence in the three assertions. By common sense, there is a long way to go before one accepts this conclusion.

1.2   The concept of probability

Although the definition of probability is well accepted by almost every statistician (keeping away some technical details), its interpretation or the sense attributed to it varies considerably. We mention here some of the more common interpretations: physical (or classical), frequentist and subjective.

Physical or classical

The probability of any event is the ratio between the number of favorable outcomes and the total number of possible experimental results. It is implicitly assumed that all the elementary events have the same chance. The concept of probability was first formulated based on those classical ideas, which are closely related with games of chance (cards, dices, coins, etc.), where the equal chance assumption is taken for granted.

The probability associated with more elaborate events would be obtained just as a consequence of the probability of the elementary events. Obviously, this interpretation is too narrow to be used in general. Besides that, how can we recognize equal chance events? Finally, the notion of chance involves some probabilistic consideration and so the argument is in some way circular.

A similar interpretation is provided by the logical viewpoint that tries to ascertain relations between events based on logical reasoning. The main question is how to translate common scientific knowledge into undisputed, objective numbers representing probability of events.

Frequentist

The probability of an event A, denoted by $Pr(A)$, is given by

$$Pr(A) = \lim_{n \to \infty} \frac{m}{n},$$

where m is the number of times that A has occurred in n identical and independent experimental trials. This interpretation intends to be objective as far as it is based only on observable quantities. However, it is worth noting that:

i) the limit cannot be understood as a mathematical limit since given $\epsilon > 0$ and $N > 0$, there could well exist an $N_0 > N$, such that $|Pr(A) - (m/N_0)| > \epsilon$. This is improbable but not impossible.

ii) the concepts of identical and independent trial are not easy to define objectively and are in essence subjective.

iii) n does not go to infinity and so there is no way to assure the existence of such limit.

The scope of the two interpretations is limited to observable events and does not correspond to the concept used by common people. The human being evaluates (explicitly or implicitly) probabilities of observable and unobservable events.

Example 1.2 *Some examples are presented below.*

1. *Consider the proposition $A = $ "will rain today". A is typically non-observable at the moment I leave home but $Pr(A)$ is a legitimate and very useful quantity to consider. If its numerical value is low, then I will decide not to take an umbrella; I will prepare myself for a nice walk back home, etc.*

2. *Let A be "John has disease X". Although A can be an observable quantity after a delicate and expensive surgery, John's doctor can take a number of actions (including the surgery itself) based on the value he ascertains for $Pr(A)$.*

3. *The proposition A is "John will get married to Mary". Once again it makes sense to think about $Pr(A)$, especially if I have some sort of personal relationship with John and/or Mary.*

In all the cases presented above the classical and frequentist interpretations do not make sense. A is always non-observable, unique and cannot be repeated under similar conditions.

Subjective

The probability of an event A is a measure of someone's degree of belief in the occurrence of A. To emphasize its subjective character, it is better to denote this probability by $\Pr(A \mid H)$ where H (for history) represents the available information set of the individual.

Example 1.3 *For example, let A be the event "it is raining in Moscow".*

1. *The easiest probability to associate with A for someone in Rio de Janeiro who does not know anything about the Moscow climate is $\Pr(A \mid H_1) = 0.5$, which is based on his body of knowledge H_1.*

2. *On the other hand, someone in St. Petersburg could have stated*

$$Pr(A \mid H_2) = \begin{cases} 0.75 & \text{, if it is raining in St. Petersburg,} \\ 0.25 & \text{, if it is not.} \end{cases}$$

   Note that, in contrast to someone in Rio, this person will typically have more information, contained in H_2.

3. *But for someone in Moscow*

$$Pr(A \mid H_3) = \begin{cases} 1 & \text{, if it is raining,} \\ 0 & \text{, otherwise,} \end{cases}$$

   because in this case, H_3 contains A!

It is worth pointing out that the values for $P(A \mid H)$ are not equal since they depend on the information H, which is different for each case. This interpretation of probability illustrated by the above example is called subjective and obeys the basic rules of probability. Note also that adopting the subjective interpretation we can associate probability for the cases unsolved by the other schools of thought. The remaining question is how to obtain its value for a given event or proposition based on a specified information set.

Probabilities can be evaluated directly or indirectly. One standard tool for direct probability evaluation can be an urn with 100 (or 1000, say) balls with two different colors: blue and red. For example, let us suppose that you want to assess the probability that the Canoas road (a very pleasant road along the mountains within the city of Rio) is not accessible because of a road accident. In the direct approach, you must compare this probability with the chance of drawing a red ball from the urn. If these probabilities were judged equal when the urn has 20 red balls, then the probability that the road is not accessible is 0.2.

For the case of indirect measurements, let us assume that we have two lottery systems. One involving the event you are interested in and the other is any direct evaluation instrument. Imagine the following lotteries:

1. Bet A: If the road is not blocked you win 5 monetary units, and otherwise you do not win anything.

2. Bet B: If the draw from the urn is red, you win 5 monetary units, otherwise nothing.

Considering the two lotteries offered to you, in which one do you prefer to bet? If you prefer A, it might be because there is a bigger chance to win the premium betting in A than in B. Now, let's do a small modification in the urn composition, to 10 red balls and 90 blue ones, so that the probability of winning bet B is 0.1. If you still prefer A, redefine again the composition of the urn and continue until you become indifferent between bets A and B.

There are also difficulties associated with these forms of probability evaluations. In the first case, the difficulty is associated with the comparison of probabilities coming from different propositions. In the second case, the difficulty is caused by the introduction of monetary units and the evaluation of their respective utilities.

1.3   Assessing subjective probabilities

There are many alternative ways to determine subjective probabilities. de Finetti (1974) provides a very useful scheme based on the notion of squared error loss function. Let A be a proposition or an event identified with the value 1 when it is true and 0 otherwise. The probability p that we associate to A is obtained minimizing the square error loss function.

$$(p - A)^2 = \begin{cases} (p - 1)^2 & \text{, if } A = 1, \\ p^2 & \text{, if } A = 0. \end{cases}$$

The basic properties of probability follow easily from this definition as will be shown below.

1. $p \in [0, 1]$

   If $p > 1$ then $p^2 > 1$ and $(p - 1)^2 > 0$. Therefore the losses are always bigger than 1 and 0, the losses obtained by making $p = 1$. A similar argument is used to show that if $p < 0$ the losses will be bigger than those evaluated with $p = 0$. Then, minimization of the square error losses imposes that $p \in [0, 1]$. Figure 1.1 arrives at the same conclusion graphically.

2. $P(\bar{A}) = 1 - P(A)$

   The possible losses associated with the specification of $P(A) = p$ and $P(\bar{A}) = q$ are A=1: $(p - 1)^2 + q^2$ and A=0: $p^2 + (q - 1)^2$.

   As we have already seen in Item 1, the possible values of (p, q) are in the unit square. In Figure 1.2 the line segments are drawn to describe

Figure 1.1 *The possible losses are given by $\overline{BD}^2$ and $\overline{CD}^2$ which are minimized if D is between B and C.*

the possible losses. The squared distance between two consecutive ver-
tices represents the losses. It is clear from the figure that the losses
are reduced by making $p + q = 1$.

3. $P(A \cap F) = P(A \mid F)P(F)$

   Define $P(A \mid F)$ as the probability of A if $F = 1$. Denoting this
   probability by p, $P(F)$ by q and $P(A \cap F)$ by r, the total loss is given
   by $(p - A)^2 F + (q - F)^2 + (r - AF)^2$. Its possible values are

$$
\begin{aligned}
\text{A=F=1} \quad &: (p - 1)^2 + (q - 1)^2 + (r - 1)^2; \\
\text{A=0, F=1} \quad &: p^2 + (q - 1)^2 + r^2; \\
\text{A=0} \quad &: q^2 + r^2.
\end{aligned}
$$

Note that (p, q, r) assume values in the unit cube. The same arguments
used in Item 2 can be developed in the cube. Minimization of the 3
losses is attained when $p = r/q$.

1.4 An example

In this section a simple example will be presented with the main intention
of anticipating many of the general questions to be discussed later in this
book. The problem to be described is extremely simple but is useful to illustrate some relevant ideas involved in the statistical reasoning. Only very
basic concepts on probability are needed for the reader to follow the classical
and Bayesian approaches to inference we will present. The interested reader is
recommended to read the excellent paper by Lindley and Phillips (1976) for
further discussion.

On his way to University one morning, one of the authors of this book was
stopped by a lady living in the neighboring Maré slumtown. She was pregnant
and anxious to know the chance of her seventh baby being male.

Initially, his reaction was to answer that the chance is $1/2$ and to continue

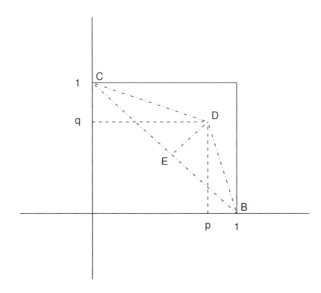

Figure 1.2 *The possible losses are given by* $\overline{BD}^2$ *and* $\overline{CD}^2$ *which are minimized if* D *is between* B *and* C. *The possible losses are given by* $\overline{BD}^2$ *when* $A = 1$ *and* $\overline{CD}^2$ *when* $A = 0$, *which are minimized if* D *is projected on* E *over the line* $p + q = 1$.

his way to work. But the lady was so disappointed with the response that he decided to proceed as a professional statistician would. He asked her some questions about her private life and she told him that her big family was composed of five boys (M) and one girl (F) and the sequence in which the babies were born was MMMMMF. The lady was also emphatic in saying that all her pregnancies were consequence of her long relationship with the same husband. In fact, her disappointment with the naive answer was now understandable.

Our problem is to calculate the chance of the 7th baby being a male, taking into consideration the specific experience of this young lady. How can we solve this problem?

Assuming that the order of the M's and F's in the outcome is not relevant to our analysis, it is enough or sufficient to take note that she had exactly five baby boys (5 M's) and one baby girl (1 F). The question about the order of the births is brought up because people usually want to know if there is any sort of abnormality in the sequence. However, it seems reasonable to assume the births to be equally distributed in probabilistic terms.

Before proceeding with an analysis, it is useful to define some quantities.

Let us denote by X_i the indicator variable of a boy in the ith child, $i = 1, ..., 7$ and let θ denote the common unknown probability of a boy, i.e., $Pr(X_i = 1|\theta) = \theta$ with $0 \leq \theta \leq 1$. Note that θ is a fixed but unknown quantity that does not exist in reality but becomes a useful summary of the situation under study.

A frequentist statistician would probably impose independence between the X_i's. In doing that the only existing link between the X_i's is provided by the value of θ. It seems reasonable at this stage to provide to the lady the value that one considers the most reasonable representation of θ. Given that he is allowed to use only the observed data in his analysis, this representative value of θ must be derived from the observations X_i, $i = 1, ..., 6$.

There are many possible ways to do that. Let us start by the probabilistic description of the data. Given the assumptions above, it is not difficult to obtain that

$$Pr(X_1 = 1, X_2 = 1, ..., X_5 = 1, X_6 = 0|\theta) = \theta^5(1 - \theta).$$

One can proceed in this choice of value for θ by finding the single value $\hat{\theta}$ that maximizes the above probability for the actual observed data. It is a simple exercise to verify that, in the above case, this value is given by $\hat{\theta} = 5/6 = 0.83$. We would then say that she has 83% chance of giving birth to a boy.

That are other ways to proceed still based only on the observed data but more assumptions are needed now. One possible assumption is that the lady had previously decided that she wanted to have only 6 children, the last one being an unwanted pregnancy. In this case, the observed data can be summarized into the number Y of M's among the 6 births. It is clear that Y has a Binomial distribution with size 6 and success probability θ, denoted by $Y \sim Bin(6, \theta)$, and that $E(Y/6|\theta) = \theta$. Using frequentist arguments, one can reason that when we are able to observe many ladies in a similar situation one would like to be correct on average. Therefore, one would estimate the value of θ as $Y/6$, the relative frequency of girls in the data. Given that the observed value of Y is 5, a reasonable estimate for θ is $\tilde{\theta} = 5/6$, coinciding with $\hat{\theta}$. There is no guarantee that the two approaches coincide in general, but it is reassuring that they did in this case.

When asking the lady if the assumption to stop at the 6th child was true, she said that her decision had to do with having had her 1st baby girl. In this case, the observed data should have been summarized by the number Z of M's she had until the 1st girl, and not by Y. (Even though the observed values of Z and Y are the same, their probability distributions are not). It is not difficult to see that Z has a Negative Binomial distribution with size 1 and success probability $1 - \theta$, denoted $Z \sim NB(1, \theta)$, and that $E(Z|\theta) = \theta/(1 - \theta)$.

Proceeding on Z with the reasoning used for Y leads to the estimation of θ by 5/6 as in the previous cases.

The main message from this part of the example for choosing a representative value for θ is that there are many possible methods, two of which were applied above and while the first one did not depend on the way the data were observed, the second one did. These issues are readdressed at greater length on Chapters 2 and 4.

Another route that can be taken is to decide whether it is reasonable to discard 1/2 as a possible value for θ. This can be done by evaluating how extreme (in discordance of the assumption $\theta = 1/2$) the observed value is. To see that, one can evaluate the probabilities that $Y \geq 5$ and $Z \geq 5$, depending on which stopping rule was used by the lady. The values of these probabilities are respectively given by 0.109 and 0.031. It is generally assumed that the cutoff point for measuring extremeness in the data is to have probabilities smaller than 0.05. It is interesting that in this case, the stopping rule has a strong effect on the decision to discard the equal probabilities assumption. This will be readdressed in a more general form in Chapter 6.

Intuition however leads to the belief that specification of the stopping rule is not relevant to solve our problem. This point can be more formally expressed in the following way: the unique relevant evidence is that in 6 births, the sex of the babies was honestly written as 1 F and 5 M's. Furthermore, these outcomes occurred in the order specified previously. This statement points to the conclusion that only the results that have effectively been observed are relevant for our analysis.

For a Bayesian statistician, the elements for the analysis are only the sequence of the observed results and a probability distribution describing the initial information about the chance of a baby being male. The experimental conditions were carefully described previously and they guarantee that a result observed in any given birth is equivalent to that obtained in any other birth. The same is true for pairs, triplets, etc, of birth. This idea is formalized by the concept of exchangeability. The sequence of births is exchangeable if the order of the sex outcomes in the births is irrelevant. In the next chapter, we will define precisely the concept of exchangeability. For our present example this means that the probability of any sequence of r M's and s F's (subject to $r + s = n$) is the same as that of any other sequence with the same number of M's and F's.

Let us return to our original problem, that is, to calculate the chance of the 7th baby born being a male based on the information gathered, namely that provided by the previous births. This probability is denoted by $Pr[X_7 =$

$1|(5,1)]$ where the pair $(5,1)$ denotes the number of births from each sex previously observed. Using basic notions of probability calculus we can obtain

$$
\begin{aligned}
Pr[X_7 = 1|(5,1)] &= \int_0^1 P[X_7 = 1, \theta|(5,1)] \, d\theta \\
&= \int_0^1 P[X_7 = 1|\theta, (5,1)] \, p(\theta|(5,1)) \, d\theta \\
&= \int_0^1 \theta \, p(\theta|(5,1)) \, d\theta \\
&= E[\theta \mid (5,1)],
\end{aligned}
$$

where the expected value is with respect to the distribution of θ given the past results. As we will see in the next chapter this is the unique possible representation for our problem if the assumption of exchangeability of the sequences of births is acceptable. One of the elements involved in the above calculation is $p(\theta|(5,1))$, which has not yet been defined. It has the interpretation, under the subjective approach, of the probability distribution of the possible values for θ after observing the data $(5,1)$.

Let us suppose that before observing the values of $(5,1)$, the subjective probability specification for θ can be represented by the density

$$
p(\theta) = k \, \theta^{a-1} (1-\theta)^{b-1} \qquad 0 \le \theta \le 1, \qquad (a, b > 0)
$$

which is a Beta distribution with parameters a and b (see list of distributions). Note that

$$
\begin{aligned}
p(\theta \mid (5,1)) &= \frac{p((5,1), \theta)}{p((5,1))} \\
&= \frac{p((5,1) \mid \theta) \, p(\theta)}{p((5,1))} \\
&\propto \theta^5 (1-\theta) \theta^{a-1} (1-\theta)^{b-1},
\end{aligned}
$$

since $p((5,1))$ does not depend on θ

$$
\propto \theta^{5+a-1} (1-\theta)^{1+b-1},
$$

where the symbol $\propto$ is to be read as *is proportional to*. The stopping rule or the sample space, in a classical language, is irrelevant because it gets incorporated into the proportionality constant. Furthermore, for any experimental result the final distribution will be a Beta. So we can complete the calculations to obtain

$$
Pr[X_7 = 1 \mid (5,1)] = E[\theta|(r,s)] = \frac{a+5}{a+b+6}.
$$

We still have a problem to be solved. What are the values of a and b?

Suppose, in the case of births of babies, that our initial opinion about the chances associated with M and F are symmetric and concentrated around 0.5. This means that the distribution of θ is symmetrically distributed with mean 0.5 and with high probability around the mean. We can choose in the family of Beta distribution that one with $a = b = 2$, for instance. With this specification, $E(\theta) = 0.5$, $P(0.4 < \theta < 0.6) = 0.296$ and the probability of the 7th birth be a boy is $7/10=0.70$.

If we have been thinking of an experiment with honest coins instead of births of babies, we could have chosen a Beta(50,50) which is symmetrically distributed but much more concentrated around 0.5 than the Beta(2,2). This is a clear representation of the fact that we know much more about coins than about the sex of a new birth, even before observing the data. Under this specification of the Beta, $Pr(0.4 < \theta < 0.6) = 0.956$ and the chance of the 7th outcome to be heads would be $55/106 = 0.519$. Evidently this result is closer to 0.5, which seems reasonable since sex of babies and coins are quite different things. In Chapter 3 we will come back to the discussion of how to specify and determine this prior distribution in any given statistical problem.

1.5   Linear algebra and probability

This is a book concerned with the study of Statistics. In order to do that, a few simple results from linear algebra and probability theory will be extensively used. We thought it might be useful to have the main ones cited here to prepare the reader with the basic results he/she will be using throughout the book. We will start with the basic results concerning densities and probability functions of collections of random variables and will then define the Multivariate Normal distribution to motivate the importance of linear algebra results about matrices and their connection with distributions.

1.5.1   Probability theory

Let $\mathbf{X} = (X_1, \ldots, X_p)$, $\mathbf{Y} = (Y_1, \ldots, Y_q)$ and $\mathbf{Z} = (Z_1, \ldots, Z_r)$ be three random vectors defined over sample spaces $\mathcal{X}$, $\mathcal{Y}$ and $\mathcal{Z}$, respectively $(p, q, r \geq 1)$. Assume for simplicity that they are all continuous with joint probability density function $p(\mathbf{x}, \mathbf{y}, \mathbf{z})$. The marginal and conditional densities will be denoted by their relevant arguments. So, for example, $p(\mathbf{x})$ denotes the marginal density of $\mathbf{X}$ and $p(\mathbf{z}|\mathbf{x}, \mathbf{y})$ denotes the conditional density of $\mathbf{Z}|\mathbf{X} = \mathbf{x}, \mathbf{Y} = \mathbf{y}$.

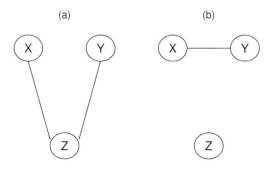

Figure 1.3 *Possible influence diagrams for three random variables: (a) X and Y are conditionally independent given Z; (b) (X, Y) is independent of Z.*

Then, the following equations hold

$$p(\mathbf{x}) = \int_{\mathcal{Y}} p(\mathbf{x}, \mathbf{y}) d\mathbf{y},$$

$$p(\mathbf{x}|\mathbf{y}) = \frac{p(\mathbf{x}, \mathbf{y})}{p(\mathbf{y})},$$

$$p(\mathbf{x}|\mathbf{z}) = \int p(\mathbf{x}, \mathbf{y}|\mathbf{z}) d\mathbf{y},$$

$$p(\mathbf{x}|\mathbf{y}, \mathbf{z}) = \frac{p(\mathbf{x}, \mathbf{y}|\mathbf{z})}{p(\mathbf{y}|\mathbf{z})}.$$

There are many possible combinations of these results but most can be derived easily from one of the above results. In fact, all results below are valid under more general settings with some components of the vectors being discrete and other continuous. The only change in the notation is the replacement of integrals by summations over the relevant parameter space.

These relations define a structure of dependence between random variables. An interesting concept is conditional independence. $\mathbf{X}$ and $\mathbf{Y}$ are said to be conditionally independent given $\mathbf{Z}$ if $p(\mathbf{x}, \mathbf{y}|\mathbf{z}) = p(\mathbf{x}|\mathbf{z}) \, p(\mathbf{y}|\mathbf{z})$. Conditional dependence structures can be graphically displayed with the use of influence diagrams. Figure 1.3 shows some possible representations involving three random variables. These diagrams can be very useful in establishing probability structures for real problems that tend to have many more than just three variables.

Important elements to describe a distribution are its mean vector and variance-covariance matrix. The mean vector $\boldsymbol{\mu}$ of a collection of random variables $\mathbf{X} = (X_1, \ldots, X_p)$ has components $\mu_i = E(X_i)$, $i = 1, \ldots, p$. The variance-covariance matrix $\boldsymbol{\Sigma}$ of $\mathbf{X}$ has elements $\sigma_{ij} = Cov(X_i, X_j)$. It is clearly a symmetric matrix since $Cov(X_i, X_j) = Cov(X_j, X_i)$, for every possible pair (i, j).

These moments are well defined for any well-defined distribution although they may not exist for some distributions. Therefore, one can evaluate the conditional mean of $X_3|X_1, X_2$ and X_5 by calculating the mean of X_3 under the conditional distribution of $X_3|X_1, X_2$ and X_5. Likewise, one can evaluate the joint marginal variance-covariance matrix of X_1, X_2 and X_5 by evaluating the variance-covariance matrix of the joint marginal distribution of (X_1, X_2, X_5).

Another useful result concerns transformation of random vectors. Let $\mathbf{X} = (X_1, \ldots, X_p)$ and $\mathbf{Y} = (Y_1, \ldots, Y_p)$ be p-dimensional random vectors defined over continuous spaces $\mathcal{X}$ and $\mathcal{Y}$, respectively. Assume further that $\mathbf{X}$ and $\mathbf{Y}$ are uniquely related by the 1-to-1 transformation $\mathbf{Y} = \mathbf{g}(\mathbf{X})$ with inverse function $\mathbf{X} = \mathbf{h}(\mathbf{Y})$ and these functions are at least differentiable. As a consequence, $\mathcal{Y} = \mathbf{g}(\mathcal{X})$ and $\mathcal{X} = \mathbf{h}(\mathcal{Y})$. Then the densities $f(\mathbf{x})$ and $p(\mathbf{y})$ are related via

$$p(\mathbf{y}) = f(\mathbf{h}(\mathbf{y})) \left| \frac{\partial \mathbf{h}(\mathbf{y})}{\partial \mathbf{y}} \right| , \mathbf{y} \in \mathcal{Y},$$

where

$$\left| \frac{\partial \mathbf{h}(\mathbf{y})}{\partial \mathbf{y}} \right|$$

is the absolute value of the Jacobian of $\mathbf{h}$, the determinant of the matrix of derivatives of $\mathbf{h}$ with respect to $\mathbf{y}$. This matrix has element (i, j) given by $\partial h_i(\mathbf{y})/\partial y_j$, $\forall (i, j)$. In the case of scalar X and Y, the relation becomes

$$f_Y(y) = f_X(h(y)) \left| \frac{\partial h(y)}{\partial y} \right| , y \in \mathcal{Y}.$$

An important distribution where some of these results can be used is the Multivariate Normal distribution with mean vector $\boldsymbol{\mu}$ and variance-covariance matrix $\boldsymbol{\Sigma}$, denoted by $N(\boldsymbol{\mu}, \boldsymbol{\Sigma})$ with density

$$(2\pi)^{-p/2}|\boldsymbol{\Sigma}|^{-1/2} \exp \left\{ -\frac{1}{2}(\mathbf{x} - \boldsymbol{\mu})'\boldsymbol{\Sigma}^{-1}(\mathbf{x} - \boldsymbol{\mu}) \right\} , \mathbf{x} \in R^p,$$

where $|\mathbf{A}|$ denotes the determinant of $\mathbf{A}$. The scalar version $(p = 1)$ of this density will simply be referred to as Normal distribution with density

$$(2\pi\sigma^2)^{-1/2} \exp \left\{ -\frac{1}{2\sigma^2}(x - \mu)^2 \right\} , \mathbf{x} \in R.$$

When $\mu = 0$ and $\sigma^2 = 1$, the distribution is referred to as standard Normal.

It can be shown that the quadratic form in the exponent of the density has a χ^2 distribution. Also, the Normal distribution is preserved under linear transformations. The Multivariate Normal distribution is completely characterized by its parameters $\boldsymbol{\mu}$ and specially $\boldsymbol{\Sigma}$. If the components are uncorrelated with $\sigma_{ij} = 0$, $\forall i \neq j$, then they are also independent. In any case, the correct understanding of the variance-covariance structure of a distribution is vital, for example, to obtain some properties such as marginal and conditional distributions (see list of exercises). In order to do that, a few basic results about matrices will be reviewed below.

There are many other distributions that will be considered in some detail in the book. They will be introduced in the text as they become needed.

1.5.2   Linear algebra

Let $\mathbf{A}$ be a real matrix of order $r \times p$, $p, r \geq 1$. Denote the matrix element in row i and column j by a_{ij}, $i = 1, \ldots, r$ and $j = 1, \ldots, p$. If $p = r$, the matrix is said to be squared of order p. Such a matrix is said to be symmetric if $a_{ij} = a_{ji}$, for every possible pair (i, j). In this case, the transpose of $\mathbf{A}$, denoted by $\mathbf{A}'$, is $\mathbf{A}' = \mathbf{A}$.

Let $\mathbf{Y} = (Y_1, \ldots, Y_r)'$ be a random vector defined by the linear transformation $\mathbf{Y} = \mathbf{c} + \mathbf{C}\,\mathbf{X}$ of another random vector $\mathbf{X} = (X_1, \ldots, X_p)'$ where $\mathbf{c}$ and $\mathbf{C}$ are a r-dimensional vector and a $r \times p$ matrix of constants. Then the expectation and variance-covariance matrix of $\mathbf{Y}$ are respectively given by $E(\mathbf{Y}) = \mathbf{c} + \mathbf{C}E(\mathbf{X})$ and $V(\mathbf{Y}) = \mathbf{C}\boldsymbol{\Sigma}\mathbf{C}'$. As an example, let $Y = \mathbf{1}_p'\mathbf{X} = \sum_i X_i$ where $\mathbf{1}_p$ is the p-dimensional vector of 1's. Then, Y will have mean $\mathbf{1}_p'E(\mathbf{X}) = \sum_i E(X_i)$ and variance $\mathbf{1}_p'V(\mathbf{X})\mathbf{1}_p = \sum_{i,j} Cov(X_i, X_j)$.

A matrix $\mathbf{A}$ is said to be positive (non-negative) definite if $\mathbf{b}'\mathbf{A}\,\mathbf{b} > (\geq)0$, for every non-null vector $\mathbf{b} = (b_1, \ldots, b_p)'$. Similarly, negative (non-positive) definite matrices can be defined by replacement of the $>$ ($\geq$) sign by the $<$ ($\leq$) sign. A trivial example of a positive definite matrix is the p-dimensional identity matrix, denoted by $\mathbf{I}_p$ with diagonal elements equal to 1 and off-diagonal elements equal to 0. It is very easy to check that $\mathbf{b}'\mathbf{I}_p\mathbf{b} = b_1^2 + \ldots + b_p^2$ which must be larger than zero because $\mathbf{b}$ is non-null.

Variance-covariance matrices are always non-negative definite and usually are positive definite matrices. To see that, assume that $\boldsymbol{\Sigma}$ is the variance-covariance matrix of $\mathbf{X} = (X_1, \ldots, X_p)$. Then, a non-null random variable $Z = \mathbf{b}'\mathbf{X}$ can be formed. This variable will have variance $V(Z) = \mathbf{b}'\boldsymbol{\Sigma}\mathbf{b}$ which is necessarily non-negative. Varying $\mathbf{b}$ over all its possible values in R^p (excluding the origin) proves the result.

Positive definiteness defines an ordering over matrices. Using the notation $\mathbf{A} > (\geq)\mathbf{0}$ for a positive (non-negative) matrix $\mathbf{A}$ allows one to denote by $\mathbf{A} > (\geq)\mathbf{B}$ the fact that the matrix $\mathbf{A} - \mathbf{B} > (\geq)\mathbf{0}$. This ordering makes sense in the context of probability. Let $\mathbf{A}$ and $\mathbf{B}$ be the variance-covariance matrices of independent p-dimensional random vectors $\mathbf{X}$ and $\mathbf{Y}$. Then $\mathbf{A} > \mathbf{B}$ implies that $V(\mathbf{b}'\mathbf{X}) - V(\mathbf{b}'\mathbf{Y}) = \mathbf{b}'\mathbf{A}\,\mathbf{b} - \mathbf{b}'\mathbf{B}\,\mathbf{b} = \mathbf{b}'(\mathbf{A} - \mathbf{B})\,\mathbf{b} > 0$, for every non-null vector $\mathbf{b}$. So, there is a sense in which matrices can be compared in magnitude and one can say that $\mathbf{A}$ is *larger* than $\mathbf{B}$.

Symmetric positive-definite matrices are said to be non-singular, because they have a non-null determinant. This implies that they have full rank and all their rows are linearly independent. Likewise, singular matrices have null determinant, which means that they do not have full rank and some of their rows can be represented as linear combinations of the other rows. Non-singular matrices also have a well-defined inverse matrix.

A squared matrix $\mathbf{A}$ of order p with p-dimensional rows $\mathbf{a}_i = (a_{i1}, \ldots, a_{ip})'$, for $i = 1, \ldots, p$ is said to be orthogonal if $\mathbf{a}_i'\mathbf{a}_j = 1$, if $i = j$, and 0, if $i \neq j$. Note that if $\mathbf{A}$ is orthogonal then $\mathbf{A}'\mathbf{A} = \mathbf{A}\,\mathbf{A}' = \mathbf{I}_p$. Therefore, orthogonal matrices can be shown to be full rank with inverse $\mathbf{A}^{-1}$ and $\mathbf{A}' = \mathbf{A}^{-1}$. There are many methods available in linear algebra for iteratively constructing an orthogonal matrix from given starting rows.

1.6   Notation

Before effectively beginning the study of Statistical Inference, it will be helpful to make some general comments. First, it is worth noting that here we will deal with regular models only; that is, the random quantities involved in our models are of the discrete or continuous type. A unifying notation will be used and additional distinction will be clear from the context. Then, if $\mathbf{X}$ is a random vector with distribution function denoted by $F(\mathbf{x})$, its probability (or density) function will be denoted by $p(\mathbf{x})$ if $\mathbf{X}$ is discrete (or continuous) and we will assume that

$$\int dF(\mathbf{x}) = \int p(\mathbf{x})d\mathbf{x}$$

independently of $\mathbf{X}$ being continuous or discrete with the integral symbol representing a sum in the discrete case. In addition, as far as the probability (or density) function is defined from the distribution of $\mathbf{X}$, we will use the notation $\mathbf{X} \sim p$ meaning that $\mathbf{X}$ has distribution p or, being more precise, a distribution whose probability (or density) function is p. Similarly, $\mathbf{X} \xrightarrow{\mathcal{D}} p$ will be used to denote that $\mathbf{X}$ converges in distribution to a random variable with density p.

In general, the observables are denoted by the capital letters of the alphabet $(X, Y, \ldots)$, as usual, and their observed values by lower case letters $(x, y, \ldots)$. Known quantities are denoted by the first letters of the alphabet $(A, B, \ldots)$, and the greek letters $(\theta, \lambda, \ldots)$ are used to describe unobservable quantities. Matrices, observed or not, will be denoted by capitals. Additionally, vectors and matrices will be distinguished from scalars denoting them in bold face. Results will generally be presented for the vector case, and whenever the specialization to the scalar case is not immediate, they will be presented again in a univariate version.

The probability of an event A is denoted by $P(A)$ or $Pr(A)$ and the distribution of X by $P(X)$. The expected value of $\mathbf{X}|\mathbf{Y}$ is denoted by $E(\mathbf{X}|\mathbf{Y})$, $E_{\mathbf{X}|\mathbf{Y}}(\mathbf{X})$ or even $E_{\mathbf{X}|\mathbf{Y}}(\mathbf{X}|\mathbf{Y})$, and the variance of $\mathbf{X}|\mathbf{Y}$ is denoted by $V(\mathbf{X}|\mathbf{Y})$, $V_{\mathbf{X}|\mathbf{Y}}(\mathbf{X})$ or even $V_{\mathbf{X}|\mathbf{Y}}(\mathbf{X}|\mathbf{Y})$. The indicator function, denoted by $I_x(A)$, assumes the value

$$
I_x(A) = \begin{cases} 1 & \text{, if } x \in A, \\ 0 & \text{, otherwise.} \end{cases}
$$

1.7   Outline of the book

The purpose of this book is to present an integrated approach of Statistical Inference at an intermediate level by discussion and comparison of the most important results of the two main schools of statistical thought: frequentist and Bayesian. With that in mind, the results are presented for multiparameter models. Whenever needed, the special case of a single parameter is presented and derivations are sometimes made at this level to help the understanding. Also, most of the examples are presented for this level. It is hoped that they will provide motivation for the usefulness of the results in the more general setting.

Presentations of results for the two main schools of thought are made in parallel as much as possible. Estimation and prediction are introduced with the Bayesian approach followed by the classical approach. The presentation of hypothesis testing and linear models goes the other way round. All chapters, including this introduction, contain a set of exercises at the end. These are included to help the student practice his/her knowledge of the material covered in the chapter. The exercises are divided according to the section of the chapter to which they refer, even though many exercises contain a few items which cover different sections. At the end of the book, a selection of exercises from all chapters have their solutions presented. We tried to spread these exercises evenly across the material contained in the chapters.

The material of the book will be briefly presented below. The book is composed of eight chapters that can broadly be divided into three parts. The first part contains the first three chapters introducing basic concepts needed for Statistics. The second part is composed of Chapters 4, 5 and 6 which discuss in an integrated way the standard topics of estimation and hypothesis testing. The final two chapters deal with other important topics of inference: prediction and linear models.

Chapter 1 consist of an introduction with the aim of providing the flavor of and intuition for the task ahead. In doing so, it has anticipated at an elementary level many of the points to be addressed later in the book.

Chapter 2 presents the main ingredients used in statistical inference. The key concepts of likelihood, sufficiency, posterior distribution, exponential family, Fisher information and exchangeability are introduced here. The issue of parameter elimination leading to marginal, conditional and profile likelihoods is presented here too.

A key element of statistical inference for the Bayesian approach is the use of prior distributions. These are separately presented and discussed in Chapter 3. Starting from an entirely subjective specification, we move on to functional form specifications where conjugate priors play a very important role. Then, non-informative priors are presented and illustrated. Finally, the structuring of a prior distribution in stages with the so-called hierarchical form is presented. A real data example is introduced and is carried through into future chapters to illustrate concepts.

Chapter 4 deals with parameter estimation where, intentionally, point and interval estimation are presented as different responses to the same summarization question, and not as two unrelated procedures. Different methods of estimation (maximum likelihood, method of moments, least squares) are presented. The classical results in estimation are shown to numerically coincide with Bayesian results obtained using vague prior distribution in many of the problems considered for the Normal observational model.

Chapter 5 deals with approximate and computationally intensive methods of inference. These results are useful when an explicit analytic treatment is not available. Maximization techniques including Newton-Raphson, Fisher scoring and EM algorithms are presented. Asymptotic theory is presented and includes the delta method and Laplace approximations. Quadrature integration rules are also presented here. Finally, methods based on stochastic simulation are presented. They include the bootstrap and its Bayesian versions, Monte Carlo integration and MCMC methods.

Chapter 6 is about hypothesis testing problems under the frequentist approach and also under the various forms of the Bayesian paradigm. Various test procedures are presented and illustrated for the models with Normal observations. Tests based on the asymptotic results of the previous chapter are also presented.

Chapter 7 deals with prediction of unknown quantities to be observed. The prediction analysis is covered from the classical and Bayesian points of view. Linear models are briefly introduced here and provide an interesting example of prediction. This chapter also includes linear Bayes methods by relating them to prediction in linear models.

Chapter 8 deals with linear models. Initially, the frequentist inference for linear models is presented, followed by the Bayesian one. Generalizations motivated with the Bayesian approach are presented leading to hierarchical and dynamic models. Also, brief introductions to generalized linear models and constrained linear models are presented.

Exercises

§1.3

1. Consider the equation $P(A \cap F) = P(A \mid F)P(F)$ in the light of de Finetti loss function setup with the 3 losses associated with events $A = 0, F = 1$, $A = 1, F = 1$ and $F = 0$ and respective probabilities p, q and r. Show that losses are all minimized when $p = r/q$.

   §1.4

2. Consider the example of the pregnant lady.

   (a) Show that proceeding on Z with the reasoning used for Y leads to the estimation of θ by 5/6.

   (b) Evaluate the probabilities that $Y \geq 5$ and $Z \geq 5$ and show that the values of these probabilities are respectively given by 0.109 and 0.031.

3. Consider again the example of the pregnant lady. Repeat the evaluation of the $P(X_7 = 1|(r, s))$ assuming now that

   (a) the observed values of (r, s) are $(3, 1)$ and $(27, 9)$ using Beta priors with parameters $(a, b) = (1, 1)$ and $(a, b) = (5, 5)$. Compare the results obtained.

(b) her next pregnancy will produce twins and the observed value of (r, s) is $(5, 1)$.

§1.5

4. Let $X|Y \sim Bin(Y, \pi)$ and let $Y \sim Pois(\lambda)$.

   (a) Show that $E(X) = E[E(X|Y)] = \lambda\pi$ and that $V(X) = E[V(X|Y)] + V[E(X|Y)] = \lambda\pi$.

   (b) Show that $X \sim Pois(\lambda\pi)$ and that $Y - X|X \sim Pois[\lambda(1 - \pi)]$.

5. Let $X|Y \sim N(0, Y^{-1})$ and $Y \sim G(a/2, b/2)$. Obtain the marginal distribution of X and the conditional distribution of $Y|X$.

6. Show that normality is preserved under linear transformations, i.e., if $\mathbf{X} \sim N(\boldsymbol{\mu}, \boldsymbol{\Sigma})$ and $\mathbf{Y} = \mathbf{c} + \mathbf{C}\ \mathbf{X}$, then $\mathbf{Y} \sim N(\mathbf{c} + \mathbf{C}\boldsymbol{\mu}, \mathbf{C}\boldsymbol{\Sigma}\mathbf{C}')$. Apply the result to obtain the marginal distribution of any sub-vector of $\mathbf{X}$.

7. Show that if

$$\begin{pmatrix} \mathbf{X} \\ \mathbf{Y} \end{pmatrix} \sim N \left[\begin{pmatrix} \boldsymbol{\mu}_X \\ \boldsymbol{\mu}_Y \end{pmatrix}, \begin{pmatrix} \boldsymbol{\Sigma}_X & \boldsymbol{\Sigma}_{XY} \\ \boldsymbol{\Sigma}'_{XY} & \boldsymbol{\Sigma}_Y \end{pmatrix} \right],$$

   then $\mathbf{X}|\mathbf{Y} \sim N(\boldsymbol{\mu}_{X|Y}, \boldsymbol{\Sigma}_{X|Y})$ where $\boldsymbol{\mu}_{X|Y} = \boldsymbol{\mu}_X + \boldsymbol{\Sigma}_{XY}\boldsymbol{\Sigma}_Y^{-1}(\mathbf{Y} - \boldsymbol{\mu}_Y)$ and $\boldsymbol{\Sigma}_{X|Y} = \boldsymbol{\Sigma}_X - \boldsymbol{\Sigma}_{XY}\boldsymbol{\Sigma}_Y^{-1}\boldsymbol{\Sigma}_{XY}$.

8. Show that if $X_1, \ldots, X_p$ are independent standard Normal variables then

$$\sum_{i=1}^{p} X_i^2 \sim \chi_p^2.$$

9. Show that if $\mathbf{X} = (X_1, \ldots, X_p)' \sim N(\boldsymbol{\mu}, \boldsymbol{\Sigma})$ and $Y = (\mathbf{X} - \boldsymbol{\mu})'\boldsymbol{\Sigma}^{-1}(\mathbf{X} - \boldsymbol{\mu})$, then $Y \sim \chi_p^2$.
   Hint: Define $\mathbf{Z} = \mathbf{A}(\mathbf{X} - \boldsymbol{\mu})$ where the matrix $\mathbf{A}$ satisfies $\mathbf{A}'\mathbf{A} = \boldsymbol{\Sigma}^{-1}$ and use the result from the previous exercise.

10. Let $\mathbf{A}$, and $\mathbf{B}$ be non-singular symmetric matrices of orders p and q and $\mathbf{C}$ a $p \times q$ matrix. Show that

   (a) $(\mathbf{A} + \mathbf{C}\mathbf{B}\mathbf{C}')^{-1} = \mathbf{A}^{-1} - \mathbf{A}^{-1}\mathbf{C}(\mathbf{C}'\mathbf{A}^{-1}\mathbf{C} + \mathbf{B}^{-1})^{-1}\mathbf{C}'\mathbf{A}^{-1}$

   (b) $\begin{pmatrix} \mathbf{A} & \mathbf{C} \\ \mathbf{C}' & \mathbf{B} \end{pmatrix}^{-1} = \begin{pmatrix} \mathbf{A}^{-1} + \mathbf{E}\mathbf{D}^{-1}\mathbf{E}' & -\mathbf{E}\mathbf{D}^{-1} \\ -\mathbf{D}^{-1}\mathbf{E}' & \mathbf{D}^{-1} \end{pmatrix}$,
   where $\mathbf{D} = \mathbf{B}^{-1} - \mathbf{C}'\mathbf{A}^{-1}\mathbf{C}$ and $\mathbf{E} = \mathbf{A}^{-1}\mathbf{C}$.

Chapter 2

Elements of inference

In this chapter, the basic concepts needed for the study of Bayesian and classical statistics will be described. In the first section, the most commonly used statistical models are presented. They will provide the basis for the presentation of most of the material of this book. Section 2.2 introduces the fundamental concept of likelihood function. Theoretically sound and operationally useful definitions of measures of information are also given in this section. The Bayesian point of view is introduced in Section 2.3. The Bayes theorem acts as the basic rule in this inferential procedure. The next section deals with the concept of exchangeability. This is a strong and useful concept as will be seen in the following chapter. Other basic concepts, such as sufficiency and exponential family, are presented in Section 2.5. Finally, in Section 2.6, the multiparametric case is presented and the main concepts are revised and extended from both the Bayesian and the classical points of view. Special attention is given to the problem of parameter elimination in order to make inference with respect to the remaining parameters.

2.1 Common statistical models

Although the nature of statistical applications is only limited by our ability to formulate probabilistic models, there are a few models that are more frequently used in Statistics. There is a number of reasons for that. First, they are more commonly found in applications. Second, they are the simplest models that can be entertained in non-trivial applications. Finally, they provide a useful starting point in the process of building up models.

The first class of models considered is a random sample from a given distribution. They are followed by the location model, the scale model and the location-scale model. An execellent complementary reading for this topic is the book of Bickel and Doksum (1977).

The most basic situation of observations is the case of an homogeneous population from a distribution F_θ, depending on the unknown quantity θ. Knowledge of the value of θ is vital for the understanding and description of this population and we would need to extract information from it to accomplish this task. Typically in this case, a random sample $X_1, \ldots, X_n$ is drawn from this population and we hope to build strategies to ascertain the value of θ from the values of the sample.

In this case, the observations $X_1, \ldots, X_n$ are independent and identically distributed (iid, in short) with common distribution F_θ. Assuming that F_θ has density or probability function f, they are probabilistically described through

$$p(x_1, \ldots, x_n | \theta) = \prod_{i=1}^{n} f(x_i | \theta).$$

Example 2.1 *Consider a series of measurements made about an unknown quantity θ. Unfortunately, measurements are made with imprecise devices which means that there are errors that should be taken into account. These errors are a result of possibly many (small) contributions and are more effectively described in terms of a probability distribution. This lead to the construction of a model in the form $X_i = \theta + e_i$, $i = 1, \ldots, n$. The e_i's represent the measurement errors involved. If the experiment is performed with care with measurement being collected independently and using the same procedures, the e_i's will form a random sample from the distribution of errors F_e. For the same reason, the X_i's will also be iid with joint density*

$$p(x_1, \ldots, x_n | \theta) = \prod_{i=1}^{n} f(x_i | \theta) = \prod_{i=1}^{n} f_e(x_i - \theta),$$

where f_e is the density of the error distribution.

Definition 2.1 *(Location model) $\boldsymbol{X}$ has a location model if a function f and a quantity $\boldsymbol{\theta}$ exist such that the distribution of $\boldsymbol{X}$ given $\boldsymbol{\theta}$ satisfies $p(\boldsymbol{x} \mid \boldsymbol{\theta}) = f(\boldsymbol{x} - \boldsymbol{\theta})$. In this case, $\boldsymbol{\theta}$ is called a location parameter. The model in Example 2.1 is a location model with location parameter $\boldsymbol{\theta} = \theta \boldsymbol{1}_n$.*

Example 2.2 *Some examples of location models are presented below.*

1. *Normal with known variance*
   In this case the density is $p(x|\theta) = (2\pi\sigma^2)^{-0.5} \exp\{-0.5\,\sigma^{-2}(x-\theta)^2\}$, which is a function of $x - \theta$.

2. *Cauchy with known scale parameter*
   The Cauchy distribution is the Student-t distribution with 1 degree of freedom. In this case, the density is $p(x|\theta) = \{\pi\sigma[1 + (x - \theta)/\sigma^2]\}^{-1}$, which is again a function of $x - \theta$.

3. *Multivariate Normal with known variance-covariance matrix*
   In this case the density is $p(\boldsymbol{x}|\boldsymbol{\theta}) = (2\pi)^{-p/2}|\boldsymbol{\Sigma}|^{-1/2} \exp\{-(\boldsymbol{x} - \boldsymbol{\theta})'\boldsymbol{\Sigma}^{-1}(\boldsymbol{x} - \boldsymbol{\theta})/2\}$, which is a function of $\boldsymbol{x} - \boldsymbol{\theta}$. Note that an iid sample from the $N(\theta, \sigma^2)$ distribution is a special case with $\boldsymbol{\theta} = \theta\boldsymbol{1}$ and $\boldsymbol{\Sigma} = \sigma^2\boldsymbol{I}$.

Definition 2.2 *(Scale model) X has a scale model if a function f and a quantity σ exist such that the distribution of X is given by*

$$p(x \mid \sigma) = \frac{1}{\sigma} f\left(\frac{x}{\sigma}\right).$$

In this case, σ is called a scale parameter.

Example 2.3 *Some examples of scale models are presented below.*

1. *Exponential with parameter θ*
   The density $p(x|\theta) = \theta \exp(-\theta x)$ is in the form $\sigma^{-1}f(x/\sigma)$ with $\theta = \sigma^{-1}$ and $f(u) = e^{-u}$.

2. *Normal with known mean θ*
   The density $p(x|\sigma^2) = (2\pi)^{-1/2}\sigma^{-1} \exp\left\{-[(x - \theta)/\sigma]^2/2\right\}$ is in the form $\sigma^{-1}f(x/\sigma)$.

Definition 2.3 *(Location and scale model) X has location and scale model if there are a function f and quantities θ and σ such that the distribution of X given (θ, σ) satisfies*

$$p(x \mid \theta, \sigma) = \frac{1}{\sigma} f\left(\frac{x - \theta}{\sigma}\right).$$

In this case, θ is called the location parameter and σ the scale parameter.

Some examples in the location-scale family are the Normal and the Cauchy distributions. Once again, the location part of the model can also be multivariate.

2.2 Likelihood-based functions

Most statistical work is based on functions that are constructed from the probabilistic description of the observations. In this section, these functions are defined and their relevance to statistical inference is briefly introduced. These functions will be heavily used in later chapters, where their importance will be fully appreciated. We start with the likelihood function and then present Fisher measures of information and the score function.

2.2.1   Likelihood function

The likelihood function of $\boldsymbol{\theta}$ is the function that associates the value $p(\mathbf{x} \mid \boldsymbol{\theta})$
to each $\boldsymbol{\theta}$. This function will be denoted by $l(\boldsymbol{\theta}; \mathbf{x})$. Other common notations
are $l_{\mathbf{x}}(\boldsymbol{\theta})$, $l(\boldsymbol{\theta} \mid \mathbf{x})$ and $l(\boldsymbol{\theta})$. It is defined as follows

$$l(\,\cdot\,; \mathbf{x}) : \boldsymbol{\Theta} \quad \rightarrow \quad R^+,$$
$$\boldsymbol{\theta} \quad \rightarrow \quad l(\boldsymbol{\theta}; \mathbf{x}) = p(\mathbf{x} \mid \boldsymbol{\theta}).$$

The likelihood function associates to each value of $\boldsymbol{\theta}$, the probability of an
observed value $\mathbf{x}$ for $\mathbf{X}$. Then, the larger l is the greater are the chances
associated to the event under consideration, using a particular value of $\boldsymbol{\theta}$.
Therefore, by fixing the value of $\mathbf{x}$ and varying $\boldsymbol{\theta}$ we observe the plausibility
(or likelihood) of each value of $\boldsymbol{\theta}$. The concept of likelihood function was discussed by Fisher, Barnard and Kalbfleisch among many others. The likelihood
function is also of fundamental importance in many theories of statistical in-
ference. Note that even though $\int_R p(\mathbf{x} \mid \boldsymbol{\theta})\, d\mathbf{x} = 1$, $\int_{\boldsymbol{\Theta}} l(\boldsymbol{\theta}; \mathbf{x})\, d\boldsymbol{\theta} = k \neq 1$, in
general, and sometimes even $k = +\infty$.

Example 2.4 *Let* $X \sim Bin(2, \theta)$. *Then*

$$p(x \mid \theta) = l(\theta; x) = \binom{2}{x} \theta^x (1-\theta)^{2-x}, \ x = 0, 1, 2\,; \ \theta \in \Theta = (0, 1)$$

$$\textit{but} \int_{\Theta} l(\theta\,; x) d\theta = \binom{2}{x} \int_0^1 \theta^x (1-\theta)^{2-x} d\theta = \binom{2}{x} B(x+1, 3-x) = \frac{1}{3} \neq 1.$$

Note that:

1. *if* $x = 1$ *then* $l(\theta\,; x = 1) = 2\theta(1-\theta)$. *The value of* θ *with highest
   likelihood or, in other words, the most likely (or probable) value of* θ
   is 1/2.

2. *if* $x = 2$ *then* $l(\theta\,; x = 2) = \theta^2$, *the most likely value of* θ *is 1.*

3. *if* $x = 0$ *then* $l(\theta\,; x = 0) = (1-\theta)^2$, *the most likely value is again 0.*

These likelihood functions are plotted in Figure 2.1.

The notion of likelihood can also be introduced from a slightly more general
perspective. This broader view will be useful in more general observation
contexts than those considered so far. These include cases where observation
are only obtained in an incomplete way.

Denoting by E an observed event and assuming the probabilistic description of E depends on an unknown quantity $\boldsymbol{\theta}$, one can define the likelihood
function of $\boldsymbol{\theta}$ based on the observation E as $l(\boldsymbol{\theta}; E) \propto Pr(E|\boldsymbol{\theta})$. If E is of
discrete nature, there is no difference with respect to the previous definition.

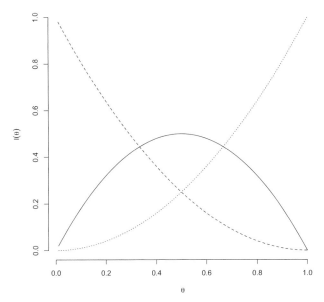

Figure 2.1 *Likelihood function of the example for different values of x. Dashed line:*
x = 0, solid line: x = 1 and dotted line: x = 3.

Example 2.5 *Let $X_1, \ldots, X_n$ be a collection of iid random variables with a common Bernoulli distribution with success probability θ. Let E be any observation of the $X_1, \ldots, X_n$ consisting on x successes. Then, $l(\theta; E) \propto Pr(E|\theta) = \theta^x (1 - \theta)^{n-x}$.*

For continuous observations, assume that any observed value x is an approximation of the real value due to rounding errors. Therefore, the observation x in fact corresponds to the observed event $E = \{x : a \leq x \leq a + \Delta\}$ for given values of a and $\Delta > 0$, which do not depend on $\boldsymbol{\theta}$. In this case, $Pr(E|\boldsymbol{\theta}) = F(a + \Delta) - F(a)$ where F is the distribution function of the observation. For typical applications, the value of Δ is very small and one can approximate $F(a + \Delta) - F(a) = p(x|\boldsymbol{\theta})\Delta$. Therefore, $l(\boldsymbol{\theta}; E) \propto p(x|\boldsymbol{\theta})$. This definition can be extended for a vector of observations with minor technical adaptations to the reasoning above to lead to $l(\boldsymbol{\theta}; E) \propto p(\mathbf{x}|\boldsymbol{\theta})$.

Example 2.6 *Let $X_1, \ldots, X_n$ be a vector of observations of independent and identically distributed (iid) random variables with a common Normal distribution with mean μ and known variance equal to 1. Let E be the observation*

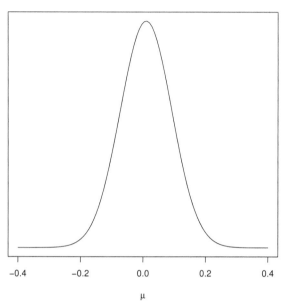

Figure 2.2 *Likelihood function for μ considering a random sample generated from a Normal distribution with mean $\mu = 0$, known variance $\sigma = 1$ and size $n = 20$.*

of the sample $X_1, \ldots, X_n$. Then,

$$l(\mu; E) \propto p(\boldsymbol{x}|\mu) \propto \exp\left\{-\frac{1}{2}\sum_{i=1}^{n}(x_i - \mu)^2\right\}.$$

Figure 2.2 presents the likelihood function for μ considering a random sample generated from a Normal distribution with mean $\mu = 0$. As expected the most likely values of μ are concentrated around the true value $\mu = 0$.

The likelihood function is also easily obtained if we consider a situation where $\boldsymbol{\theta}$ is a parameter vector. In this case, perhaps the most appropriate method for visualizing the likelihood function is plotting its contours, as we shall see in the next example.

Example 2.7 *Let $\boldsymbol{\theta} = (\mu, \sigma^2)$. Let $X_1, \ldots, X_n$ be a vector of observations of iid random variables with a common Normal distribution with mean μ and unknown variance σ^2. Let E be any observation of the $X_1, \ldots, X_n$. Then,*

$$l(\boldsymbol{\theta}; E) = l(\mu, \sigma^2; E) \propto p(\boldsymbol{x}|\mu, \sigma) \propto \sigma^{-n} \exp\left\{-\frac{1}{2\sigma^2}\sum_{i=1}^{n}(x_i - \mu)^2\right\}.$$

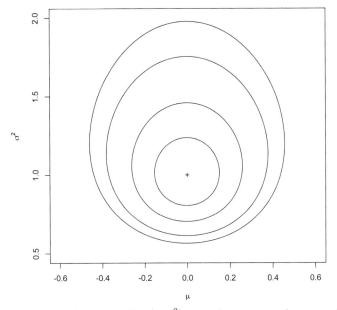

Figure 2.3 *Likelihood function for (μ, σ^2) considering a random sample generated from a Normal distribution with mean $\mu = 0$ and variance $\sigma^2 = 1$ with size $n = 20$.*

Figure 2.3 presents the likelihood function contours considering a random sample generated from a Normal distribution with mean $\mu = 0$ and variance $\sigma^2 = 1$ with size $n = 20$. The most likely values of $\boldsymbol{\theta} = (\mu, \sigma^2)$ are around $(0, 1)$.

The likelihood function leads to the Likelihood Principle, which states that all the information provided by the experiment $\mathbf{X}$ is summarized by the likelihood function. This principle draws a clear line that separates inference schools. On the same side lie the Bayesian and likelihood-based frequentist approaches, which do not violate this principle. On the other side lie the frequentist approaches that are based on the probabilities implied by the sampling distribution of $\mathbf{X}$. In this way, they take into consideration all the possible values of $\mathbf{X}$, in clear violation of the Likelihood Principle.

2.2.2 Fisher information

We have already mentioned that understanding and measuring information is one of the key aspects of the statistical activity. In this section, the most commonly accepted measures of information are introduced. They have important connections with the notion of sufficiency, to be defined later in this chapter, and will prove to be very useful. In fact, they consist of a series of related measures generally known as Fisher information measures.

Definition 2.4 *Let X be a random vector with probability (density) function $p(x \mid \theta)$. The expected Fisher information measure of θ provided by X is defined as*

$$I(\theta) = E_{X|\theta}\left[-\frac{\partial^2 \log p(X \mid \theta)}{\partial \theta^2} \right].$$

If $\boldsymbol{\theta} = (\theta_1, \ldots, \theta_p)$ is a parametric vector, then the Fisher expected information matrix of $\boldsymbol{\theta}$ provided by X can be defined as

$$\boldsymbol{I}(\boldsymbol{\theta}) = E_{X|\theta}\left[-\frac{\partial^2 \log p(X \mid \boldsymbol{\theta})}{\partial \boldsymbol{\theta}\, \partial \boldsymbol{\theta}'} \right],$$

with elements $I_{ij}(\boldsymbol{\theta})$ given by

$$I_{ij}(\boldsymbol{\theta}) = E_{X|\theta}\left[-\frac{\partial^2 \log p(X \mid \boldsymbol{\theta})}{\partial \theta_i \partial \theta_j} \right], \quad i, j = 1, \ldots, p.$$

The information measure defined this way is given by the mean value of the curvature of the likelihood. The larger this curvature is, the larger is the information content summarized in the likelihood function and so the larger will $I(\boldsymbol{\theta})$ be. Since the curvature is expected to be negative, the information value is taken as minus the curvature. The expectation is taken with respect to the sample distribution. The observed Fisher information corresponds to minus the second derivative of the log likelihood, that is,

$$J_{\mathbf{x}}(\boldsymbol{\theta}) = -\frac{\partial^2 \log p(\mathbf{x} \mid \boldsymbol{\theta})}{\partial \boldsymbol{\theta}\partial \boldsymbol{\theta}}'$$

and is interpreted as a local measure of the information content while its expected value, the expected Fisher information, is a global measure. Both $J_{\mathbf{x}}(\boldsymbol{\theta})$ and $I(\boldsymbol{\theta})$ will be used later on in connection with Bayesian and frequentist estimation. The motivation for these definitions will be clear from the following example.

Example 2.8 *Let $X \sim N(\theta, \sigma^2)$, with σ^2 unknown. It is easy to obtain $I(\theta) = J_x(\theta) = \sigma^{-2}$, the sampling precision. Then, the observed and expected Fisher information measures with respect to θ obtained from the observation X coincide with the precision (inverse variance) that we can identify with information.*

There are many properties that can be derived from the Fisher information. One of the most useful ones concerns the additivity of the information with respect to independent observations or, more generally, sources of information.

Lemma 2.1 *Let $\boldsymbol{X} = (X_1, \ldots, X_n)$ be a collection of independent random variables with distribution $p_i(x \mid \boldsymbol{\theta})$, $i = 1, \ldots, n$. Let J_x and J_{x_i} be the observed information measures provided by $\boldsymbol{X}$ and X_i, $i = 1, \ldots, n$, respectively. Let I and I_i be the expected information measures provided by $\boldsymbol{X}$ and X_i, $i = 1, \ldots, n$, respectively. Then,*

$$J_x(\boldsymbol{\theta}) = \sum_{i=1}^{n} J_{x_i}(\boldsymbol{\theta}) \ and \ I(\boldsymbol{\theta}) = \sum_{i=1}^{n} I_i(\boldsymbol{\theta}).$$

Proof $p(\mathbf{X} \mid \boldsymbol{\theta}) = \prod_{i=1}^{n} p_i(X_i \mid \boldsymbol{\theta})$ and therefore $\log p(\mathbf{X} \mid \boldsymbol{\theta}) = \sum_{i=1}^{n} \log p_i(X_i \mid \boldsymbol{\theta})$. Then,

$$-\frac{\partial^2 \log p(\mathbf{X} \mid \boldsymbol{\theta})}{\partial \boldsymbol{\theta} \partial \boldsymbol{\theta}'} = -\sum_{i=1}^{n} \frac{\partial^2 \log p_i(X_i \mid \boldsymbol{\theta})}{\partial \boldsymbol{\theta} \partial \boldsymbol{\theta}'},$$

which proves the result about observed information. Taking expectation with respect to $\mathbf{X} \mid \boldsymbol{\theta}$ on both sides, gives

$$
\begin{aligned}
I(\boldsymbol{\theta}) &= E\left[-\sum_{i=1}^{n} \frac{\partial^2 \log p_i(X_i \mid \boldsymbol{\theta})}{\partial \boldsymbol{\theta} \partial \boldsymbol{\theta}'} \mid \boldsymbol{\theta} \right] \\
&= \sum_{i=1}^{n} E\left[-\frac{\partial^2 \log p_i(X_i \mid \boldsymbol{\theta})}{\partial \boldsymbol{\theta} \partial \boldsymbol{\theta}'} \mid \boldsymbol{\theta} \right] \\
&= \sum_{i=1}^{n} I_i(\boldsymbol{\theta}).
\end{aligned}
$$

$\square$

Lemma 2.1 states that the total information obtained from independent observations is the sum of the information of the individuals observations. This provides further intuition about the appropriateness of the Fisher measures as actual summaries of information.

Another important statistic involved in the study of the likelihood function is the score function.

Definition 2.5 *The score function of X, denoted by $U(X; \theta)$, is defined as*

$$U(X; \theta) = \frac{\partial \log p(X \mid \theta)}{\partial \theta}.$$

In the case of a parametric vector $\boldsymbol{\theta}$, the score function is also a vector $\boldsymbol{U}(X; \boldsymbol{\theta}) = \partial \log p(X \mid \boldsymbol{\theta})/\partial \boldsymbol{\theta}$ with components $U_i(X; \boldsymbol{\theta}) = \partial \log p(X \mid \boldsymbol{\theta})/\partial \theta_i$, $i = 1, \ldots, p$.

The score function is very relevant for statistical inference as will be shown in the next chapters. The following lemma shows an alternative way to obtain the Fisher information based on the score function.

Lemma 2.2 *Under certain regularity conditions,*

$$I(\theta) = E_{X|\theta}[U^2(X; \theta)].$$

In the case of a vector parameter $\boldsymbol{\theta}$, the result becomes

$$\boldsymbol{I}(\boldsymbol{\theta}) = E_{X|\theta}[\, \boldsymbol{U}(\boldsymbol{X}; \boldsymbol{\theta}) \, \boldsymbol{U}'(\boldsymbol{X}; \boldsymbol{\theta}) \,].$$

Proof Using the equality $\int p(\mathbf{X} \mid \boldsymbol{\theta})d\mathbf{X} = 1$ and differentiating both sides with respect to $\boldsymbol{\theta}$, it follows, after interchanging the integration and differentiation operators, that

$$
\begin{aligned}
0 = \int \frac{\partial p(\mathbf{X} \mid \boldsymbol{\theta})}{\partial \boldsymbol{\theta}} \, d\mathbf{X} &= \int \frac{1}{p(\mathbf{X} \mid \boldsymbol{\theta})} \frac{\partial p(\mathbf{X} \mid \boldsymbol{\theta})}{\partial \boldsymbol{\theta}} p(\mathbf{X} \mid \boldsymbol{\theta}) \, d\mathbf{X} \\
&= \int \frac{\partial \log p(\mathbf{X} \mid \boldsymbol{\theta})}{\partial \boldsymbol{\theta}} p(\mathbf{X} \mid \boldsymbol{\theta}) \, d\mathbf{X}.
\end{aligned}
$$

Therefore, the score function has expected value equal to a zero vector. Differentiating again with respect to $\boldsymbol{\theta}$ and interchanging integration and differentiation, we have

$$
\begin{aligned}
0 &= \int \frac{\partial \log p(\mathbf{X}|\boldsymbol{\theta})}{\partial \boldsymbol{\theta}} \left[\frac{\partial p(\mathbf{X}|\boldsymbol{\theta})}{\partial \boldsymbol{\theta}} \right]' p(\mathbf{X}|\boldsymbol{\theta})d\mathbf{X} + \int \frac{\partial^2 \log p(\mathbf{X}|\boldsymbol{\theta})}{\partial \boldsymbol{\theta}\, \partial \boldsymbol{\theta}'} p(\mathbf{X}|\boldsymbol{\theta}) \, d\mathbf{X} \\
&= \int \left[\frac{\partial \log p(\mathbf{X} \mid \boldsymbol{\theta})}{\partial \boldsymbol{\theta}} \right] \left[\frac{\partial \log p(\mathbf{X} \mid \boldsymbol{\theta})}{\partial \boldsymbol{\theta}} \right]' p(\mathbf{X} \mid \boldsymbol{\theta}) \, d\mathbf{X} - I(\boldsymbol{\theta}).
\end{aligned}
$$

The result follows straightforwardly.   □

Although we shall not go into the technical details of the regularity conditions, the main reasons for their presence is to ensure that differentiation of the likelihood can be performed over the entire parameter space and integration and differentiation can be interchanged.

Finally, assume that $\boldsymbol{\theta}_1$ and $\boldsymbol{\theta}_2$ are vectors containing distinct components of the parameter $\boldsymbol{\theta}$. The parameters $\boldsymbol{\theta}_1$ and $\boldsymbol{\theta}_2$ are said to be orthogonal if the portion of the Fisher information matrix related to them is in block diagonal form. In this case, it may be said that there is orthogonality between $\boldsymbol{\theta}_1$ and $\boldsymbol{\theta}_2$. Usually the definition is applied to the expected information matrix but it may also be used for the observed information matrix.

2.3   Bayes theorem

We have seen that a (parametric) statistical inference problem can be stated as having an unknown, unobserved quantity of interest $\boldsymbol{\theta}$ assuming values in a set denoted by $\boldsymbol{\Theta}$. $\boldsymbol{\theta}$ can be a scalar, a vector or a matrix. Up until now, the only relevant source of inference was provided by the probabilistic description of the observations. In this section, we will formalize the use of other sources of information in statistical inference. This will define the Bayesian approach to inference.

Let H (for history) denote the initial available information about some parameter of interest. Assume further that this initial information is expressed in probabilistic terms. It can then be summarized through $p(\boldsymbol{\theta} \mid H)$ and, if the information content of H is enough for our inferential purpose, this is all that is needed. In this case the description of our uncertainty about $\boldsymbol{\theta}$ is complete.

Depending on the relevance of the question with which we are involved, H may not be sufficient and, in this case, it must be augmented. The main tool used in this case is experimentation. Assume a vector of random quantities $\mathbf{X}$ related with $\boldsymbol{\theta}$ can be observed, thus providing further information about $\boldsymbol{\theta}$. (If $\mathbf{X}$ is not random then a functional relationship relating it with $\boldsymbol{\theta}$ should exist. We can then evaluate the value of $\boldsymbol{\theta}$ and the problem is trivially solved.) Before observing $\mathbf{X}$, we should know the sampling distribution of $\mathbf{X}$ given by $p(\mathbf{X} \mid \boldsymbol{\theta}, H)$, where the dependence on $\boldsymbol{\theta}$, central to our argument, is clearly stated. After observing the value of $\mathbf{X}$, the amount of information we have about $\boldsymbol{\theta}$ has increased from H to $H^* = H \cap \{\mathbf{X} = \mathbf{x}\}$. In fact, H^* is a subset of H (a refinement on H was performed).

Now the information about $\boldsymbol{\theta}$ is summarized by $p(\boldsymbol{\theta} \mid \mathbf{x}, H)$, and the only remaining question left is how to pass from $p(\boldsymbol{\theta} \mid H)$ to $p(\boldsymbol{\theta} \mid \mathbf{x}, H)$. From Section 1.4, one can write

$$p(\boldsymbol{\theta} \mid \mathbf{x}, H) = \frac{p(\boldsymbol{\theta}, \mathbf{x} \mid H)}{p(\mathbf{x} \mid H)} = \frac{p(\mathbf{x} \mid \boldsymbol{\theta}, H) p(\boldsymbol{\theta} \mid H)}{p(\mathbf{x} \mid H)},$$

where

$$p(\mathbf{x} \mid H) = \int_{\boldsymbol{\Theta}} p(\mathbf{x}, \boldsymbol{\theta} \mid H) \, d\boldsymbol{\theta}.$$

The result presented above is known as Bayes theorem. This theorem was introduced by Thomas Bayes in a 1763 paper published after his death. As we can see the function in the denominator does not depend on $\boldsymbol{\theta}$ and so, as far as the quantity of interest $\boldsymbol{\theta}$ is concerned, it is just a constant. Therefore, Bayes theorem can be rewritten in its more usual form

$$p(\boldsymbol{\theta} \mid \mathbf{x}) \propto p(\mathbf{x} \mid \boldsymbol{\theta}) \, p(\boldsymbol{\theta}).$$

The dependence on H is dropped, for simplicity of notation, since it is a common factor to all the terms. Nevertheless, it should not be forgotten. The above formula is valid for discrete and continuous, scalar, vector and matrix quantities. The theorem provides a rule for updating probabilities about $\boldsymbol{\theta}$, starting from $p(\boldsymbol{\theta})$ and leading to $p(\boldsymbol{\theta} \mid \mathbf{x})$. This is the reason the above distributions are called prior and posterior distributions, respectively.

To recover the removed constant in the former equation, it is enough to notice that densities must integrate to 1 and to rewrite it as

$$p(\boldsymbol{\theta} \mid \mathbf{x}) = k\, p(\mathbf{x} \mid \boldsymbol{\theta})\, p(\boldsymbol{\theta}),$$

where

$$1 = \int_{\Theta} p(\boldsymbol{\theta} \mid \mathbf{x})\, d\boldsymbol{\theta} = k \int_{\Theta} p(\mathbf{x} \mid \boldsymbol{\theta})\, p(\boldsymbol{\theta}) d\boldsymbol{\theta}$$

and hence

$$
\begin{aligned}
k^{-1} = p(\mathbf{x}|H) &= \int_{\Theta} p(\mathbf{x} \mid \boldsymbol{\theta}) p(\boldsymbol{\theta})\, d\boldsymbol{\theta} \\
&= E_{\boldsymbol{\theta}}[p(\mathbf{x} \mid \boldsymbol{\theta})].
\end{aligned}
$$

This is the predictive (or marginal) distribution of $\mathbf{X}$. As before, after removing the dependence on H, it can be denoted by $p(\mathbf{x})$. This is the expected distribution of $\mathbf{X}$ (under the prior) and it behaves like a prediction for a given H. So, before observing $\mathbf{X}$ it is useful to verify the prior adequacy through the predictions it provides for $\mathbf{X}$. After observing $\mathbf{X}$, it serves to test the model as a whole. An observed value of $\mathbf{X}$ with a low predictive probability is an indication that the stated model is not providing good forecasts. This is an evidence that something unexpected has happened. Either the model must be revised or an aberrant observation has occurred.

2.3.1   Prediction

Another relevant aspect that follows from the calculations presented above is that we obtain an automatic way to make predictions for future observations. If we want to predict $\mathbf{Y}$, whose probabilistic description is $P(\mathbf{Y} \mid \boldsymbol{\theta})$, we have

$$
\begin{aligned}
p(\mathbf{y} \mid \mathbf{x}) &= \int_{\Theta} p(\mathbf{y}, \boldsymbol{\theta} \mid \mathbf{x}) d\boldsymbol{\theta} \\
&= \int_{\Theta} p(\mathbf{y} \mid \boldsymbol{\theta}, \mathbf{x}) p(\boldsymbol{\theta} \mid \mathbf{x}) d\boldsymbol{\theta} \\
&= \int_{\Theta} p(\mathbf{y} \mid \boldsymbol{\theta}) p(\boldsymbol{\theta} \mid \mathbf{x}) d\boldsymbol{\theta},
\end{aligned}
$$

where the last equality follows from the independence between $\mathbf{X}$ and $\mathbf{Y}$, once $\boldsymbol{\theta}$ is given. This conditional independence assumption is typically present in many statistical problems. Also, it follows from the last equation that

$$p(\mathbf{y}|\mathbf{x}) = E_{\boldsymbol{\theta}|\mathbf{x}}[p(\mathbf{y} \mid \boldsymbol{\theta})].$$

It is always useful to concentrate on prediction rather than on estimation because the former is verifiable. The reason for the difference is that $\mathbf{Y}$ is observable and $\boldsymbol{\theta}$ is not. This concept can be further explored by reading the books of Aitchison and Dunsmore (1975) and Geisser (1993).

Example 2.9 *John goes to the doctor claiming some discomfort. The doctor is led to believe that he may have disease A. He then takes some standard procedures for this case: he examines John, carefully observes the symptoms and prescribes routine laboratory exams.*

Let θ be the unknown quantity of interest indicating whether John has disease A or not. The doctor assumes that $P(\theta = 1|H) = 0.7$. H in this case contains the information John gave him and all other relevant knowledge he has obtained from former patients. To improve the evidence about the illness, the doctor asks John to undertake an exam. Exam X is related with θ and provides an uncertain result, of the positive/negative type, with the following probability distribution

$$\begin{cases} P(X = 1 \mid \theta = 0) = 0.40, & \textit{positive test without disease,} \\ P(X = 1 \mid \theta = 1) = 0.95, & \textit{positive test with disease.} \end{cases}$$

Suppose that John goes through the exam and that its result is $X = 1$. So, for the doctor, the probability that John has the disease is

$$\begin{aligned} P(\theta = 1 \mid X = 1) \quad &\propto \quad l(\theta = 1\,; X = 1)P(\theta = 1) \\ &\propto \quad (0.95)(0.7) = 0.665 \end{aligned}$$

and the probability that he has not the disease is

$$\begin{aligned} P(\theta = 0 \mid X = 1) \quad &\propto \quad l(\theta = 0\,; X = 1)P(\theta = 0) \\ &\propto \quad (0.40)(0.30) = 0.120. \end{aligned}$$

The normalizing constant, such that the total probability adds 1, is calculated so that $k(0.665) + k(0.120) = 1$ and $k = 1/0.785$. Consequently,

$$\begin{aligned} P(\theta = 1 \mid X = 1) &= 0.665/0.785 = 0.847 \ \textit{and} \\ P(\theta = 0 \mid X = 1) &= 0.120/0.785 = 0.153. \end{aligned}$$

For the doctor, the information $X = 1$ increases the probability that John has disease A from 0.7 to 0.847. This is not too much since the probability that the test would give a positive result even if John were not ill was reasonably high. So the doctor decides to ask John to undertake another test, Y, again of the positive/negative type, where

$$\begin{cases} P(Y = 1 \mid \theta = 1) = 0.99, \\ P(Y = 1 \mid \theta = 0) = 0.04. \end{cases}$$

Note that the probability of this new test yielding a positive result given that John doesn't have the illness is very small. Although this test might be more expensive, its results are more accurate.

The posterior distribution of θ given X, $P(\theta \mid X)$ will be the prior distribution for the Y test. Before observing the result of test Y, it is useful to ask ourselves what will the predictive distribution be; that is, what are the values of $P(Y = y \mid X = 1)$, for $y = 0, 1$. As we have already seen, in the discrete case,

$$p(Y = y \mid X = x) = \sum_{j=0}^{1} p(Y = y \mid \theta = j)\, p(\theta = j \mid X = x)$$

and so

$$P(Y = 1 \mid X = 1) = (0.04)(0.153) + (0.99)(0.847) = 0.845 \ and$$
$$P(Y = 0 \mid X = 1) = 1 - P(Y = 1 \mid X = 1) = 0.155.$$

Let us suppose now that John undertook test Y and the observed result was $Y = 0$. This is a reasonably unexpected result as the doctor only gave it a 15.5% chance. He should rethink the model based on this result. In particular, he might want to ask himself:

1. *Did 0.7 adequately reflect his $P(\theta = 1|H)$?*

2. *Is test X really so unreliable? Is the sample distribution of X correct?*

3. *Is test Y so powerful?*

4. *Have the tests been carried out properly?*

In any case, the doctor has to re-evaluate the chances of $\theta = 1$ considering that he now also knows that $Y = 0$, in addition to the knowledge that $X = 1$. By Bayes theorem

$$P(\theta = 1 \mid X = 1, Y = 0) \quad \propto \quad l(\theta = 1; Y = 0)P(\theta = 1 \mid X = 1)$$
$$\propto \quad (0.01)(0.847) = 0.008 \ and$$
$$P(\theta = 0 \mid X = 1, Y = 0) \quad \propto \quad l(\theta = 0; Y = 0)P(\theta = 0 \mid X = 1)$$
$$\propto \quad (0.96)(0.155) = 0.147.$$

Note that now $P(\theta \mid X = 1)$ is working as the prior distribution for the experiment Y. The normalizing constant is $1/(0.008 + 0.147) = 1/0.155$ and

$$P(\theta = 1 \mid Y = 0, X = 1) = 0.008/0.155 = 0.052 \text{ and}$$
$$P(\theta = 0 \mid Y = 0, X = 1) = 0.147/0.155 = 0.948.$$

Summarizing the doctor's findings chronologically,

$$P(\theta = 1) = \begin{cases} 0.7 & , \text{ before the tests } X \text{ and } Y, \\ 0.847 & , \text{ after } X \text{ and before } Y, \\ 0.055 & , \text{ after } X \text{ and } Y. \end{cases}$$

The doctor can then decide his course of action as these updating operations take place.

2.3.2 Sequential nature of the Bayes theorem

Bayes theorem is nothing more than a rule for updating probabilities. From an experimental result $\mathbf{X}_1$ with probability distribution $p_1(\mathbf{x}_1 \mid \boldsymbol{\theta})$ (and consequently, likelihood $l_1(\boldsymbol{\theta}; \mathbf{x}_1)$), it follows that

$$p(\boldsymbol{\theta} \mid \mathbf{x}_1) \propto l_1(\boldsymbol{\theta}; x_1)p(\boldsymbol{\theta}).$$

After observing another experimental result $\mathbf{X}_2$ with probability $p_2(\mathbf{x}_2 \mid \boldsymbol{\theta})$ not depending on $\mathbf{X}_1$, we have

$$p(\boldsymbol{\theta} \mid \mathbf{x}_2, \mathbf{x}_1) \propto l_2(\boldsymbol{\theta}; \mathbf{x}_2)p(\boldsymbol{\theta} \mid \mathbf{x}_1)$$
$$\propto l_2(\boldsymbol{\theta}; \mathbf{x}_2)l_1(\boldsymbol{\theta}; \mathbf{x}_1)p(\boldsymbol{\theta}).$$

Repeating this procedure n times, after observing $\mathbf{X}_3, \ldots, \mathbf{X}_n$ related to $\boldsymbol{\theta}$ through the observational distributions $p_i(\mathbf{x}_i \mid \boldsymbol{\theta})$, for $i = 3, \ldots, n$, we get

$$p(\boldsymbol{\theta} \mid \mathbf{x}_n, \ldots, \mathbf{x}_1) \propto l_n(\boldsymbol{\theta}; \mathbf{x}_n)p(\boldsymbol{\theta} \mid \mathbf{x}_{n-1}, \ldots, \mathbf{x}_1)$$

or alternatively

$$p(\boldsymbol{\theta} \mid \mathbf{x}_n, \mathbf{x}_{n-1}, \ldots, \mathbf{x}_1) \propto \left[\prod_{i=1}^{n} l_i(\boldsymbol{\theta}; \mathbf{x}_i) \right] p(\boldsymbol{\theta}),$$

and it is not difficult to see that the order in which the observations are processed by the theorem is irrelevant. In fact, the observations can be processed on one same batch, on an individual basis or through disjoint subgroups. The final result is always the same as long as the observations are conditionally (on

$\boldsymbol{\theta}$) independent. The sequential nature of Bayesian inference has been deeply explored by Lindley (1965).

Another basic result corresponds to the case of Normal observations with unknown mean, which is used in many practical situations. It shows that if the mean is described by a Normal prior distribution, the posterior distribution will also be a Normal distribution.

Theorem 2.1 *Normal prior and observation. Let $\theta \sim N(\mu, \tau^2)$ and $X \mid \theta \sim N(\theta, \sigma^2)$, with σ^2 known. Therefore, the posterior distribution of θ is $(\theta \mid X = x) \sim N(\mu_1, \tau_1^2)$ where*

$$\mu_1 = \frac{\tau^{-2}\mu + \sigma^{-2}x}{\tau^{-2} + \sigma^{-2}} \quad and \quad \tau_1^{-2} = \tau^{-2} + \sigma^{-2}.$$

Proof From Bayes theorem, it follows that

$$
\begin{aligned}
p(\theta \mid x) &\propto l(\theta; x)p(\theta) \\
&\propto \exp\left\{-\frac{1}{2\sigma^2}(x-\theta)^2 - \frac{1}{2\tau^2}(\theta-\mu)^2\right\} \\
&\propto \exp\left\{-\frac{\theta^2}{2\sigma^2} - \frac{\theta^2}{2\tau^2} + \frac{x\theta}{\sigma^2} + \frac{\mu\theta}{\tau^2}\right\} \\
&= \exp\left\{-\frac{\theta^2}{2}\left(\frac{1}{\sigma^2} + \frac{1}{\tau^2}\right) + \theta\left(\frac{x}{\sigma^2} + \frac{\mu}{\tau^2}\right)\right\},
\end{aligned}
$$

where in the first step all constants involved were included in the proportionality factor. Now let $\tau_1^2 = (\tau^{-2} + \sigma^{-2})^{-1}$ and $\mu_1 = (\sigma^{-2}x + \mu\tau^{-2})\tau_1^2$. Substituting into the above expression gives

$$
\begin{aligned}
p(\theta \mid x) &\propto \exp\left\{-\frac{\theta^2}{2\tau_1^2} + \frac{\theta\mu_1}{\tau_1^2}\right\} \\
&\propto \exp\left\{-\frac{1}{2\tau_1^2}(\theta-\mu_1)^2\right\} \\
&\propto \frac{1}{\sqrt{2\pi\tau_1^2}}\exp\left\{-\frac{1}{2\tau_1^2}(\theta-\mu_1)^2\right\}.
\end{aligned}
$$

It is easy to recognize that the last term in the above expression corresponds to a Normal density. Therefore, the last proportionality constant is equal to 1 and $\theta \mid x \sim N(\mu_1, \tau_1^2)$.

$\square$

Defining the precision as the reciprocal of the variance, it follows from the theorem that the posterior precision is the sum of the prior and likelihood precisions and does not depend on x. Interpreting the precision as a measure of information and defining $w = \tau^{-2}/(\tau^{-2} + \sigma^{-2}) \in (0,1)$, w measures the relative information contained in the prior distribution with respect to the total information (prior plus likelihood). Then one can write

$$\mu_1 = w\mu + (1 - w)x,$$

which is the weighted mean of prior and likelihood means.

Example 2.10 *(Box & Tiao, 1992) Two physicists, A and B, want to determine the value of a physical constant θ. Physicist A with large experience in the area specifies his prior as $\theta \sim N(900, (20)^2)$. On the other side, physicist B, not so experienced on the subject, states a more uncertain prior $\theta \sim N(800, (80)^2)$. It is easy to obtain that for physicist A, $Pr(\theta \in (860, 940)) \doteq 0.95$ and for physicist B, $Pr(\theta \in (640, 960)) \doteq 0.95$. Both physicists agree to make an evaluation of θ, denoted by X, using a calibrated device with sampling distribution $X \mid \theta \sim N(\theta, (40)^2)$. After observing the value $X = 850$, the posterior distributions of θ can be obtained using Theorem 2.1 and the values stated above as*

    1. physicist A: $(\theta \mid X = 850) \sim N(890, (17.9)^2)$,

    2. physicist B: $(\theta \mid X = 850) \sim N(840, (35.7)^2)$.

    The inference procedure of the two physicists is summarized in Figure 2.4. It is worth noting that due to the different initial uncertainties, the same experiment provides very little additional information for A, although the uncertainty of B has been substantially reduced. Identifying precision with information we have that the information about θ for physicist A increases from 0.0025 to 0.00312 since the likelihood precision was 0.000625 (an increase of 25%). For physicist B, it increases from 0.000156 to 0.000781 (an increase of 400%).

2.4   Exchangeability

Exchangeability is a very important concept introduced by de Finetti (1937). It is weaker than independence but it is just as useful, as will be shown below.

Definition 2.6 *Let $\mathcal{K} = \{k_1, \ldots, k_n\}$ be a permutation of $\{1, \ldots, n\}$. Random quantities $X_1, \ldots, X_n$ are exchangeable if the $n!$ permutations $(X_{k_1}, \ldots, X_{k_n})$ have the same n-dimensional probability distribution. An infinite sequence of random quantities is exchangeable if any finite sub-sequence is exchangeable.*

    One immediate consequence of exchangeability is that all marginal distributions must be the same. To see this, consider any two distinct permutations $\mathcal{K}$ and $\mathcal{K}'$ of an exchangeable sequence of random variables, that therefore

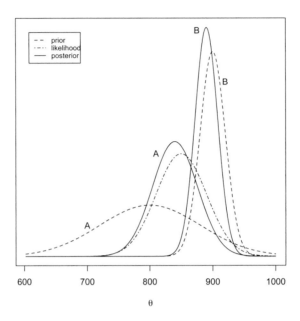

Figure 2.4 *Prior and posterior densities and likelihood for θ for the physicist's example.*

must have the same probability. Let k_1 and k'_1 be the first index of the two permutations, respectively. If both sides of this equality are integrated with respect to all the components but the first, one gets that the marginal distribution of X_{k_1} and $X_{k'_1}$ must be the same. Since one is free to choose the values of k_1 and k'_1, this means that all marginal distributions are equal.

A sequence (finite or not) of iid random quantities is trivially exchangeable, although the reciprocal is not true, in general.

Example 2.11 *Two examples are presented below.*

1. *Consider an urn with m balls, r with number 1 and $m - r$ with number 0. Balls are drawn from the urn, one at time, without replacement. Let X_k denote the number associated with the k-th ball selected. Then, $X_1, \ldots, X_n$, $n \leq m$ is an exchangeable sequence, but the X_i's are not independent.*

2. *Let $X_1, X_2, \ldots$ be a sequence of Bernoulli trials with unknown success probability θ. The classical assumption is that the X_k's are iid. For a Bayesian, if θ is unknown, the information content of the j-th observation can modify his/her belief about the distribution of X_k. The ex-*

ample of the previous chapter illustrated this point. If the experiments are judged similar in some sense, the hypothesis of exchangeability is acceptable, while marginal independence is not.

The relevance of the concept of exchangeability is due to the fact that, although it is based on weak assumptions, it allows one to state a very powerful result, known as the representation theorem of de Finetti. The theorem is stated here without proof; that can be found in de Finetti (1937). A simple outline of the proof of this representation theorem, worth reading, is found in Heath and Sudderth (1976).

Theorem 2.2 *To all infinite sequences of exchangeable random quantities $\{X_n, n = 1, 2, \ldots\}$ assuming values $\{0, 1\}$, there corresponds a distribution F in (0,1) such that for all n and $k \leq n$,*

$$P[(k, n-k)] = \int_0^1 \theta^k (1-\theta)^{n-k} dF(\theta),$$

where $(k, n-k)$ denotes the event that k of the X_i's are 1 and the other $n-k$ of the X_i's are 0. Note that, due to the exchangeability assumption, any sequence with k 1's and $n-k$ 0's also admits a representation according to the above theorem.

The importance of the theorem is that it provides further backing for the Bayesian approach. If one is willing to consider a collection of $0-1$ observations as exchangeable, then one is prepared to rephrase one's model into a sampling Bernoulli model with success probability θ that itself is random with probability distribution F. The theorem, however, does not say anything about the distribution F.

Example 2.12 *Some possible specifications for F are*

1. *a degenerate distribution: if $P(\theta = \theta_0) = 1$, for some θ_0, then*

$$P[(k, n-k)] = \theta_0^k (1-\theta_0)^{n-k}.$$

2. *a discrete distribution: if $P(\theta = \theta_i) = p_i$, if $\theta = \theta_i$, $i = 1, \ldots, s$, with $p_i \geq 0$ and $\sum p_i = 1$, then*

$$P[(k, n-k)] = \sum_{i=1}^{s} p_i \theta_i^k (1-\theta_i)^{n-k}.$$

3. *a continuous distribution: if $\theta \sim Beta(a, b)$, then*

$$
\begin{aligned}
P[(k, n - k)] &= \int_0^1 \theta^k (1 - \theta)^{n-k} \frac{\theta^{a-1}(1 - \theta)^{b-1}}{B(a, b)} d\theta \\
&= \frac{1}{B(a, b)} \int_0^1 \theta^{a+k-1}(1 - \theta)^{b+n-k-1} d\theta \\
&= \frac{B(a + k, b + n - k)}{B(a, b)}.
\end{aligned}
$$

The exchangeability concept has already been extended to many other distributions with the inclusion of some additional hypotheses; see Bernardo and Smith (1994) for a review of the main results. The definition is the same as presented previously with removal of the constraint imposed on the sample space. In particular, if we introduce the hypothesis of symmetry of the distributions and invariance under linear transformations, it can be shown that the joint density of any finite sub-sequence is given by

$$
p(x_1, \ldots, x_n) = \int_0^\infty \int_{-\infty}^\infty \prod p_N(x_i; \theta, \sigma^2) dF(\theta, \sigma^2),
$$

where θ is a quantity varying in R, σ a quantity assuming values in R^+, $p_N(\cdot | a, b)$ is the density of an $N(a, b)$ distribution and F is a distribution function in $R \times R^+$. Exchangeability, along with invariance now, lead to a representation where a Normal sampling distribution is obtained with parameters having some prior distribution F. It is worth noting that F once again is not specified.

Another useful extension, well explored in the last years, is the concept of partial exchangeability. In this case the exchangeability holds only under certain conditions. For example, we can define some groups of variables where exchangeability is valid only within each group. This concept can be extended to more general cases than group classification and is formalized as follows.

Definition 2.7 *Let $\{X_i, i = 1, 2, \ldots, n\}$ be any sequence of random quantities and $\mathcal{K}$ be any permutation of $\{1, 2, \ldots, n\}$. We say that $\mathbf{X}$ is partially exchangeable if there are quantities $\{Z_i, i = 1, 2, \ldots, n\}$ such that the distribution of $(\mathbf{X} \mid \mathbf{Z})$ is the same as that of $(\mathbf{X}_\mathcal{K} \mid \mathbf{Z}_\mathcal{K})$, for any permutation $\mathcal{K}$, where for any vector $\mathbf{c} = (c_1, \ldots, c_n)$, $\mathbf{c}_\mathcal{K} = (c_{k_1}, \ldots, c_{k_n})$.*

The main idea behind this definition is that when the indexes of the X_i's are permuted, the resulting vector will have the same conditional distribution, as long as the same permutation is applied to the Z_i's indexes. This is clearly a weaker concept than exchangeability. The case $Z_i = 1$, $\forall i$, corresponds to the exchangeability definition. Another interesting and less trivial case is when there are s groups and each Z_i takes on a value in $\{1, \ldots, s\}$ to identify the group to which the observation X_i belongs. In this case, one has exchangeability within each group but not globally.

The extension of this concept to countable sequences is immediate. The notion of partial exchangeability will be returned to in Chapter 3 when the related concept of hierarchical prior is introduced and in Chapters 4 and 8 when inference for hierarchical models is discussed.

2.5   Sufficiency and exponential family

As we have said previously, one of the main goals of statistics is to summarize information. An important aspect is to know if, given a set of observations $\mathbf{X}$ sampled to get information about a parameter of interest $\boldsymbol{\theta}$, there are statistics, i.e., functions of the observations $\mathbf{X}$ that summarize all the information contained in $\mathbf{X}$.

Definition 2.8 *(classical) Let $\mathbf{X}$ be a random quantity. Then, the statistic $\mathbf{T} = \mathbf{T}(\mathbf{X})$ is sufficient for the parameter $\boldsymbol{\theta}$ if the distribution of $\mathbf{X}|\mathbf{T} = \mathbf{t}$ does not depend on $\boldsymbol{\theta}$, for all possible values of $\mathbf{t}$. If $\mathbf{X}$ has density (or probability) function $p(\boldsymbol{x} \mid \boldsymbol{\theta})$ then*

$$p(\boldsymbol{x} \mid \boldsymbol{t}, \boldsymbol{\theta}) = p(\boldsymbol{x} \mid \boldsymbol{t}).$$

The definition states that given $\mathbf{T}$, $\mathbf{X}$ does not bring any additional information about $\boldsymbol{\theta}$. From the Bayesian point of view this means that $\mathbf{X}$ and $\boldsymbol{\theta}$ are independent conditionally on $\mathbf{T}$. The main point of the definition is that, after observing $\mathbf{T}$, one can forget the rest of the data if one is only interested in gathering information about $\boldsymbol{\theta}$. The concept of sufficient statistics was introduced by Fisher and was studied by Lehman, Scheffe and Bahadur in the 50's, as pointed out in DeGroot (1970).

The strength of the definition lies in the possibility of finding sufficient statistics of a smaller dimension than data $\mathbf{X}$, thus implying savings in information storage. In some cases, it is possible to find sufficient statistics with fixed dimension independent of the sample size. In these cases the dimension reduction and consequent storage savings can be huge if large sample sizes are considered.

Theorem 2.3 *If $\mathbf{T} = \mathbf{T}(\mathbf{X})$ is a sufficient statistic for $\boldsymbol{\theta}$, then*

$$p(\boldsymbol{\theta} \mid \boldsymbol{x}) = p(\boldsymbol{\theta} \mid \boldsymbol{t}), \ \text{for all priors } p(\boldsymbol{\theta}).$$

Proof $p(\mathbf{x} \mid \boldsymbol{\theta}) = p(\mathbf{x}, \mathbf{t} \mid \boldsymbol{\theta})$, if $\mathbf{t} = \mathbf{T}(\mathbf{x})$ and 0, if $\mathbf{t} \neq \mathbf{T}(\mathbf{x})$. So,

$$
\begin{aligned}
p(\mathbf{x} \mid \boldsymbol{\theta}) &= p(\mathbf{x} \mid \mathbf{t}, \boldsymbol{\theta}) p(\mathbf{t} \mid \boldsymbol{\theta}) \\
&= p(\mathbf{x} \mid \mathbf{t}) p(\mathbf{t} \mid \boldsymbol{\theta}), \text{ by the definition of sufficiency.}
\end{aligned}
$$

But, by Bayes theorem,

$$
\begin{aligned}
p(\boldsymbol{\theta} \mid \mathbf{x}) \quad &\propto \quad p(\mathbf{x} \mid \boldsymbol{\theta})p(\boldsymbol{\theta}) \\
&= \quad p(\mathbf{x} \mid \mathbf{t})p(\mathbf{t} \mid \boldsymbol{\theta})p(\boldsymbol{\theta}) \\
&\propto \quad p(\mathbf{t} \mid \boldsymbol{\theta})p(\boldsymbol{\theta}), \text{ since } p(\mathbf{x} \mid \mathbf{t}) \text{ does not depend on } \theta \\
&\propto \quad p(\boldsymbol{\theta} \mid \mathbf{t}).
\end{aligned}
$$

Then $p(\boldsymbol{\theta} \mid \mathbf{x}) = k\, p(\boldsymbol{\theta} \mid \mathbf{t})$, for some $k > 0$.
    Additionally,

$$
1 = \int_{\Theta} p(\boldsymbol{\theta} \mid \mathbf{x})d\boldsymbol{\theta} = k \int_{\Theta} p(\boldsymbol{\theta} \mid \mathbf{t})d\boldsymbol{\theta} = k,
$$

and so $p(\boldsymbol{\theta} \mid \mathbf{x}) = p(\boldsymbol{\theta} \mid \mathbf{t})$. $\square$

This theorem leads to a possible Bayesian definition of sufficiency below.

Definition 2.9 *(Bayesian) The statistic $\boldsymbol{T}(\boldsymbol{X})$ is sufficient for $\boldsymbol{\theta}$ if there is a function f such that*

$$
p(\boldsymbol{\theta} \mid \boldsymbol{x}) \propto f(\boldsymbol{\theta}, \boldsymbol{t}).
$$

A useful insight into sufficiency is gained through the notion of partitions of the sample space. Let $\mathbf{T} = \mathbf{T}(\mathbf{X})$ be a p-dimensional statistic and $A_{\mathbf{t}} = \{\mathbf{x} : \mathbf{T}(\mathbf{x}) = \mathbf{t}\}$. The collection of sets $\{A_{\mathbf{t}} : \mathbf{t} \in R^p\} = \{A_{\mathbf{t}}\}$ is a partition if $A_{\mathbf{t}} \cap A_{\mathbf{t}'} = \emptyset$, $\forall \mathbf{t}, \mathbf{t}' \in R^p$ and $\cup_{\mathbf{t}} A_{\mathbf{t}} = \mathcal{S}$, the sample space. A partition induced by a sufficient statistic is called a sufficient partition.

Example 2.13 *Assume X_1, X_2, X_3 to be independent, identically distributed $Ber(\theta)$ and let $T = X_1 + X_2 + X_3$. The partition of the sample space induced by the statistics T has components $A_0 = \{(0,0,0)\}$, $A_1 = \{(1,0,0),(0,1,0),(0,0,1)\}$, $A_2 = \{(1,1,0),(1,0,1),(0,1,1)\}$ and $A_3 = \{(1,1,1)\}$.*

The equivalence between the classical and Bayesian definitions follows easily from the theorem presented below for the classical definition of sufficiency.

Theorem 2.4 *(Neyman's factorization criterion) The statistics $\boldsymbol{T}$ is sufficient for $\boldsymbol{\theta}$ if and only if*

$$
p(\boldsymbol{x} \mid \boldsymbol{\theta}) = f(\boldsymbol{t}, \boldsymbol{\theta})\, g(\boldsymbol{x}),
$$

where f and g are non-negative functions.

Proof ($\Longrightarrow$) We have already seen that $p(\mathbf{x} \mid \boldsymbol{\theta}) = p(\mathbf{x} \mid \mathbf{t})p(\mathbf{t} \mid \boldsymbol{\theta})$. Then it is enough to define $g(\mathbf{x}) = p(\mathbf{x} \mid \mathbf{t}) = p(\mathbf{x} \mid \mathbf{T}(\mathbf{x}))$ and $f(\mathbf{t}, \boldsymbol{\theta}) = p(\mathbf{t} \mid \boldsymbol{\theta})$, completing this part of the proof.

($\Longleftarrow$) Conversely, we have that $p(\mathbf{x} \mid \boldsymbol{\theta}) = f(\mathbf{t}, \boldsymbol{\theta})g(\mathbf{x})$. Defining $A_{\mathbf{t}} = \{\mathbf{x} : \mathbf{T}(\mathbf{x}) = \mathbf{t}\}$ and assuming $\mathbf{X}$ continuous, the density function of $\mathbf{T} \mid \boldsymbol{\theta}$ is

$$p(\mathbf{t} \mid \boldsymbol{\theta}) = \int_{A_{\mathbf{t}}} p(\mathbf{x} \mid \boldsymbol{\theta})d\mathbf{x} = f(\mathbf{t}, \boldsymbol{\theta}) \int_{A_{\mathbf{t}}} g(\mathbf{x})d\mathbf{x} = f(\mathbf{t}, \boldsymbol{\theta})G(\mathbf{x}),$$

for some function G and so, $f(\mathbf{t}, \boldsymbol{\theta}) = p(\mathbf{t} \mid \boldsymbol{\theta})/G(\mathbf{x})$. Similar calculations follow in the discrete case.

On the other hand, from the hypothesis of the theorem, $f(\mathbf{t}, \boldsymbol{\theta}) = p(\mathbf{x} \mid \boldsymbol{\theta})/g(\mathbf{x})$. Equating the two forms for $f(\mathbf{t}, \boldsymbol{\theta})$ leads to

$$\frac{p(\mathbf{x} \mid \boldsymbol{\theta})}{p(\mathbf{t} \mid \boldsymbol{\theta})} = \frac{g(\mathbf{x})}{G(\mathbf{x})}.$$

Since $p(\mathbf{x} \mid \mathbf{t}, \boldsymbol{\theta}) = p(\mathbf{x} \mid \boldsymbol{\theta})/p(\mathbf{t} \mid \boldsymbol{\theta})$, then

$$p(\mathbf{x} \mid \mathbf{t}, \boldsymbol{\theta}) = \frac{g(\mathbf{x})}{G(\mathbf{x})} = p(\mathbf{x} \mid \mathbf{t})$$

since the ratio g/G does not depend on $\boldsymbol{\theta}$. Thus, $\mathbf{T}$ is sufficient for $\boldsymbol{\theta}$. □

Neyman's factorization criterion is the tool usually applied for identification of sufficient statistics.

We can now show that the two concepts of sufficiency are equivalent, since

1. (classical $\Longrightarrow$ Bayesian) Follows trivially from Theorem 2.4.

2. (Bayesian $\Longrightarrow$ classical)

$$p(\boldsymbol{\theta} \mid \mathbf{x}) = \frac{p(\mathbf{x} \mid \boldsymbol{\theta})p(\boldsymbol{\theta})}{p(\mathbf{x})} = f(\mathbf{t}, \boldsymbol{\theta})p(\boldsymbol{\theta}), \text{ by hypothesis.}$$

So, $p(\mathbf{x} \mid \boldsymbol{\theta}) = f(\mathbf{t}, \boldsymbol{\theta})p(\mathbf{x})$ which, by the factorization criterion, is equivalent to saying that $\mathbf{T}$ is a sufficient statistic.

Definition 2.10 *Suppose that $\mathbf{X}$ has density $p(\boldsymbol{x} \mid \boldsymbol{\theta})$, then $\boldsymbol{T}(\boldsymbol{X})$ is an ancillary statistic for $\boldsymbol{\theta}$ if $p(\boldsymbol{t} \mid \boldsymbol{\theta}) = p(\boldsymbol{t})$.*

In this case, $\mathbf{T}$ does not provide any information for $\boldsymbol{\theta}$ although it is a function of $\mathbf{X}$, which is related with $\boldsymbol{\theta}$. Ancillarity may be understood as an antonym of sufficiency.

Sufficiency is a basic concept in classical statistic even though it is not so relevant for the Bayesian approach. Also, from the applied point of view this is also not a very useful concept since even small perturbations in the model can imply the loss of sufficiency. Nevertheless, it is a concept that highlights the importance of data summarization and as such becomes relevant in other seemingly unrelated areas where parsimony may be called into play.

Example 2.14 *Two examples are provided below.*

1. *Let* $\boldsymbol{X} = (X_1, \ldots, X_n)$ *be observations with values 0 or 1, where* $P(X_i = 1 \mid \theta) = \theta$.

$$p(\boldsymbol{x} \mid \theta) = \theta^t (1 - \theta)^{n-t}, \quad \text{with } t = \sum_{i=1}^n x_i.$$

From the factorization criterion it follows that $T(\boldsymbol{X}) = \sum_{i=1}^n X_i$ *is a sufficient statistic for* θ. *In this case, it is also possible to conclude straightforwardly, from the definition of sufficiency and using some combinatorial arguments, that* $T(\boldsymbol{X})$ *is a sufficient statistic since* $p(\boldsymbol{x} \mid T(\boldsymbol{x}) = t) = [\binom{n}{t}]^{-1}$, *which does not depend on* θ.

2. *Let* $X_1, X_2, \ldots, X_n$ *be iid conditional on* $\boldsymbol{\theta}$ *with common density* $p(x_i \mid \boldsymbol{\theta})$. *Then*

$$p(x_1, \ldots, x_n \mid \boldsymbol{\theta}) = \prod_{i=1}^n p(x_i \mid \boldsymbol{\theta}).$$

The order statistics are defined as $Y_1 = X_{(1)} = \min_i X_i$, $Y_2 = X_{(2)} =$ *second smallest sample value,* $\ldots$, $Y_n = X_{(n)} = \max_i X_i$. *The order of the terms does not alter the product and to each* x_i *corresponds a unique* y_i *(assuming continuity)*

$$\prod_{i=1}^n p(x_i \mid \boldsymbol{\theta}) \propto \prod_{i=1}^n p(y_i \mid \boldsymbol{\theta}).$$

Then, with $g(\boldsymbol{x}) = 1$, $\boldsymbol{t} = (y_1, \ldots, y_n)$ *and* $f(\boldsymbol{t}, \boldsymbol{\theta}) = \prod_{i=1}^n p(y_i \mid \boldsymbol{\theta})$, *the factorization criterion holds and* $\boldsymbol{T} = (Y_1, \ldots, Y_n)$ *is a sufficient statistic for* $\boldsymbol{\theta}$. *Note that the dimension of* $\boldsymbol{T}$ *depends upon the sample size. In this case no dimensionality reduction was achieved and the definition becomes deprived of its strength. It is also clear that the sample* $\boldsymbol{X}$ *itself is trivially a sufficient statistic for* $\boldsymbol{\theta}$.

The application of the sufficiency concept developed above is not necessarily useful. It is only relevant when the dimension of the sufficient statistic is significantly smaller than the sample size. An interesting question is the concern on how to obtain a sufficient statistic with maximal reduction of the sample data. Such a statistic is known in the literature as a minimal sufficient statistic.

Definition 2.11 *Let* $\boldsymbol{X} \sim p(\boldsymbol{x} \mid \boldsymbol{\theta})$. *The statistic* $\boldsymbol{T}(\boldsymbol{X})$ *is a minimal sufficient statistic for* $\boldsymbol{\theta}$ *if it is a sufficient statistic for* $\boldsymbol{\theta}$ *and a function of every other sufficient statistic for* $\boldsymbol{\theta}$.

If $\mathbf{S}(\mathbf{X})$ is a statistic obtained as a bijective function of a sufficient statistic $\mathbf{T}(\mathbf{X})$, then $\mathbf{S}$ is also a sufficient statistic. On the other hand, the minimal sufficient statistic is unique, apart from bijective transformation of itself.

Definition 2.12 *Two elements $\boldsymbol{x}$ and $\boldsymbol{y}$ of the sample space are information equivalent if and only if the ratio $p(\boldsymbol{x}|\boldsymbol{\theta})/p(\boldsymbol{y}|\boldsymbol{\theta})$ or, equivalently, $l(\boldsymbol{\theta};\boldsymbol{x})/l(\boldsymbol{\theta};\boldsymbol{y})$ does not depend on $\boldsymbol{\theta}$.*

Information equivalence defines an equivalence relation of the elements of the sample space. Therefore, it defines a partition of the sample space. This partition is called a minimal sufficient partition. It can be shown that a sufficient statistic is minimal if and only if the partition it defines on the sample space is minimal sufficient.

Example 2.15 *Let $X_1, \ldots, X_n$ be iid Poisson variables with mean λ, and define $T(\boldsymbol{X}) = \sum_{i=1}^{n} X_i$. Then,*

$$p(\boldsymbol{x}|\lambda) = \prod_{i=1}^{n} p(x_i|\lambda) = \prod_{i=1}^{n} e^{-\lambda} \frac{\lambda^{x_i}}{x_i!} = e^{-n\lambda} \frac{\lambda^{T(\boldsymbol{x})}}{\prod_i x_i!}.$$

Therefore, $p(\boldsymbol{x}|\lambda)/p(\boldsymbol{y}|\lambda) = \lambda^{T(\boldsymbol{x})-T(\boldsymbol{y})} \prod_i (y_i!)/x_i!)$, which does not depend on λ if and only if $T(\boldsymbol{x}) = T(\boldsymbol{y})$. Hence, $T(\boldsymbol{X})$ is a minimal sufficient statistic for λ.

Another interesting question is whether there are families of distributions admitting fixed dimension sufficient statistics for $\boldsymbol{\theta}$. Fortunately, for a large class of distributions, the dimension of the sufficient statistic $\mathbf{T}$ is equal to the number of parameters. Maximum summarization is obtained when we have one sufficient statistics for each parameter. Subject to some weak regularity conditions, all distributions with the dimension of $\mathbf{T}$ equal to the number of the parameters belonging to the exponential family.

Definition 2.13 *The family of distributions with probability (density) function $p(x \mid \boldsymbol{\theta})$ belongs to the exponential family with r parameters if $p(x \mid \boldsymbol{\theta})$ can be written as*

$$p(x \mid \boldsymbol{\theta}) = a(x) \exp\left\{ \sum_{j=1}^{r} U_j(x)\phi_j(\boldsymbol{\theta}) + b(\boldsymbol{\theta}) \right\}, \; x \in \mathcal{X} \subset R$$

and $\mathcal{X}$ does not depend on $\boldsymbol{\theta}$.

By the factorization criterion, $U_1(X), \ldots, U_r(X)$ are sufficient statistics for $\boldsymbol{\theta}$ (when a single X is observed). For a size n sample $\mathbf{X}$ we have

$$p(\mathbf{x} \mid \boldsymbol{\theta}) = \left[\prod_{i=1}^{n} a(x_i)\right] \exp\left\{ \sum_{j=1}^{r} \left[\sum_{i=1}^{n} U_j(x_i)\right] \phi_j(\boldsymbol{\theta}) + nb(\boldsymbol{\theta}) \right\},$$

which belongs to the exponential family too, with $a(\mathbf{x}) = \prod_{i=1}^{n} a(x_i)$ and $U_j(\mathbf{X}) = \sum_{i=1}^{n} U_j(X_i)$, $j = 1, \ldots, r$. So, $\mathbf{T} = (T_1, \ldots, T_r)$ with $T_j = U_j(\mathbf{X})$, for $j = 1, 2, \ldots, r$, is a sufficient statistic for $\boldsymbol{\theta}$.

The exponential family is very rich including most of the distributions more commonly used in statistics. Among the most important distributions not included in this family are the uniform distribution (which also has sufficient statistics with dimension not depending on the sample size) and the Student-t distribution (which does not). Darmois, Koopman and Pitman have independently showed that among families satisfying some regularity conditions, a sufficient statistic of fixed dimension will only exist for the exponential family.

Example 2.16 *Some special cases are provided below.*

1. *Bernoulli(θ)*

$$
\begin{aligned}
p(x \mid \theta) &= \theta^x (1-\theta)^{1-x} I_x(\{0,1\}), \\
&= \exp\left\{ x \log\left(\frac{\theta}{1-\theta} \right) + \log(1-\theta) \right\} I_x(\{0,1\}).
\end{aligned}
$$

For a sample $\mathbf{x}$ we have

$$
\begin{aligned}
p(\mathbf{x} \mid \theta) &= \prod_{i=1}^{n} \theta^{x_i} (1-\theta)^{1-x_i} I_{x_i}(\{0,1\}) \\
&= \exp\left\{ \sum_{i=1}^{n} x_i \log\left(\frac{\theta}{1-\theta} \right) + n \log(1-\theta) \right\} I_x(\{0,1\}^n).
\end{aligned}
$$

Then, the Bernoulli distribution belongs to the one parameter exponential family with $a(\boldsymbol{x}) = I_x(\{0,1\}^n)$, $b(\theta) = n \log(1-\theta)$, $\phi(\theta) = \log[\theta/(1-\theta)]$ and $U(\boldsymbol{X}) = \sum_{i=1}^{n} X_i$. So, U is a sufficient statistic for θ as we have seen previously.

2. *Poisson(λ)*

$$
\begin{aligned}
p(x \mid \lambda) &= \frac{e^{-\lambda} \lambda^x}{x!} I_x(\{0,1,\ldots\}), \text{ which in the exponential form is} \\
&= \frac{1}{x!} \exp\{-\lambda + x \log \lambda\} I_x(\{0,1,\ldots\}).
\end{aligned}
$$

For a sample $\mathbf{x}$ we have

$$
p(\boldsymbol{x} \mid \lambda) = \frac{1}{\prod_{i=1}^{n} x_i!} \exp\left\{ \sum_{i=1}^{n} x_i \log \lambda - n\lambda \right\} I_x(\{0,1,\ldots\}^n).
$$

So, the Poisson distribution belongs to the one parameter exponential family with $a(\boldsymbol{x}) = I_x(\{0,1,\ldots\}^n)/\prod_{i=1}^{n} x_i!$, $b(\theta) = -n\lambda$, $\phi(\theta) = \log \lambda$ and $U(\boldsymbol{X}) = \sum_{i=1}^{n} X_i$. Then U is sufficient for λ.

3. $Normal(\mu, \sigma^2)$

$$p(x \mid \mu, \sigma^2) = \frac{1}{\sqrt{2\pi}\sigma} \exp\left\{-\frac{(x-\mu)^2}{2\sigma^2}\right\}$$

$$= \frac{1}{\sqrt{2\pi}\sigma} \exp\left\{\frac{\mu}{\sigma^2}x - \frac{1}{2\sigma^2}x^2 - \frac{\mu^2}{2\sigma^2}\right\}$$

$$= \frac{1}{\sqrt{2\pi}} \exp\left\{\frac{\mu}{\sigma^2}x - \frac{1}{2\sigma^2}x^2 - \frac{\mu^2}{2\sigma^2} - \frac{1}{2}\log\sigma^2\right\}.$$

For a sample $\boldsymbol{x}$, it follows

$$p(\boldsymbol{x} \mid \mu, \sigma^2) = \frac{1}{(2\pi)^{n/2}} \exp\left\{\frac{\mu}{\sigma^2}\sum_{i=1}^{n}x_i - \frac{1}{2\sigma^2}\sum_{i=1}^{n}x_i^2 - \frac{n}{2}\left(\frac{\mu^2}{\sigma^2} + \log\sigma^2\right)\right\}.$$

Then, the Normal distribution is a member of the exponential family with a bidimensional parameter $\boldsymbol{\theta} = (\mu, \sigma^2)$, $a(\boldsymbol{x}) = (2\pi)^{-n/2}$, $b(\boldsymbol{\theta}) = -(n/2)[(\mu^2/\sigma^2) + \log\sigma^2]$, $\phi_1(\boldsymbol{\theta}) = \mu^2/\sigma^2$, $\phi_2(\boldsymbol{\theta}) = -1/2\sigma^2$, $U_1(\boldsymbol{X}) = \sum_{i=1}^{n} X_i$ and $U_2(\boldsymbol{X}) = \sum_{i=1}^{n} X_i^2$. So, $\boldsymbol{U} = (U_1, U_2)$ is sufficient for (μ, σ^2).

2.6 Parameter elimination

Sometimes models need to be developed with the inclusion of various parameters, many of which are included regardless of our particular interest on them. These parameters are often included to describe relevant aspects of the reality we are modeling, even though they are not related to our main concern about the problem. To simplify the discussion, the parametric vector can be broadly split into two sub-vectors: $\boldsymbol{\theta}$ containing the parameters of interest and $\boldsymbol{\phi}$ containing the components that despite being present in the model are not our immediate concern. The first sub-vector is called the parameter of interest and the second one is known as the nuisance parameter. Usually, it is our desire to eliminate $\boldsymbol{\phi}$ from the analysis as soon as possible, in order to concentrate efforts on $\boldsymbol{\theta}$.

2.6.1 Bayesian approach

From the Bayesian point of view, this is done using probability rules. After observing $\boldsymbol{X} = \mathbf{x}$, the posterior distribution $p(\boldsymbol{\theta}, \boldsymbol{\phi} \mid \mathbf{x})$ is obtained by Bayes theorem. Thus we can calculate the following quantities:

1. the marginal posterior distributions of $\boldsymbol{\theta}$ and $\boldsymbol{\phi}$ are defined as

$$p(\boldsymbol{\theta} \mid \mathbf{x}) = \int_{\Phi} p(\boldsymbol{\theta}, \boldsymbol{\phi} \mid \mathbf{x})d\boldsymbol{\phi} \text{ and } p(\boldsymbol{\phi} \mid \mathbf{x}) = \int_{\Theta} p(\boldsymbol{\theta}, \boldsymbol{\phi} \mid \mathbf{x})d\boldsymbol{\theta},$$

where $\boldsymbol{\Theta}$ and $\boldsymbol{\Phi}$ are the respective parameter spaces of $\boldsymbol{\theta}$ and ϕ.

2. the conditional posterior distributions of θ and ϕ are defined as

$$p(\boldsymbol{\theta} \mid \phi, \mathbf{x}) = \frac{p(\boldsymbol{\theta}, \phi \mid \mathbf{x})}{p(\phi \mid \mathbf{x})} \propto p(\boldsymbol{\theta}, \phi \mid \mathbf{x})$$

and

$$p(\phi \mid \boldsymbol{\theta}, \mathbf{x}) = \frac{p(\boldsymbol{\theta}, \phi \mid \mathbf{x})}{p(\boldsymbol{\theta} \mid \mathbf{x})} \propto p(\boldsymbol{\theta}, \phi \mid \mathbf{x}),$$

where these conditional distributions are well defined if the corresponding denominators are non-zero, that is, for values of $\boldsymbol{\theta}$ and ϕ with strictly positive marginal posterior density. The above calculations use the fact that the terms in the denominator do not depend on the quantity of interest and can be included in the proportionality constant.

3. the likelihood function of $(\boldsymbol{\theta}, \phi)$ is defined as $l(\boldsymbol{\theta}, \phi; \mathbf{x}) = p(\mathbf{x} \mid \boldsymbol{\theta}, \phi)$. In a similar way, one can define the marginal likelihood functions for $\boldsymbol{\theta}$ and ϕ as

$$\begin{aligned} l(\boldsymbol{\theta}; \mathbf{x}) = p(\mathbf{x} \mid \boldsymbol{\theta}) &= \int_{\boldsymbol{\Phi}} p(\mathbf{x}, \phi \mid \boldsymbol{\theta}) d\phi \\ &= \int_{\boldsymbol{\Phi}} p(\mathbf{x} \mid \phi, \boldsymbol{\theta}) p(\phi \mid \boldsymbol{\theta}) d\phi \end{aligned}$$

and similarly

$$l(\phi; \mathbf{x}) = \int_{\boldsymbol{\Theta}} p(\mathbf{x} \mid \boldsymbol{\theta}, \phi) p(\boldsymbol{\theta} \mid \phi) d\boldsymbol{\theta}.$$

Example 2.17 *Let* $\boldsymbol{X} = (X_1, \ldots, X_n)$ *be a random sample from an* $N(\theta, \sigma^2)$ *and define* $\phi = \sigma^{-2}$. *The parameter vector* (θ, ϕ) *is unknown and we assume that the main interest lies in the mean of the population. The precision* ϕ *is essentially a nuisance parameter that we would like to eliminate from the analysis.*

Suppose that the prior distribution is such that $n_0 \sigma_0^2 \phi \sim \chi_{n_0}^2$ *or, equivalently,* $\phi \sim G(n_0/2, n_0 \sigma_0^2/2)$ *and* ϕ *is independent of* θ *a priori. In these conditions,*

$$p(\phi \mid \theta) = p(\phi) \propto \phi^{\frac{n_0}{2} - 1} \exp\left\{-\frac{n_0 \sigma_0^2}{2}\phi\right\},$$

while the likelihood function for θ *and* ϕ *is given by*

$$p(\boldsymbol{x} \mid \theta, \phi) \propto \phi^{n/2} \exp\left\{-\frac{\phi}{2}\sum_{i=1}^{n}(x_i - \theta)^2\right\},$$

where $\sum_{i=1}^{n}(x_i - \theta)^2$ can be written as

$$\sum_{i=1}^{n}[(x_i - \bar{x}) + (\bar{x} - \theta)]^2 = \sum_{i=1}^{n}(x_i - \bar{x})^2 + n(\bar{x} - \theta)^2 = n[s^2 + (\bar{x} - \theta)^2],$$

where $s^2 = \sum_{i=1}^{n}(x_i - \bar{x})^2/n$.

Therefore, the marginal likelihood of θ is

$$
\begin{aligned}
l(\theta; \boldsymbol{x}) &= \int_{\Phi} p(\boldsymbol{x} \mid \phi, \theta) p(\phi \mid \theta) d\phi \\
&\propto \int_{0}^{\infty} \phi^{\frac{n_0}{2}-1} \exp\left\{-\frac{\phi}{2}n_0\sigma_0^2\right\} \phi^{\frac{n}{2}} \exp\left\{-\frac{\phi}{2}[ns^2 + n(\bar{x} - \theta)^2]\right\} d\phi \\
&= \int_{0}^{\infty} \phi^{\frac{n+n_0}{2}-1} \exp\left\{-\frac{\phi}{2}[ns^2 + n(\bar{x} - \theta)^2 + n_0\sigma_0^2]\right\} d\phi \\
&= \frac{\Gamma[(n+n_0)/2]}{\{[(ns^2 + n(\bar{x} - \theta)^2 + n_0\sigma_0^2]/2\}^{(n+n_0)/2}} \\
&= \frac{\Gamma[(n+n_0)/2]}{[(ns^2 + n_0\sigma_0^2)/2]^{(n+n_0)/2}} \left[1 + \frac{n(\bar{x} - \theta)^2}{ns^2 + n_0\sigma_0^2}\right]^{-\frac{n+n_0}{2}}.
\end{aligned}
$$

Assume that there is no prior information and let $n_0 \to 0$. This is technically not possible since it implies a degenerate prior distribution. This discussion will be deferred to the next chapter and we shall proceed with the assumption. Then,

$$
\begin{aligned}
l(\theta; \boldsymbol{x}) &\rightarrow k\left[1 + \frac{(\bar{x} - \theta)^2}{s^2}\right]^{-n/2} \\
&= k\left[1 + \frac{T^2(\boldsymbol{x}, \theta)}{n-1}\right]^{-[(n-1)+1]/2}, \quad \text{where } T(\boldsymbol{X}, \theta) = \frac{\bar{X} - \theta}{\sqrt{\frac{s^2}{n-1}}}.
\end{aligned}
$$

Interpreting the likelihood as proportional to the probability density function of T, it follows that $T \sim t_{n-1}(0, 1)$.

The sampling distribution of $\bar{X}$, the minimal sufficient statistic for θ, is used in the frequentist inference. This distribution depends on σ^2. Substituting it with the estimator S^2, the minimal sufficient statistic for ϕ, leads to $T(\boldsymbol{X}, \theta)$ with a $t_{n-1}(0, 1)$ sampling distribution as will be seen in Chapter 4. Therefore, if the adopted prior distribution for θ is proportional to a constant then the Bayesian and classical results will agree numerically at least, although theoretically they were obtained using different arguments.

Example 2.18 *Let $\boldsymbol{X} = (X_1, \ldots, X_n)$ be a random sample from an $N(\mu, \sigma^2)$ and define $\phi = \mu$. Let the precision parameter σ^2 be the parameter of interest and the mean μ the nuisance parameter. Suppose that the prior distribution* *for μ and σ^2 is given by $p(\mu, \sigma^2) = 1/\sigma$. In these conditions we have the joint posterior density given by*

$$p(\mu, \sigma^2 | \boldsymbol{x}) \quad \propto \quad p(\boldsymbol{x} | \mu, \sigma^2) \, p(\mu, \sigma^2)$$

$$\propto \quad \frac{1}{(2\pi\sigma^2)^{n/2}} \exp\left\{ -\frac{n}{2\sigma^2}(\mu - \bar{x})^2 - \frac{1}{2\sigma^2} \sum_{i=1}^{n}(x_i - \bar{x})^2 \right\} \frac{1}{\sigma}.$$

Then the marginal posterior, obtained by integrating μ out, is given by

$$p(\sigma^2 | \boldsymbol{x}) \propto (\sigma^2)^{-(n-1)/2} \exp\left\{ -\frac{1}{2\sigma^2} \sum_{i=1}^{n}(x_i - \bar{x})^2 \right\},$$

which is an Inverse Gamma distribution with parameters $\alpha = (n-2)/2$ and $\beta = \sum_{i=1}^{n}(x_i - \bar{x})^2/2$.

Suppose now that $p(\mu | \sigma^2) = 1$ and that $p(\sigma^2) = \frac{1}{\sigma}$. Then, the marginal likelihood for σ^2 is given by

$$\begin{aligned} l(\sigma^2) \quad &= \quad \int_{\infty}^{\infty} p(\boldsymbol{x} | \mu, \sigma^2) \, \pi(\mu | \sigma^2) d\mu \\ &= \quad \int_{\infty}^{\infty} \prod_{i=1}^{n} \frac{1}{\sqrt{2\pi\sigma^2}} \exp\left\{ -\frac{1}{2\sigma^2}(x_i - \mu)^2 \right\} \\ &= \quad \frac{1}{(2\pi\sigma^2)^{n/2}\sqrt{n}} \exp\left\{ -\frac{1}{2\sigma^2} \sum_{i=1}^{n}(x_i - \bar{x})^2 \right\}. \end{aligned}$$

Figure 2.5 presents the marginal posterior and the marginal likelihood for σ^2 considering a random sample generated from a Normal distribution with mean $\mu = 1$ and variance $\sigma^2 = 1$ with size $n = 20$.

In this context, the marginal posterior distribution for σ^2, obtained via the marginal likelihood, is given by

$$p(\sigma^2 | x) \propto l(\sigma^2) \, p(\sigma^2) \propto (\sigma^2)^{-(n-1)/2} \exp\left\{ -\frac{1}{2\sigma^2} \sum_{i=1}^{n}(x_i - \bar{x})^2 \right\},$$

which is also an Inverse Gamma distribution with parameters $\alpha = (n-2)/2$ and $\beta = n \sum_{i=1}^{n}(x_i - \bar{x})^2/2$.

In other words, the marginal posterior distribution of σ^2 obtained via the marginal likelihood for σ^2 coincides with the marginal posterior distribution of σ^2 obtained by marginalization of the joint posterior distribution with respect to μ, since the same prior distributions were considered.

We observe that, conceptually, there is no difficulty to define and obtain any of the above quantities although sometimes there are difficulties in analytically solving these integrals. The same however is not true for classical

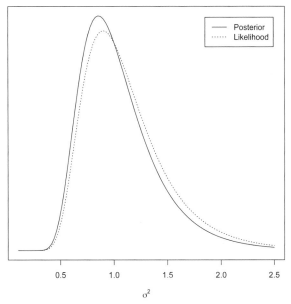

Figure 2.5 *Marginal posterior and marginal likelihood for σ^2 considering a random sample generated from a Normal distribution with mean $\mu = 1$ and variance $\sigma^2 = 1$ with size $n = 20$.*

or frequentist inference. Some special rules must be stated leading to ad-hoc procedures to solve the stated problem.

2.6.2 Classical approach

In classical inference efforts are geared towards an appropriate definition of marginal likelihood. This is also a relevant problem for the frequentist statistician and has received much research attention, beginning with the work of Bartlett in the 30's and further developed by Kalbfleisch and Sprott in the 70's, as discussed in Cox and Hinkley (1974), Berger, Liseo and Wolpert (1999) and Brazzale, Davison and Reid (2007). Many proposals to express the marginal likelihood $l(\theta; \mathbf{x})$ are based on substituting ϕ in the (total) likelihood with some particular value ϕ_0. Often, ϕ_0 is taken as the value that maximizes the likelihood, possibly leading to the more straightforward procedure proposal, the profile likelihood, which is defined as follows.

Definition 2.14 *Let $\hat{\phi}(\theta)$ be the value of ϕ that maximizes $l(\theta, \phi; \mathbf{x})$ for any* *given* $\boldsymbol{\theta}$. *The relative or profile likelihood for* $\boldsymbol{\theta}$ *is defined as*

$$l_P(\boldsymbol{\theta}; \boldsymbol{x}) = \max_{\phi} l(\boldsymbol{\theta}, \phi; \boldsymbol{x}) = l(\boldsymbol{\theta}, \hat{\phi}(\boldsymbol{\theta}); \boldsymbol{x}).$$

Example 2.19 *(Example 2.18 continued) The likelihood for* (ϕ, σ^2) *is given by*

$$l(\phi, \sigma^2; \boldsymbol{x}) = (2\pi\sigma^2)^{-n/2} \exp\left\{ -\frac{1}{2\sigma^2} \sum_{i=1}^{n} (x_i - \phi)^2 \right\},$$

where σ^2 *is the parameter of interest and* $\phi = \mu$ *the nuisance one. Let* $\hat{\phi}(\sigma^2)$ *be the value that maximizes* $l(\phi, \sigma^2; \boldsymbol{x})$ *with respect to* ϕ *for a given* σ^2. *Then the profile likelihood for* σ^2 *is given by*

$$l_P(\sigma^2; \boldsymbol{x}) = (2\pi\sigma^2)^{-n/2} \exp\left\{ -\frac{1}{2\sigma^2} \sum_{i=1}^{n} (x_i - \hat{\phi}(\sigma^2))^2 \right\}.$$

An alternative procedure to eliminate the nuisance parameter is the use of integration, leading to the definition of the integrated likelihood outlined below.

Definition 2.15 *Let* ϕ *the nuisance parameter and* $q(\phi)$ *a non-negative weight function be defined in* ϕ *such that* $\int_{\phi} |q(\phi)| d\phi < \infty$. *Then, the integrated likelihood for* θ *is defined as*

$$\tilde{l}_q(\theta) \propto \int_{\Phi} f(\boldsymbol{x}|\theta, \phi) q(\phi) d\phi = f_q(\boldsymbol{x}|\theta).$$

A special case of the integrated likelihood is the uniform integrated likelihood, which is obtained by assuming that $q(\phi) = 1$, giving the same weight for all ϕ. Such procedure is equivalent to assuming that $q(\phi)$ is the density of a Uniform distribution in the interval $(0, \infty)$; that is, $q(\phi) = I_{\phi}(R^+)$. From the Bayesian point of view, the function $q(\phi)$ can be identified as the prior distribution of $\phi|\theta$, given by $p(\phi|\theta)$.

Example 2.20 *(Example 2.18 continued) Considering* $q(\mu) = 1$, *the uniform integrated likelihood for* σ^2 *is equal to the marginal likelihood for* σ^2 *and is given by*

$$\tilde{l}_q(\sigma^2) = \frac{1}{2\pi^{(n-1)/2}(\sigma^2)^{n/2}} \exp\left\{ -\frac{1}{2\sigma^2} \sum_{i=1}^{n} (x_i - \bar{x})^2 \right\}.$$

Figure 2.6 presents the uniform integrated, profile and marginal likelihood for σ^2 considering a random sample generated from a Normal distribution. Note that they all coincide, apart from multiplicative constants.

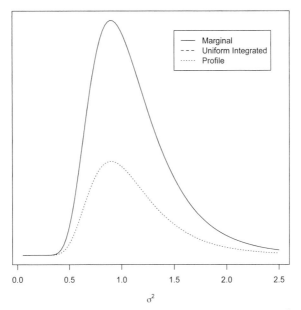

Figure 2.6 *Uniform integrated, profile and marginal likelihoods for σ^2 considering a random sample generated from a Normal distribution with mean $\mu = 0$ and variance $\sigma^2 = 1$ with size $n = 20$, assuming $q(\mu) = 1$. The Uniform integrated and marginal likelihoods are overlapping.*

It is possible to find forms and variations of the likelihood function other than those defined above. A special likelihood function is obtained when it is possible to factor it as

$$l(\theta, \phi) = l_1(\theta; \mathbf{x}) \, l_2(\phi; \mathbf{x}).$$

In this case, only $l_1(\theta; \mathbf{x})$ is used for inference on the parameter of interest θ. Unfortunately, this decomposition is rarely encountered in practice.

Also, in some situations, the sufficient statistics have conditional distributions that depend on θ, but not ϕ. And the original likelihood can be rewritten as a product of two functions, say, l_1 and l_2, one depending only on θ and the other depending on θ and ϕ, leading to the conditional likelihood defined as

$$l(\theta, \phi; \mathbf{x}) = l_1(\theta; \mathbf{x}) l_2(\theta, \phi; \mathbf{x}).$$

In this case, $l_1(\theta|\mathbf{x})$ may be used for inference about the parameter of interest θ. Note that there is loss of information about θ when $l_2(\theta, \phi|\mathbf{x})$ is not incorporated into the estimation process.

There are other frequentist definitions for the marginal and conditional likelihood functions. Suppose that the vector $\mathbf{X}$, or some one to one transformation of it, is partitioned into $(\mathbf{T}, \mathbf{U})$, with joint density given by

$$p(\mathbf{t}, \mathbf{u} \mid \boldsymbol{\theta}, \boldsymbol{\phi}) = p(\mathbf{t} \mid \boldsymbol{\theta}, \boldsymbol{\phi})\, p(\mathbf{u} \mid \mathbf{t}, \boldsymbol{\theta}, \boldsymbol{\phi}).$$

The likelihood function of $\boldsymbol{\theta}, \boldsymbol{\phi}$ is given by the left hand side of the equation above and the right hand side terms can also be written in likelihood terms as

$$l(\boldsymbol{\theta}, \boldsymbol{\phi}; \mathbf{t}, \mathbf{u}) = l(\boldsymbol{\theta}, \boldsymbol{\phi}; \mathbf{t})\, l(\boldsymbol{\theta}, \boldsymbol{\phi}; \mathbf{u} \mid \mathbf{t}).$$

The first term on the right hand side is called the marginal likelihood and the second, the conditional likelihood. These forms of likelihood are useful when some of the parametric components can be eliminated. For example, if $\mathbf{T}$ were such that $l(\boldsymbol{\theta}, \boldsymbol{\phi}; \mathbf{t}) = l(\boldsymbol{\theta}; \mathbf{t})$ only this term could be used to make inference about $\boldsymbol{\theta}$. For conditional likelihood, this form is related with sufficient statistics because if $\mathbf{T}$ is sufficient for $\boldsymbol{\phi}$, with $\boldsymbol{\theta}$ fixed, then $l(\boldsymbol{\theta}, \boldsymbol{\phi}; \mathbf{u} \mid \mathbf{t}) = l(\boldsymbol{\theta}; \mathbf{u} \mid \mathbf{t})$. Again only this term is used to make inference about $\boldsymbol{\theta}$. The question is how much information is lost when ignoring the other part of the likelihood.

Some authors suggest corrections in the profile likelihood expression taking into consideration measures of information associated with ϕ (Barndorff-Nielsen, 1983; Cox and Reid, 1987; Severini, 1998). For instance, a local measure of the information, the observed Fisher information, may be considered. In this context, the adjusted profile likelihood is given by

$$l_{AP}(\theta; \mathbf{x}) = l_P(\theta; \mathbf{x}) |J_{\phi\phi}(\theta, \hat{\phi}(\theta))|^{1/2},$$

where $J_{\phi\phi}(\theta, \hat{\phi}(\theta))$ is the (θ, θ) element of the the observed Fisher information $J(\theta, \phi)$ evaluated for fixed θ at the maximum likelihood estimate of ϕ for fixed θ given by $\hat{\phi}(\theta)$. The adjusted profile likelihood $l_{AP}(\theta; \mathbf{x})$ was proposed under the assumption of orthogonality of θ and ϕ, which is not always possible to be attained.

Then, a second correction in the profile likelihood is given by the modified profile likelihood defined as

$$l_{MP}(\theta; \mathbf{x}) = l_P(\theta; \mathbf{x}) |J_{\phi\phi}(\theta, \hat{\phi}(\theta))|^{1/2} \left| \frac{\partial \hat{\phi}(\theta)}{\partial \hat{\phi}} \right|^{-1}.$$

Several approximations to $l_{MP}(\theta; \mathbf{x})$ were proposed in order to simplify its evaluation, since it is difficulty to obtain $|\partial\hat{\phi}(\theta)/\partial\hat{\phi}|$. According to the approximation proposed by Severini (1998), we can rewrite $l_{MP}(\theta; \mathbf{x})$ as

$$l_{MP}(\theta; \mathbf{x}) = l_P(\theta; \mathbf{x}) |J_{\phi\phi}(\theta, \hat{\phi}(\theta))|^{1/2} |\hat{J}(\theta, \hat{\phi}(\theta); \hat{\theta}, \hat{\phi})|^{-1},$$

where

$$\hat{J}(\theta, \hat{\phi}(\theta); \hat{\theta}, \hat{\phi}) = \sum_{i=1}^{n} \ell_\phi^{(i)}(\theta, \hat{\phi}(\theta)) \; \ell_\phi^{(i)}(\hat{\theta}, \hat{\phi})',$$

and $\ell_\phi^{(i)}(.)$ is the first derivative of the likelihood function for the i-th observation in the sample. If $\hat{\phi}(\theta) = \hat{\phi}$ for all θ, then $l_{AP}(\theta; \mathbf{x}) = l_{MP}(\theta; \mathbf{x})$.

Example 2.21 *Let* $\mathbf{X} = (X_1, \ldots, X_n)$ *be a random sample from an* $N(\mu, \sigma^2)$ *distribution. Assume that the precision parameter* $\phi = 1/\sigma^2$ *is a nuisance parameter that we would like to eliminate from the analysis. The adjusted profile likelihood for* μ *is given by*

$$l_{AP}(\mu) = \left(\frac{\hat{\phi}(\mu)}{2\pi} \right)^{n/2} \exp\left\{ -\frac{\hat{\phi}(\mu)}{2} \sum_{i=1}^{n} (x_i - \mu)^2 \right\} \left| J_{\phi\phi}(\mu, \hat{\phi}(\mu)) \right|^{-1/2},$$

where $J_{\phi\phi}(\mu, \hat{\phi}(\mu)) = -\dfrac{n\hat{\phi}^2}{2} + \hat{\phi}^3 \sum_{i=1}^{n}(x_i - \mu)^2$ *and* $\hat{\phi}(\mu) = \dfrac{n}{\sum_{i=1}^{n}(x_i - \mu)^2}.$
   Then,

$$l_{MP}(\mu) = \left(\frac{\hat{\phi}(\mu)}{2\pi} \right)^{n/2} \exp\left\{ -\frac{\hat{\phi}(\mu)}{2} \sum_{i=1}^{n} (x_i - \mu)^2 \right\} \frac{\left| J_{\phi\phi}(\mu, \hat{\phi}(\mu)) \right|^{1/2}}{\left| I(\mu, \hat{\phi}(\mu); \hat{\mu}, \hat{\phi}) \right|},$$

where $I(\mu, \hat{\phi}(\mu); \hat{\mu}, \hat{\phi}) = \sum_{i=1}^{n} \ell_\phi^{(i)}(\mu, \hat{\phi}(\mu)) \ell_\phi^{(i)}(\hat{\mu}, \hat{\phi})'$ *and* $\ell_\phi^{(i)}(\mu, \phi) = -\dfrac{\phi}{2} + \phi^2 (x_i - \mu)^2/2.$

   Figure 2.7 presents the profile, the adjusted profile and the modified profile likelihoods for μ considering a random sample generated from a Normal distribution.

Exercises

§2.1

1. For each of the following distributions verify if the model is location, scale or location-scale.

   (a) $t_\alpha(\mu, \sigma^2)$, α known;
   (b) Pareto (x_0, α), with α fixed, with density $p(x|x_0) = \alpha x_0^\alpha / x^{1+\alpha}$, $x > x_0$, $(\alpha, x_0 > 0)$;
   (c) Uniform distribution on $(\theta - 1, \theta + 1)$;
   (d) Uniform distribution on $(-\theta, \theta)$.

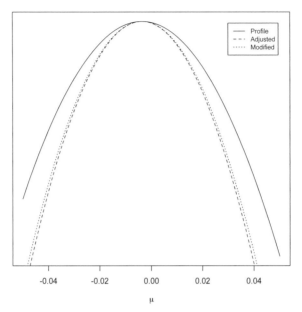

Figure 2.7 *Profile, adjusted profile and marginal profile likelihoods for* σ^2 *consider-ing a random sample generated from a Normal distribution with mean* $\mu = 0$ *and variance* $\sigma^2 = 0.01$ *with size* $n = 10$.

§2.2

2. Let X have sampling density $p(x|\theta)$ with $\theta \in \Theta \subset R$. Prove or give a counterexample to the assertion that if the sampling distribution $p(x|\theta)$ has a unique maximum then the likelihood $l(\theta; x)$ also has a unique maximum. Generalize the result to the case of vector of observations $\mathbf{X}$ and also for a parameter vector $\boldsymbol{\theta}$.

3. A situation that usually occurs in lifetime or survival analysis is having observations that are censored because of time and/or cost restrictions on the experiment. One common situation occurs when the experiment is run only until time $T > 0$. If the observational unit is observed until time T, it is uncensored but if it is censored then all one can extract from the experiment is that the lifetime of this unit is larger than T. Assuming that a random sample $X_1, \ldots, X_n$ from a density $f(x|\theta)$ is observed, show that

(a) the likelihood is

$$l(\theta) = \prod_{i=1}^{n} [f(x_i|\theta)]^{1-\chi_i} [1 - F(T|\theta)]^{\chi_i},$$

where F is the distribution function of the observations and χ_i is the censoring indicator, $i = 1, \ldots, n$, taking values 0, when failure is observed, and 1, when censoring takes place.

(b) in the case that $f(x|\theta) = \theta e^{-\theta x}$, $l(\theta) = \theta^m e^{-\theta U}$ where $m \leq n$ is the number of uncensored observations and $U = (n - m)T + \sum_i (1 - \chi_i)x_i$ is the total time on test.

4. Let $X_1, \ldots, X_n$ be iid random quantities from the Weibull distribution, denoted by $Wei(\alpha, \beta)$ $(\alpha, \beta > 0)$, with

$$p(x|\alpha, \beta) = \beta\alpha x^{\alpha-1} \exp(-\beta x^\alpha), \text{ for } x > 0.$$

(a) Obtain the likelihood function, the score function and the observed and expected Fisher information matrix for the pair of parameters (α, β).

(b) The Weibull distribution is sometimes parameterized in terms of α and $\theta = 1/\beta^\alpha$. Repeat item (a) for the pair of parameters (α, θ).

§2.3

5. Return to the example about John's illness and consider the same distributions used in the example.

(a) Which test result makes the doctor more certain about John's illness? Why?

(b) The test X is applied and provides the result $X = 1$. Suppose the doctor is not satisfied with the available evidence and decides to ask for another replication of test X and again the result is $X = 1$. What is now the probability that John has disease A?

(c) What is the minimum number of repetitions of the test X which allow the doctor to assure that John has the disease with 99.9% probability. What are the results of these replications that guarantee this?

6. Suppose that $X \mid \theta \sim N(\theta, 1)$ (for example, X is a measurement of a physical constant θ made with an instrument with variance 1). The prior distribution for θ elicited by physician A corresponds to a $N(5, 1)$ distribution and physician B elicits a $N(15, 1)$ distribution. The value $X = 6$ was observed.

(a) Which prior distribution fits the data better?

(b) What kind of comparison can be made between the two physicians?

7. Classify the following assertions as TRUE or FALSE, briefly justifying your answer.

   (a) The posterior distribution is always more precise than the prior because it is based on more information.

   (b) When X_2 is observed after X_1, the prior distribution before observing X_2 has to be necessarily the posterior distribution after observing X_1.

   (c) The predictive density is the prior expected value of the sampling distribution.

   (d) The smaller is the prior information, the bigger is the influence of the sample in the posterior distribution.

8. A test to verify if a driver is driving drunk has 0.8 chance of being correct, that is, to provide a positive result when in fact the driver has a high level of alcohol in his/her blood or negative result when it is below the acceptable limit. A second test is only applied to the suspected cases. This never fails if the driver is not drunk, but has only a 10% chance of error with drunk drivers. If 25% of all the drivers stopped by the police are *above the limit*, calculate the following:

   (a) the proportion of drivers stopped that have to be submitted to a second test;

   (b) the posterior probability that a driver really has the high level of alcohol a in his blood informed by the two tests;

   (c) the proportion of drivers who will be submitted only to the first test.

9. (DeGroot, 1970, pg. 152) The random variables $X_1, \ldots, X_k$ are such that $k - 1$ have probability function h and one has probability function g. X_j will have the probability function g with probability α_j, $j = 1, \ldots, k$, where $\alpha_j > 0$, $\forall j$ and $\sum_{j=1}^{k} \alpha_j = 1$. What is the probability that X_1 has probability function g given that:

   (a) $X_1 = x$ was observed?

   (b) $X_i = x, i \neq 1$ was observed?

10. Let $X \sim U[\theta - 1, \theta + 1]$ and assume that the prior is $p(\theta) \propto \theta^{-1}$, $\theta > 0$.

   (a) Show that $p(\theta|x) = c\theta^{-1} I_\theta(x-1, x+1)$, where $c^{-1} = \log[(x+1)/(x-1)]$, $x > 1$.

   (b) Obtain the mean, mode and median of the posterior distribution.

11. Let $X \mid \theta, \mu \sim N(0, \sigma^2), \sigma^2$ known and $\theta \mid \mu \sim N(\mu, \tau^2), \tau^2$ known and $\mu \sim N(0, 1)$. Obtain the following distributions:

(a) $(\theta \mid x, \mu)$;

(b) $(\mu \mid x)$;

(c) $(\theta \mid x)$.

12. Let $(X|\theta) \sim N(\theta, 1)$ be observed. Suppose that your prior is such that θ is $N(\mu, 1)$ or $N(-\mu, 1)$ with equal probabilities. Write the prior distribution and find the posterior after observing $X = x$. Show that

$$\mu' = E(\theta \mid x) = \frac{x}{2} + \frac{\mu}{2} \frac{1 - \exp(-\mu x)}{1 + \exp(-\mu x)}$$

and draw a graph of μ' as a function of x.

13. The standard Cauchy density function is given by $p(x|\theta) = (1/\pi)\{1/[1 + (x - \theta)^2]\}$. It is similar to the $N(\theta, 1)$ density and may be used in its place in many applications.

(a) Obtain the modal equation (the first order condition to the maximum) of the posterior supposing that the prior is $p(\theta) = 1/\pi(1 + \theta^2)$.

(b) Solve it for $x = 0$ and $x = 3$.

(c) Compare with the results obtained assuming that $(x|\theta) \sim N(\theta, 1)$ and $\theta \sim N(0, 1)$.

14. Assume that an observation vector $\mathbf{X}$ has multivariate Normal distribution, introduced in Chapter 1, with mean vector $\boldsymbol{\mu}$ and variance-covariance matrix $\boldsymbol{\Sigma}$. Assuming that $\boldsymbol{\Sigma}$ is known and the prior distribution is $\boldsymbol{\mu} \sim N(\boldsymbol{\mu}_0, \mathbf{B}_0)$, obtain the posterior distribution for $\boldsymbol{\mu}$.

§2.4

15. Let $\mathbf{X} = (X_1, \ldots, X_n)$ be an exchangeable sample of $0 - 1$ observations. Show that

(a) $E(X_i) = E(X_j), \forall i, j = 1, \ldots, n$;

(b) $V(X_i) = V(X_j), \forall i, j = 1, \ldots, n$;

(c) $Cov(X_i, X_j) = Cov(X_k, X_l), \forall i, j, k, l = 1, \ldots, n$.

16. Let $\mathbf{X} = (X_1, \ldots, X_n)$ be an exchangeable sample of $0 - 1$ observations and $T = \sum_{i=1}^{n} X_i$. Show that

(a) $P(T = t) = \int_0^1 \binom{n}{t}\theta^t(1 - \theta)^{n-t} p(\theta) \, d\theta, \quad t = 1, \ldots, n$;

(b) $E(T) = n E(\theta)$.

Hint: use the definition of $E(T)$ and exchange the summation and integration signs.

17. Let $\theta_1, \ldots, \theta_k$ be the probability that patients $I_1, \ldots, I_k$ have disease B. After summarizing all the available information the doctor concludes that

   (a) the patients can be divided in two groups, the 1st containing the patients $I_1, \ldots, I_j$, $j < k$ and the 2nd with patients $I_{j+1}, \ldots, I_k$.

   (b) the patients in the same group are similar, that is to say they are indistinguishable with respect to B.

   (c) there is no relevant information relating these two groups.

   Use the idea of partial exchangeability to construct a prior distribution for $\boldsymbol{\theta} = (\theta_1, \ldots, \theta_k)$. If instead of (c), there was information relating the 2 groups, what modifications would this information imply in the prior for $\boldsymbol{\theta}$?

§2.5

18. Let $\boldsymbol{X} = (X_1, \ldots, X_n)$ be a random sample from the $U(\theta_1, \theta_2)$ distribution, with density

$$p(x \mid \theta_1, \theta_2) = \frac{1}{\theta_2 - \theta_1}, \quad \theta_1 \leq x \leq \theta_2.$$

   Let $\boldsymbol{T}(\boldsymbol{X}) = (X_{(1)}, X_{(n)})$, obtain its joint distribution and show that it is a sufficient statistic for $\boldsymbol{\theta} = (\theta_1, \theta_2)$.

19. Let $\boldsymbol{X}$ be a random sample from $P(X \mid \boldsymbol{\theta})$. Show that if $\boldsymbol{T} = \boldsymbol{T}(\boldsymbol{X})$ is a sufficient statistic for $\boldsymbol{\theta}$ and $\boldsymbol{S}(\boldsymbol{X})$ is 1-to-1 function of $\boldsymbol{T}$, then $\boldsymbol{S}(\boldsymbol{X})$ is also a sufficient statistic for $\boldsymbol{\theta}$.

20. Let $X_1, \ldots, X_n$ be a random sample from $P(X \mid \theta_1, \theta_2)$. Show that if T_1 is sufficient for θ_1 when θ_2 is known and T_2 is sufficient for θ_2 when θ_1 is known, then $T = (T_1, T_2)$ is sufficient for $\theta = (\theta_1, \theta_2)$.

21. Verify if the following distributions belong to the exponential family. If so, determine the functions a, b, u and ϕ.

   (a) $Bin(n, \theta)$, n known;

   (b) $Exp(\theta)$;

   (c) $G(\alpha, \beta)$;

   (d) $Beta(\alpha, \beta)$;

   (e) $N(\boldsymbol{\mu}, \boldsymbol{\Sigma})$, $\boldsymbol{\Sigma}$ known.

22. Which of the following distributions families are members of the exponential family? Also, obtain the minimal sufficient statistic for those belonging to the exponential family.

   (a) $p(x \mid \theta) = 1/9$, $x \in \{0.1 + \theta, \ldots, 0.9 + \theta\}$;

(b) the family of $N(\theta, \theta^2)$ distributions;

(c) the family of $N(\theta, \theta)$ distributions, with $\theta > 0$;

(d) $p(x \mid \theta) = 2(x + \theta)/(1 + 2\theta)$, $x \in (0, 1)$, $\theta > 0$;

(e) the distribution family of $X \mid X \neq 0$ where $X \sim Bin(n, \theta)$;

(f) $f(x \mid \theta) = \theta/(1 + x)^{1+\theta}$, $x \in R^+$;

(g) $f(x \mid \theta) = \theta^x \log \theta/(\theta - 1)$, $x \in (0, 1)$;

(h) $f(x \mid \theta) = (1/2) \exp(-|x - \theta|)$, $x \in R$.

23. Let $X_1, \ldots, X_n$ be a random sample from the $N(\mu, \sigma^2)$, with σ^2 unknown. Show, using the classical definition, that the sample mean $\overline{X}$ is a sufficient statistic for μ.

   Hint: It is enough to show that $E(X \mid \overline{X})$ and $V(X \mid \overline{X})$ are not functions of μ. Why?

24. Let (X_1, X_2, X_3) be a vector with probability function

$$\frac{n!}{\prod_{i=1}^{3} x_i!} \prod_{i=1}^{3} p_i^{x_i} \quad , x_i \geq 0, \ x_1 + x_2 + x_3 = n$$

   where $p_1 = \theta^2$, $p_2 = 2\theta(1 - \theta)$, $p_3 = (1 - \theta)^2$ and $0 \leq \theta \leq 1$.

   (a) Verify if this distribution belongs to the exponential family with k parameters. If this is true, what is the value of k?

   (b) Obtain the minimal sufficient statistic for θ.

25. Using the same notation adopted for the one parameter exponential family,

   (a) Show that

$$E[U(X)] = -\frac{b'(\theta)}{\phi'(\theta)} \text{ and } V[U(X)] = \frac{b'(\theta)\phi''(\theta) - \phi'(\theta)b''(\theta)}{[\phi'(\theta)]^3}.$$

   Hint: From the relationship $\int p(x \mid \theta)dx = 1$, differentiate both sides with respect to θ.

   (b) Verify that the result in (a) is correct for the case where $X \sim Exp(\theta)$ by direct evaluation of $E[U(X)]$ and $V[U(X)]$.

26. Show that information equivalence defines an equivalence relation of the elements of the sample space.

27. Let X_1, X_2, X_3 be iid Bernoulli variables with success probability θ and define $T = T(\mathbf{X}) = \sum_{i=1}^{n} X_i$, $T_1 = X_1$ and $\mathbf{T}_2 = (T, T_1)$. Note that the sample space is $S = \{0, 1\}^3$.

   (a) Obtain the partitions induced by T, T_1 and $\mathbf{T}_2$.

(b) Show that $\mathbf{T}_2$ is a sufficient statistic.

(c) Prove that T is a minimal sufficient statistic for θ but $\mathbf{T}_2$ isn't by showing that T induces a minimal sufficient partition on $\mathcal{S}$ but $\mathbf{T}_2$ does not.

28. Consider a sample $\mathbf{X} = (X_1, \ldots, X_n)$ from a common density $p(x|\boldsymbol{\theta})$ and let $\mathbf{T}$ be the vector of order statistics from the sample.

(a) Prove that the sample $\mathbf{X}$ is always sufficient for $\boldsymbol{\theta}$.

(b) Obtain the factor $g(\mathbf{x})$ in the factorization criterion for the sufficient statistic $\mathbf{T}$.

29. Consider a sample $\mathbf{X} = (X_1, \ldots, X_n)$ from a uniform distribution on the interval $[\theta_1, \theta_2]$, $\theta_1 < \theta_2$, so that $\boldsymbol{\theta} = (\theta_1, \theta_2)$.

(a) Show that this distribution does not belong to the exponential family.

(b) Obtain a sufficient statistic of fixed size.

(c) Specialize the results above for the cases that θ_1 is known and θ_2 is known.

30. Consider a sample $\mathbf{X} = (X_1, \ldots, X_n)$ from a common density in the form

$$p(x|\mu, \sigma^2, \nu) = \frac{\Gamma[(\nu + 1)/2]}{\Gamma(\nu/2)\sqrt{\pi\nu\sigma^2}} \left[1 + \frac{(x - \mu)^2}{\nu\sigma^2}\right]^{-(\nu+1)/2}.$$

In this case, X is said to follow a Student-t distribution with ν degrees of freedom, location parameter μ and scale parameter σ^2, denoted $t_\nu(\mu, \sigma^2)$, and $\boldsymbol{\theta} = (\nu, \mu, \sigma^2)$.

(a) Show that this distribution does not belong to the exponential family.

(b) Show that it is not possible to obtain a sufficient statistic for $\boldsymbol{\theta}$ of fixed size.

(c) Show that the results above are retained even for the cases when some of the components of $\boldsymbol{\theta}$ are known.

§2.6

31. Let X and Y be independent random variables Poisson distributed with parameters θ and ϕ, respectively, and suppose that the prior distribution is $p(\theta, \phi) = p(\theta)p(\phi) \propto k$. Let $\psi = \theta/(\theta + \phi)$ and $\xi = \theta + \phi$ be a parametric transformation of (θ, ϕ).

(a) Obtain the prior for (ψ, ξ).

(b) Show that $\psi \mid x, y \sim Beta(x + 1, y + 1)$ and $\xi \mid x, y \sim G(x + y + 2, 1)$ are independent.

(c) Show that the conditional distribution X given $X + Y$ depends only on ψ, that is, $p(x \mid x + y, \psi, \xi) = p(x \mid x + y, \psi)$, and that the distribution of $X + Y$ depends only on ξ.

(d) Show that $X + Y$ is a sufficient statistic for ξ, X is a sufficient statistic for ψ, given $X + Y$, and that $(X, X + Y)$ is a sufficient statistic for (ψ, ξ).

(e) Obtain the marginal likelihoods of ψ and ξ.

(f) To make inference for ξ a statistician decides to use the fact presented in item (d). Show that the posterior is identical to that obtained in (b). Does it mean that $X + Y$ does not contain information about ψ?

32. Suppose that X has density $f(x \mid \boldsymbol{\theta})$ where $\boldsymbol{\theta} = (\theta_1, \theta_2, \theta_3)$ and the prior for θ is built up as $p(\boldsymbol{\theta}) = g(\theta_1, \theta_2 \mid \theta_3) h(\theta_3)$ where g and h are densities. Obtain the marginal likelihood $f(x \mid \theta_2, \theta_3)$ as function of f, g and h.

33. Let $\boldsymbol{X} = (\boldsymbol{X}_1, \boldsymbol{X}_2)$ where $\boldsymbol{X}_1 = (X_{11}, \ldots, X_{1m})$ and $\boldsymbol{X}_2 = (X_{21}, \ldots, X_{2n})$ are samples from the $Exp(\theta_1)$ and $Exp(\theta_2)$ distributions, respectively. Suppose that the independent $G(a_i, b_i)$ priors are assumed, $i = 1, 2$ and define $\psi = \theta_1/\theta_2$.

(a) Obtain the distribution of (θ_1, θ_2) given $\boldsymbol{X} = \boldsymbol{x}$.

(b) Repeat item (a), assuming now that $a_i, b_i \to 0$, $i = 1, 2$.

(c) Using the posterior obtained in item (b), show that

$$\frac{\bar{x}_1}{\bar{x}_2} \psi \mid \boldsymbol{X} \sim F(2m, 2n), \text{ where } \bar{x}_1 = \frac{\sum x_{1j}}{m} \text{ and } \bar{x}_2 = \frac{\sum x_{2j}}{n}.$$

Hint: Complete the transformation with $\psi_1 = \theta_2$.

34. Let (X_1, X_2, X_3) be a random vector with trinomial distribution with parameter $\boldsymbol{\theta} = (\theta_1, \theta_2, \theta_3)$ where $\theta_3 = 1 - \theta_1 - \theta_2$ and assume that the prior for $\boldsymbol{\theta}$ is constant.

(a) Define $\lambda = \theta_1/(\theta_1 + \theta_2)$ and $\psi = \theta_1 + \theta_2$ and obtain their priors.

(b) Obtain the marginal likelihood of ψ.

(c) Show that $X_1 + X_2$ is a sufficient statistic for ψ.

35. Consider a sample $\boldsymbol{X} = (X_1, \ldots, X_n)$ from a common density

$$p(x \mid \theta, \phi) = \frac{\theta^\phi}{\Gamma(\phi)} x^{\phi-1} e^{-\theta x}, \text{ for } x > 0.$$

(a) Let θ be the interest parameter and ϕ the nuisance parameter. Determine the profile and uniform integrated likelihood functions for θ.

(b) Determine the adjusted profile likelihood function for θ.

(c) Let $\theta|\phi$ be with prior distribution with density given by

$$d(\theta \mid \phi_0, \theta_0) = \frac{\theta_0^{\phi_0}}{\Gamma(\theta_0)} \theta^{\theta_0-1} e^{-\phi_0\theta}, \text{ for } \theta > 0.$$

with known ϕ_0 and θ_0. Find the marginal likelihood function and the marginal posterior density of θ.

(d) Generate a random sample with size $n = 10$ from $p(x; \theta, \phi)$ and construct the plot of the profile, adjusted and modified likelihoods.

36. A machine emits particles following a Poisson process with mean intensity of λ particles per time unit. Each particle generates a $N(0,1)$ signal. A signal detector registers the particles. Unfortunately, the detector registers only the positive signals (making it impossible to observe the number n of emitted particles).

(a) Obtain the distribution of $k|n$ where k is the number of particles registered.

(b) Show that the likelihood $L(\theta, \lambda)$ based on the observation of just one registered signal $(k = 1)$ assuming the value $x_1 > 0$ during an unit interval is given by

$$\phi(x_1 - \theta)\lambda\Phi(\theta) \exp\{-\lambda\Phi(\theta)\},$$

where ϕ and Φ are the density and the cumulative distribution function of the standard Normal distribution.

Hint: Obtain the joint distribution of (x_1, k, n) and eliminate n by integration.

(c) Obtain the profile likelihood of θ, that is, $l(\theta, \hat{\lambda}(\theta))$ where $\hat{\lambda}(\theta)$ maximizes the likelihood of λ assuming θ known.

(d) Assuming that the prior is $p(\theta, \lambda) \propto k$, obtain the marginal likelihood of θ.

37. Derive the expressions for the adjusted and the modified profile likelihood functions for μ obtained from a random sample of size n drawn from a $N(\mu, \sigma^2)$ population.

Chapter 3

Prior distribution

In this chapter, different specification forms of the prior distribution will be discussed. Apart from the interpretation of probability, this is the only novelty introduced by the Bayesian analysis, with respect to the frequentist approach. It can be seen as an element implied from exchangeability by de Finetti's representation theorem. It is determined in a subjective way, although it is not forbidden to use past experimental data to set it. The only requirement is that this distribution should represent the knowledge about θ before observing the results of the new experiment. In this chapter, alternative forms of assessing the prior distribution will be discussed. In Section 3.1 entirely subjective methods for direct assessment of the prior will be presented. An indirect approach, via functional forms, is discussed in Section 3.2. The parameters of those functional forms, denominated *hyperparameters*, must be specified in correspondence with the subjective information available. The conjugate distribution will be introduced and the most common families will be presented in Section 3.3. The concept of reference prior and different forms of building up non-informative priors will be presented in Section 3.4. Finally, hierarchical prior specification will be discussed in Section 3.5.

3.1 Entirely subjective specification

Let θ be an unknown quantity and consider its possible values. If it is discrete, a prior probability for each possible value of θ can be evaluated directly. Also one may use some auxiliary tools, such as lotteries or roulettes, as described in Chapter 1. de Finetti (1974) characterizes subjective probability through the consideration of betting and scoring rules.

The continuous case is slightly more complicated. Some suggestions follow.

1. *The histogram approach*: First, the range of values of θ is divided into intervals, and prior probabilities for θ belonging to each interval are

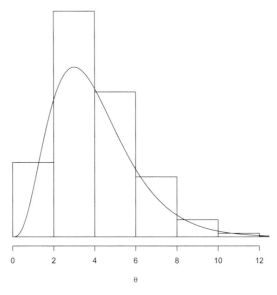

Figure 3.1 *Histogram representing (subjective) probabilities of the intervals I_1, I_2, I_3, I_4, I_5 and I_6 with a fitted density.*

specified, as in the discrete case. Hence, a histogram for θ is built up and a smooth curve can be fitted to obtain the prior density of θ. Note that the number of intervals involved is arbitrarily chosen. Although the probability in the tails of the prior distribution is often very small, they can influence the subsequent inference. This is a relevant aspect in prior elicitation that deserves some caution. Figure 3.1 shows an elicitation exercise for a positive quantity θ.

2. *The distribution function approach*: First, let us define percentiles. q_α is the $100\alpha\%$ percentile (α quantile) of X if $P(X \leq q_\alpha) = \alpha$, $\alpha \in [0, 1]$. The median of X, denoted by m, is the 50% percentile (0.5 quantile); that is, $P(X \leq m) = 0.5$. The collection of all percentiles of X describes the distribution function of X. In this approach, some percentiles are subjectively assessed, as in the discrete case, and later, a smooth curve is fitted to the distribution function of θ. Figure 3.2 is a graphical representation of this process. This approach is used less

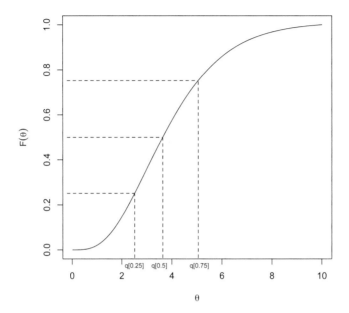

Figure 3.2: *Distribution function fitted to the quartiles $q_{0.25}$, $q_{0.5}$ and $q_{0.75}$.*

than the previous one because it is easier to identify a distribution through its density than through its distribution function.

3. *Relative likelihood approach*: The procedure is similar to the histogram approach but instead of intervals it evaluates the relative chances of isolated points. Even though $Pr(\theta = \theta_0) = 0$, $\forall \theta_0 \in \Theta$, probabilities can be calculated and compared because

$$Pr(\theta = \theta_0 \mid \theta = \theta_0 \text{ or } \theta = \theta_1) = \frac{p(\theta_0)}{p(\theta_0) + p(\theta_1)},$$

where $p(\theta)$ is the prior density of θ. Then a set of values proportional to the prior density of θ can be evaluated. For example, if $\theta = 2$ is three times more probable than $\theta = 1$ and $\theta = 3$ is twice more likely than $\theta = 1$, then we have $p(2) = 3p(1) = 1.5p(3)$. Again a smooth curve can be fitted to these points. One problem still outstanding is the evaluation of the normalization constant. Note that every density must integrate to 1 and, by construction, this curve does not necessarily satisfy this requirement.

These concepts are well described by Berger (1985).

3.1.1 Aspects on elicitation of expert beliefs

This subject has experienced great development during the last decades. Some introductory aspects of the elicitation of expert knowledge is presented in this section and a more in-depth review of the area can be found in Garthwaite, Kadane and O'Hagan (2005) and references therein.

Elicitation is the process of evaluation of the knowledge or beliefs of someone about an uncertain quantity. This task involves two main actors: the *facilitator* and the *expert*. The former helps the expert to describe his knowledge through a probability function. The expert is simply someone whose opinion we want to obtain. The effort to attempt elicitation is not only worthwhile to make inference, in particular to calculate the posterior distribution, but to making decisions too. The elicitation process involves the following stages: problem setup, elicitation of specific summaries, fitting of a joint probability distribution and assessment of the adequacy of the elicitation procedure. Usually the following summaries are desired in the elicitation process: measures of central location (mean, mode or median), measures of dispersion (variances, coefficient of variation, interquartile range) and credibility intervals.

Sometimes the expert identifies points that correspond to specified percentiles of his subjective distribution, as described in Section 3.1. The bisection method is often used. It starts by asking first the value of θ that splits the probability into two equal halves, i.e., this value is the median. Later, for values below the median, it asks the value of θ that splits the probability into two equal halves, i.e., this value is the first quartile. Finally the same task is performed above the median. The elicitation task is completed by converting the expert information into a probability distribution, actually a prior distribution to be updated to a posterior distribution. Alternative methods to quantify subjective opinion of an expert about a parameter are described in Garthwaite, Kadane and O'Hagan (2005).

In what follows we will concentrate on the estimation of π, the parameter of a Bernoulli process. This is one of the problems that has attracted substantial attention in the recent literature. In the example presented below we concentrate on the quantile method.

Example 3.1 *A large elicitation study is developed to assess the* impact of a treatment for osteoporosis *with a particular drug. In this context, the expert beliefs are essential, for instance, to determine the sample size necessary to run a clinical trial to investigate this drug. The main actors involved in this* *elicitation study are a medical doctor at the university hospital and a senior statistician. The parameter of interest is the chance that a patient will not have any sort of fracture in the next five years. After describing the nature of the parameter of interest and summarizing the elicitation method, the expert assesses his initial quartiles as $\pi_{0.25} = .47$, $\pi_{0.5} = 0.61$ and $\pi_{0.75} = 0.74$.*

Fitting a Beta distribution with only those initial judgments may not be satisfactory since the tail probability could be too heavy. In order to overcome this unpleasant situation the expert is asked to provide extra information, in the form of the tail quantiles $\pi_{0.10} = 0.35$ and $\pi_{0.90} = 0.83$. A Beta distribution with unknown hyperparameters $a, b > 0$ is then fitted by the facilitator to these percentiles. The hyperparameters are obtained via a least square procedure: $min_{a,b} \sum_{i=1}^{5}(F(\pi_i; a, b) - p_i)^2$, where $\pi = (0.35, 0.47, 0.61, 0.74, 0.83)$, $p = (0.10, 0.25, 0.50, 0.75, 0.90)$ and $F(\cdot; a, b)$ is the $Beta(a, b)$ distribution function. The values obtained are $\hat{a} = 2.1$ and $\hat{b} = 1.4$ providing the following summary: $\hat{\pi}_{0.25} = 0.43, \hat{\pi}_{0.50} = 0.62, \hat{\pi}_{0.75} = 0.79$ which are consistent with the values initially assessed even though $P[\pi < 0.35] = 0.26$, which seems too high when compared against the expected value of 10%. After thinking about these figures, the expert decides to make adjustments to the fitted values of a and b, while keeping the proportionality of $3 : 2$. Figure 3.3 shows the fitted Beta distribution with parameters $\hat{\hat{a}} = 3.6$ and $\hat{\hat{b}} = 2.4$, with fitted quantiles $\hat{\hat{\pi}} = (.35, .47, .61, .74, .84)$, where the $P(\pi < .35) = 0.10$ and so forth.

3.2 Specification through functional forms

The prior knowledge about θ can be used to specify a prior density with a particular functional form. A parametric family of densities can, for instance, be defined. Although very often this family can make the analysis easier, one must be careful and make sure that the chosen density really represents the available information. For example, we could make the following assumptions about θ:

- θ is symmetrically distributed with respect to the mode;
- its density decays fast (say, exponentially) when far away from the mode;
- intervals far from the mode have irrelevant probabilities.

These considerations can characterize, at least approximately, a Normal distribution with parameters, generically called hyperparameters, determined in correspondence with the information expressed in H. These ideas may be put in a more general framework and have led to a systematic approach of

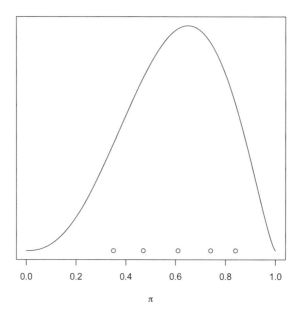

Figure 3.3 *Elicited Beta(3.6, 2.4), with 1st decil (.35), quartiles (.47, .61, .74) and 9th decil (.84) indicated by small dots.*

determination of prior distributions. The most relevant case corresponds to the conjugate family of distributions.

We have seen in Theorem 2.1 that if the observational distribution is $(X \mid \theta) \sim N(\theta, \sigma^2)$ and the prior is $\theta \sim N(\mu, \tau^2)$, then the posterior distribution is also Normal, with mean μ_1 and variance τ_1^2. So, if we start with a Normal prior, we end up with a Normal posterior. The main advantage of this approach is the ease of the resulting analysis. Among other things, this allows for the possibility of exploring the sequential aspect of the Bayesian paradigm. Every new Normal observation that is obtained only leads to changes in the parameters of the (new) posterior distribution. No new analytic calculations are required.

Definition 3.1 *Let $\mathcal{F} = \{ p(x|\theta), \theta \in \Theta\}$ be a family of a sampling or observational distributions. A class $\mathcal{P}$ of distributions is said to be a conjugate family (with respect) to $\mathcal{F}$ if $\forall p \in \mathcal{F}$ and $p(\theta) \in \mathcal{P}$ then $p(\theta \mid x) \in \mathcal{P}$.*

Thus, we can say that the class of Normal distributions is a conjugate family with respect to the class of Normal (sampling) distributions. Some caution is necessary when using the notion of conjugacy:

- The class $\mathcal{P}$ can be very broad

  For example, take $\mathcal{P} = \{$ all distributions $\}$ and $\mathcal{F}$ to be any family of sampling distributions. It is easy to see that $\mathcal{P}$ is conjugate with respect to $\mathcal{F}$ since any posterior will be a member of $\mathcal{P}$. In this context the definition of conjugacy does not have any practical appeal and is useless.

- The class $\mathcal{P}$ can be very narrow

  Suppose, for example, that

  $$\mathcal{P} = \{p \ : \ p(\theta = \theta_0) = 1\,, \theta_0 \in A\},$$

  for some non-null set A. This means that $\mathcal{P}$ consist only of distributions concentrated on a single point. Whatever the information provided by the sample, the posterior distribution would be the same as the prior because if we know, a priori, that $\theta = \theta_0$ with certainty, nothing will remove this certainty. That is,

  $$p(\theta|x) \propto l(\theta)\,p(\theta) = \begin{cases} l(\theta) \times 1 & , \text{if } \theta = \theta_0, \\ l(\theta) \times 0 & , \text{if } \theta \neq \theta_0. \end{cases}$$

  Then it follows that

  $$p(\theta \mid x) = \begin{cases} k \times l(\theta) & , \text{if } \theta = \theta_0, \\ 0 & , \text{if } \theta \neq \theta_0. \end{cases}$$

  As $\int p(\theta \mid x)\,d\theta = 1$, we must have that $p(\theta \mid x) = 1$ if and only if (iff, in short) $\theta = \theta_0$. Hence, $\mathcal{P}$ is conjugate to any distribution family and again the definitions are not helpful.

The last consideration illustrates, in an extreme situation, another very important aspect of prior specification. When a null probability is given to a particular sub-set of the possible values of θ, no observed information will change this specification, even if later it is proved to be inadequate. In order to avoid this obviously incoherent statement, it is strongly recommended that the statistician always associates non-null prior probability to every possible value of θ, even if some of them are judged to be very unlikely. Dennis Lindley refers to this recommendation as the Cromwell's rule.

Therefore, the class $\mathcal{P}$ must be broad enough to ensure elicitation of the convenient prior and, at the same time, restricted enough in order that the

definition be useful. A general procedure for obtaining conjugate prior families
is illustrated in the following example.

Example 3.2 *(Bernoulli trials) Let $X_i|\theta \sim Ber(\theta)$, $i = 1, \ldots, n$. The joint
sampling density is*

$$p(\boldsymbol{x} \mid \theta) = \theta^t (1-\theta)^{n-t} \ where \ t = \sum_{i=1}^{n} x_i, \quad x_i = 0, 1, \ i = 1, \ldots, n$$

*defining a class of distribution parameterized by $\theta \in (0,1)$. From Bayes theo-
rem, we know that the posterior density of θ given $\boldsymbol{x}$ is given by*

$$\begin{aligned} p(\theta \mid \boldsymbol{x}) \quad &\propto \quad p(\boldsymbol{x} \mid \theta) \, p(\theta) \\ &\propto \quad \theta^t (1-\theta)^{n-t} \, p(\theta). \end{aligned}$$

It is worth pointing out that $p(\theta)$ and $p(\theta \mid \mathbf{x})$ are related through the
likelihood function. The conjugate prior can then be obtained by mimicking
the kernel of the likelihood function. In Example 3.2, the likelihood kernel is of
the form $\theta^a (1-\theta)^b$. This is also the kernel of the Beta family of distributions,
introduced in Chapter 1. Taking the prior distribution as a $Beta(\alpha, \beta)$ and
combining with the likelihood, the posterior distribution is

$$\begin{aligned} p(\theta \mid \mathbf{x}) \quad &\propto \quad \theta^t (1-\theta)^{n-t} \, \theta^{\alpha-1} (1-\theta)^{\beta-1} \\ &\propto \quad \theta^{\alpha+t-1} (1-\theta)^{\beta+n-t-1}. \end{aligned}$$

Therefore, $\theta \mid \mathbf{x} \sim Beta(\alpha + t, \beta + n - t)$ which belongs to the same family of distributions used for the prior. So, the Beta family is conjugate to
the Bernoulli sampling model. The proportionality constant for the posterior
density above is given by $1/B(\alpha + t, \beta + n - t)$. It is not difficult to show
that the same is true for Binomial, geometric and negative Binomial sampling
distributions.

We can now discuss the setting of the conjugate prior family from a practical point of view for the general case of any given random sample. Let
$\mathbf{X} = (X_1, \ldots, X_n)$ be a sample from $p(x \mid \theta)$, $\theta \in \Theta$ and consider the density
function of $\mathbf{X}$, $p(\mathbf{x} \mid \theta)$. The family $\mathcal{P}$ is said to be closed under multiplication
or sampling if for all $p_1, p_2 \in \mathcal{P}$, there is a k such that $k p_1 p_2 \in \mathcal{P}$.

Example 3.3 *Let $\mathcal{P}$ be the class of Gamma distributions. If $p_i, i = 1, 2$ denote
the Gamma densities with parameters (a_i, b_i), $i = 1, 2$ then*

$$\begin{aligned} p_1 \times p_2 \quad &= \quad \frac{b_1^{a_1}}{\Gamma(a_1)} x^{a_1-1} e^{-b_1 x} \frac{b_2^{a_2}}{\Gamma(a_2)} x^{a_2-1} e^{-b_2 x} \ for \ x > 0, \\ &\propto \quad x^{a_1+a_2-2} e^{-(b_1+b_2)x}, \ for \ x > 0, \end{aligned}$$

which is proportional to another Gamma density with parameters $a_1 + a_2 - 1$ and $b_1 + b_2$. It can be shown that the same result is true for the class of Beta distributions.

Closure under multiplication is very important in the search for conjugate families. If the kernel of the likelihood can be identified with the kernel (of a member) of a given family of distributions and this family is closed under multiplication then prior and posterior will necessarily belong to the same family and conjugacy is obtained. The concept of conjugacy was formalized by Raiffa and Schlaifer (1961). They also studied many of the families that are presented in the next section.

With the definition of closure under sampling in hand, it becomes easy to specify a procedure to determine the conjugate class. It consists of

1. identification of the class $\mathcal{P}$ of distribution for θ whose members are proportional to $l(\theta; \mathbf{x})$, and

2. verification if $\mathcal{P}$ is closed under sampling.

If, in addition, for any given likelihood $l(\theta; \mathbf{x})$ obtained from a family $\mathcal{F}$, there exists a constant k defining a density p as $p(\theta) = k\, l(\theta; \mathbf{x})$, then the family $\mathcal{P}$ of all such densities p is said to be a natural conjugate family with respect to the sampling model with likelihood function l.

Example 3.4 *(Example 3.2 continued) Setting $k^{-1} = B(t + 1, n - t + 1)$ implies that*

$$k\, l(\theta; \boldsymbol{x}) = \frac{1}{B(t + 1, n - t + 1)} \theta^t (1 - \theta)^{n-t},$$

which has the form of a $Beta(t + 1, n - t + 1)$ density. Therefore, the class of Beta distributions with integer parameters is a natural conjugate family to the Bernoulli sampling model. Nothing substantial is lost however if this class is enlarged to the class of all Beta distributions, including all positive values for the parameters. This new class, strictly speaking, is not a natural conjugate family anymore. Nevertheless it keeps the essence of the definition and is used in practice.

Natural conjugate families are especially useful because an objective meaning can be attributed to the hyperparameters involved. Revisiting the above example, suppose that n_0 hypothetical (or not) trials were previously made, with t_0 successes. Then, the likelihood l^* of this hypothetical experiment would be $l^*(\theta) \propto \theta^{t_0}(1 - \theta)^{n_0 - t_0}$. If our (subjective) prior information is equivalent to that provided by the experiment described above, the prior for θ will be a Beta with hyperparameters $t_0 + 1$ and $n_0 - t_0 + 1$.

3.3   Conjugacy with the exponential family

The one-parameter exponential family includes many of the most common probability distributions. An essential characteristic of this family is that there exists a sufficient statistic with fixed dimension. The conjugate class $\mathcal{P}$ to the one-parameter exponential family is easy to characterize. Following the reasoning behind natural conjugacy, it is not difficult to see that members of this class have density

$$p(\theta) \propto \exp\{\alpha\phi(\theta) + \beta b(\theta)\}$$

and so

$$p(\theta \mid x) \propto \exp\left\{[\alpha + u(x)]\phi(\theta) + [\beta + 1]b(\theta)\right\}.$$

Denoting the constant involved in the definition of $p(\theta)$ by $k(\alpha, \beta)$, the constant associated to $p(\theta \mid x)$ will be $k(\alpha + u(x), \beta + 1)$. Using k as defined above it is easy to obtain $p(x)$ without explicitly calculating $\int p(x \mid \theta)p(\theta)\,d\theta$. From the equation $p(x)\,p(\theta \mid x) = p(x \mid \theta)p(\theta)$, it follows that

$$p(x) = \frac{p(x \mid \theta)p(\theta)}{p(\theta \mid x)}.$$

Substituting the densities previously obtained we get

$$p(x) = \frac{a(x)\exp\{u(x)\phi(\theta) + b(\theta)\}k(\alpha, \beta)\exp\{\alpha\phi(\theta) + \beta b(\theta)\}}{k(\alpha + u(x), \beta + 1)\exp\{[\alpha + u(x)]\phi(\theta) + [\beta + 1]b(\theta)\}}$$

and after some simplification we arrive at

$$p(x) = \frac{a(x)k(\alpha, \beta)}{k(\alpha + u(x), \beta + 1)}.$$

A straightforward extension of the Bernoulli example is the Binomial model. In that case, it follows that

$$
\begin{aligned}
p(x) &= \frac{\binom{n}{x}\theta^x(1-\theta)^{n-x}\, B^{-1}(\alpha, \beta)\theta^{\alpha-1}(1-\theta)^{\beta-1}}{B^{-1}(\alpha + x, \beta + n - x)\theta^{\alpha+x-1}(1-\theta)^{\beta+n-x-1}} \\[2mm]
&= \binom{n}{x}\frac{B(\alpha + x, \beta + n - x)}{B(\alpha, \beta)}, \quad \text{for } x = 0, 1, \ldots, n,\; n \geq 1.
\end{aligned}
$$

This is the probability function of the Beta-Binomial distribution.

In general, from a sample of size n of the exponential family we obtain the joint density

$$p(\mathbf{x} \mid \theta) = \left[\prod_{i=1}^{n} a(x_i)\right]\exp\left\{\left[\sum_{i=1}^{n} u(x_i)\right]\phi(\theta) + nb(\theta)\right\}.$$

The use of a conjugate prior $p(\theta) = k(\alpha, \beta) \exp\{\alpha\phi(\theta) + \beta b(\theta)\}$ leads to the posterior density

$$
p(\theta \mid \mathbf{x}) = k\left(\alpha + \sum_{i=1}^{n} u(x_i), \beta + n\right)
$$
$$
\times \exp\left\{\left[\alpha + \sum_{i=1}^{n} u(x_i)\right]\phi(\theta) + [\beta + n]b(\theta)\right\}
$$

and the marginal or predictive distribution is

$$
p(\mathbf{x}) = \frac{[\prod a(x_i)]\, k(\alpha, \beta)}{k\left(\alpha + \sum u(x_i), \beta + n\right)}.
$$

The main members of the exponential family will be presented in the following subsections. The results obtained previously will be applied to these particular cases and the resulting conjugate families obtained. Some of these families were presented previously.

3.3.1   Binomial distribution

The family of Beta distributions is conjugate to the Binomial (or Bernoulli) model as we have shown previously.

3.3.2   Normal distribution with known variance

Theorem 2.1 stated that the Normal distribution family is conjugate to the Normal model, based on a single observation. For the case of a sample of size n, we have already seen that

$$
l(\theta; \mathbf{x}) \propto \exp\left\{-\frac{n}{2\sigma^2}(\bar{x} - \theta)^2\right\},
$$

where the terms involving σ^2 were incorporated into the proportionality constant. So, the likelihood above has the same form as that based on a single observation x, substituting x by $\bar{x}$ and σ^2 by σ^2/n. Another way to say this is to note that $\bar{X}$ is a sufficient statistic for θ and so the likelihood based on the observed value of $\bar{X}$, which is distributed as $N(\theta, \sigma^2/n)$, is proportional to likelihood obtained with the individual observations $\mathbf{X}$. Therefore, the result presented in Theorem 2.1 is still true, with the substitutions mentioned above, i.e., the posterior distribution of θ given $\mathbf{x}$ is $N(\mu_1, \tau_1^2)$, with

$$
\mu_1 = \frac{n\sigma^{-2}\bar{x} + \tau^{-2}\mu}{n\sigma^{-2} + \tau^{-2}} \quad \text{and} \quad \tau_1^{-2} = n\sigma^{-2} + \tau^{-2}.
$$

3.3.3   Poisson distribution

Suppose that $\mathbf{X} = (X_1, \ldots, X_n)$ is a random sample from the Poisson distribution with parameter θ, denoted $Pois(\theta)$. Its joint probability function is

$$p(\mathbf{x} \mid \theta) = \prod_{i=1}^{n} p(x_i \mid \theta) = \prod_{i=1}^{n} \frac{e^{-\theta}\theta^{x_i}}{x_i!}$$

and the likelihood function assumes the form

$$l(\theta \mid \mathbf{x}) \propto e^{-n\theta}\theta^{\Sigma x_i}.$$

Its kernel has the form $\theta^a e^{-b\theta}$ characterizing a Gamma family of distributions. We have already seen in the previous section that the Gamma family is closed under sampling. Then the conjugate prior distribution of θ will be $\theta \sim G(\alpha, \beta)$. The posterior density will be

$$p(\theta \mid \mathbf{x}) \propto \theta^{\alpha + \Sigma x_i - 1} \, exp\{-(\beta + n)\,\theta\},$$

corresponding to the $G(\alpha + \Sigma x_i, \beta + n)$ density. The calculation of the predictive distribution, using the method previously described, is left as an exercise.

3.3.4   Exponential distribution

Suppose that $\mathbf{X} = (X_1, \ldots, X_n)$ is a random sample from the Exponential distribution with parameter θ, denoted by $Exp(\theta)$. Its joint density function is

$$p(\mathbf{x} \mid \theta) = \theta^n \, \exp\left\{-\theta \sum_{i=1}^{n} x_i\right\}.$$

The form of the likelihood allows recognition of the kernel of the Gamma family as a conjugate distribution for θ. Assuming a $G(\alpha, \beta)$ prior, the posterior will have the form

$$p(\theta \mid \mathbf{x}) \quad \propto \quad \theta^n \, \exp\left\{-\theta \sum_{i=1}^{n} x_i\right\} \theta^{\alpha-1} \exp\{-\beta\theta\}$$

$$\propto \quad \theta^{\alpha+n-1} \, \exp\left\{-(\beta + \sum x_i)\theta\right\},$$

which is the density of a $G(\alpha + n, \beta + \sum x_i)$ distribution.

3.3.5   Multinomial distribution

Denote by $\mathbf{X} = (X_1, \ldots, X_p)$ and $\boldsymbol{\theta} = (\theta_1, \ldots, \theta_p)$, respectively, the number of observed cases in and the probabilities associated with each of p categories in a sample of size n and the following constraints are true: $\sum_{i=1}^{p} X_i = n$ and $\sum_{i=1}^{p} \theta_i = 1$. $\mathbf{X}$ is said to have a multinomial distribution with parameters n and $\theta_1, \ldots, \theta_p$. The joint probability function of the p counts $\mathbf{X}$ is

$$p(\mathbf{x} \mid \boldsymbol{\theta}) = \frac{n!}{\prod_{i=1}^{p} x_i!} \prod_{i=1}^{p} \theta_i^{x_i}.$$

It is not difficult to show that this distribution also belongs to the exponential family. The likelihood function for $\boldsymbol{\theta}$ is $l(\boldsymbol{\theta}) \propto \prod \theta_i^{x_i}$. Its kernel is the same as the kernel of the density of a Dirichlet distribution. The Dirichlet family with integer parameters $a_1, \ldots, a_p$ is a natural conjugate with respect to the multinomial sampling distribution. Again, little is lost by extending natural conjugacy over all Dirichlet distributions.

The posterior distribution will then be

$$p(\boldsymbol{\theta} \mid \mathbf{x}) \propto \left[\prod_{i=1}^{p} \theta_i^{x_i} \right] \left[\prod_{i=1}^{p} \theta_i^{a_i - 1} \right] = \prod_{i-1}^{p} \theta_i^{x_i + a_i - 1}$$

and, as anticipated, this posterior is also a Dirichlet distribution with parameters $a_1 + x_1, \ldots, a_p + x_p$ which is denoted by $(\boldsymbol{\theta} \mid \mathbf{x}) \sim D(a_1 + x_1, \ldots, a_p + x_p)$. From the above results about the Dirichlet distribution, it is not difficult to obtain the proportionality constant as

$$\frac{\Gamma(a + n)}{\prod_{i=1}^{p} \Gamma(a_i + x_i)}.$$

This conjugate analysis generalizes the analysis for Binomial samples with Beta priors, obtained when $p = 2$.

3.3.6 Normal distribution with known mean and unknown variance

Let $\mathbf{X} = (X_1, \ldots, X_n)$ be a random sample of size n from the $N(\theta, \sigma^2)$, θ known, and $\phi = \sigma^{-2}$. In this case the joint density function will be

$$l(\phi; \mathbf{x}) = p(\mathbf{x} \mid \theta, \phi) \propto \phi^{n/2} \exp\left\{ -\frac{\phi}{2} n s_0^2 \right\} \text{ where } s_0^2 = \frac{1}{n} \sum (x_i - \theta)^2.$$

The conjugate prior may have the kernel of $l(\phi; \mathbf{x})$, which is in the Gamma distribution form. As the Gamma family is closed under sampling, we can consider a $G(n_0/2, n_0\sigma_0^2/2)$ prior distribution or, equivalently, that $n_0\sigma_0^2\phi$ has a χ^2 distribution with n_0 degrees of freedom. The posterior distribution of ϕ is obtained using Bayes theorem as

$$
\begin{aligned}
p(\phi \mid \mathbf{x}) &\propto l(\phi; \mathbf{x}) p(\phi) \\
&\propto \phi^{n/2} \exp\{ -n\phi s_0^2/2 \} \phi^{(n_0/2) - 1} \exp\{ -n_0\phi\sigma_0^2/2 \} \\
&= \phi^{[(n_0 + n)/2] - 1} \exp\{ -(n_0\sigma_0^2 + n s_0^2)\phi/2 \}.
\end{aligned}
$$

The above expression corresponds to the kernel of the Gamma distribution as expected. Therefore,

$$\phi \mid \mathbf{x} \sim G\left(\frac{n_0 + n}{2}, \frac{n_0\sigma_0^2 + ns_0^2}{2}\right)$$

or, equivalently, $(n_0\sigma_0^2 + ns_0^2)\phi \mid \mathbf{x} \sim \chi^2_{n_0+n}$. Then it follows that the Gamma or the χ^2 family of distributions is conjugate with respect to the Normal sampling model with θ known and σ^2 unknown.

3.3.7 Normal distribution with unknown mean and variance

The conjugate prior distribution for (θ, ϕ) will be presented in two stages. First, the following conditional distribution of θ given ϕ will be considered

$$\theta \mid \phi \sim N[\mu_0, (c_0\phi)^{-1}]$$

and the marginal prior for ϕ is as previously stated, that is,

$$n_0\sigma_0^2\phi \sim \chi^2_{n_0} \text{ or } \phi \sim G\left(\frac{n_0}{2}, \frac{n_0\sigma_0^2}{2}\right),$$

where (n_0, σ_0^2) and (μ_0, c_0) are obtained from the initial information H. This distribution is usually called Normal-Gamma or Normal-χ^2 with parameters $(\mu_0, c_0, n_0, \sigma_0^2)$ and joint density given by

$$
\begin{aligned}
p(\theta, \phi) &= p(\theta|\phi)p(\phi) \\
&\propto \phi^{1/2} \exp\left\{-\frac{c_0\phi}{2}(\theta - \mu_0)^2\right\} \phi^{n_0/2-1} \exp\left(-\frac{n_0\sigma_0^2\phi}{2}\right) \\
&= \phi^{(n_0+1)/2-1} \exp\left\{-\frac{\phi}{2}\left[n_0\sigma_0^2 + c_0(\theta - \mu_0)^2\right]\right\}.
\end{aligned}
$$

The marginal prior distribution of θ can be obtained by integration, using the result

$$\int_0^\infty \phi^{a-1}e^{-b\phi}\, d\phi = \frac{\Gamma(a)}{b^a}.$$

Application of the above result gives

$$
\begin{aligned}
p(\theta) &\propto \int_0^\infty \phi^{(n_0+1)/2-1} \exp\left\{-\frac{\phi}{2}\left[n_0\sigma_0^2 + c_0(\theta - \mu_0)^2\right]\right\} d\phi \\
&= \frac{\Gamma[(n_0+1)/2]}{\{[n_0\sigma_0^2 + c_0(\theta - \mu_0)^2]/2\}^{(n_0+1)/2}} \\
&\propto [n_0\sigma_0^2 + c_0(\theta - \mu_0)^2]^{-(n_0+1)/2},
\end{aligned}
$$

since the $\Gamma(\cdot)$ term does not depend on θ. Then, we obtain

$$p(\theta) \propto \left[1 + \frac{(\theta - \mu_0)^2}{n_0(\sigma_0^2/c_0)}\right]^{-\frac{n_0+1}{2}},$$

which is the kernel of the Student-t distribution with n_0 degrees of freedom, location parameter μ_0 and scale parameter σ_0^2/c_0, denoted by $t_{n_0}(\mu_0, \sigma_0^2/c_0)$.

The conditional distribution of $\phi \mid \theta$ can be obtained from the joint distribution of (θ, ϕ) and is a $G\{(n_0+1)/2, [n_0\sigma_0^2 + c_0(\theta - \mu_0)^2]/2\}$ or, equivalently, $[n_0\sigma_0^2 + c_0(\theta - \mu_0)^2]\phi \mid \theta \sim \chi^2_{2(n_0+1)}$.

The joint distribution of a random sample $\mathbf{X} = (X_1, \ldots, X_n)$ is

$$
\begin{aligned}
p(\mathbf{x} \mid \theta, \phi) &= \prod_{i=1}^{n} \phi^{1/2} \exp\left\{-\frac{\phi}{2}(x_i - \theta)^2\right\} \\
&\propto \phi^{n/2} \exp\left\{-\frac{\phi}{2}\left[ns^2 + n(\bar{x} - \theta)^2\right]\right\}
\end{aligned}
$$

where $s^2 = \frac{1}{n}\sum(x_i - \bar{x})^2$, as we have seen in Section 2.6. The above expression has the same kernel as the Normal-Gamma density for (θ, ϕ). Next, it is necessary to check if the Normal-Gamma family is closed under sampling. It is not difficult to verify it is. The posterior distribution will then be

$$
\begin{aligned}
p(\theta, \phi | \mathbf{x}) &\propto p(\mathbf{x} \mid \theta, \phi)p(\theta, \phi) \\
&\propto \phi^{[(n+n_0+1)/2]-1} \\
&\quad \times \exp\left\{-\frac{\phi}{2}\left[n_0\sigma_0^2 + ns^2 + c_0(\theta - \mu_0)^2 + n(\bar{x} - \theta)^2\right]\right\}.
\end{aligned}
$$

It can be shown (the proof is left as an exercise) that

$$c_0(\theta - \mu_0)^2 + n(\theta - \bar{x})^2 = (c_0 + n)(\theta - \mu_1)^2 + \frac{c_0 n}{c_0 + n}(\mu_0 - \bar{x})^2,$$

where $\mu_1 = (c_0\mu + n\bar{x})/(c_0 + n)$. Thus, it follows that the posterior density for (θ, ϕ) is proportional to

$$
p(\theta, \phi | \mathbf{x}) \propto \phi^{[(n+n_0+1)/2]-1}
$$
$$
\times \exp\left\{-\frac{\phi}{2}\left[n_0\sigma_0^2 + ns^2 + \frac{c_0 n}{c_0 + n}(\mu_0 - \bar{x})^2 + (c_0 + n)(\theta - \mu_1)^2\right]\right\}.
$$

Therefore, the joint posterior for $(\theta, \phi \mid \mathbf{x})$ is Normal-Gamma with parameters $(\mu_1, c_1, n_1, \sigma_1^2)$ given by $\mu_1 = (c_0\mu_0 + n\bar{x})/(c_0 + n)$, $c_1 = c_0 + n$, $n_1 = n_0 + n$ and $n_1\sigma_1^2 = n_0\sigma_0^2 + ns^2 + [c_0 n/(c_0 + n)](\mu_0 - \bar{x})^2$.

Prior and posterior distributions are members of the same family. So, the Normal-Gamma family is conjugate with respect to the Normal sampling model when θ and σ^2 are both unknown. Table 3.1 summarizes the distributions involved in the Bayesian analysis of the Normal models with unknown mean and variance.

Table 3.1 *Distributions Involved in the Bayesian Analysis of the Normal Models with Unknown Mean and Variance.*

	Prior	Posterior
$\theta \mid \phi$	$N(\mu_0, (c_0\phi)^{-1})$	$N(\mu_1, (c_1\phi)^{-1})$
ϕ	$n_0\sigma_0^2\phi \sim \chi_{n_0}^2$	$n_1\sigma_1^2\phi \sim \chi_{n_1}^2$
θ	$t_{n_0}(\mu_0, \sigma_0^2/c_0)$	$t_{n_1}(\mu_1, \sigma_1^2/c_1)$
$\phi \mid \theta$	$[n_0\sigma_0^2+c_0(\theta-\mu_0)^2)]\phi \sim \chi_{n_0+1}^2$	$[n_1\sigma_1^2+c_1(\theta-\mu_1)^2)]\phi \sim \chi_{n_1+1}^2$

3.4   Non-informative priors

Many statisticians show concern about the nature of the prior distribution. This is mainly due to an influence from the frequentist point of view. They typically maintain that the prior distribution is arbitrary and alters the conclusions about the statistical problem at hand. Therefore, they argue that prior information is not acceptable for use in a scientific context. Besides, they may desire an inference based only on the data, completely objective, instead of considering a priori subjectivity as regarded in previous sections. In these cases, only the data should be responsible for the information contained in the posterior distribution.

In this context, this section is devoted to the concept of non-informative or reference prior, as an attempt to reconcile the above reasoning with the Bayesian point of view. The idea behind these priors comes from the desire to make statistical inference based on a minimum (or absence, if possible) of subjective prior information. This minimum is clearly a relative concept and should take in consideration, for example, the sample information content.

Another context where the concept of a reference prior may be useful was outlined in the example of the two physicists in Chapter 2. Let us suppose that two scientists have strong and divergent prior opinions about an unknown quantity and that is not possible to reconcile these initial opinions. This is a situation where it may be necessary to produce a *neutral* analysis, introducing a referential standpoint. Another plausible justification to want a reference analysis is the usual expectation that the evidence from the experiment is stronger than the prior.

Initially, uniform priors were proposed to represent situations where little or no initial information is available or where, even if it is available, we do not wish to use it. The rationale behind this proposition is not to give preference to any value of θ. That is, $p(\theta) \propto k$ for θ varying in a given sub-set of the real line means that none of the particular values of θ is preferred (Bayes, 1763). This choice brings some difficulties with it. The first one is that $p(\theta)$ is not a proper distribution if the range of values of θ is unbounded. This means that $\int p(\theta)\, d\theta \to \infty$, which goes against the basic rules of probability. Also, if $\phi = \phi(\theta)$ is a one-to-one transformation of θ and if θ is uniformly distributed, then by the theorem of variable transformation, the density of ϕ is

$$p(\phi) = p(\theta(\phi)) \left| \frac{d\theta}{d\phi} \right| \propto \left| \frac{d\theta}{d\phi} \right|,$$

which is only constant if ϕ is defined by a linear transformation. However, the same assumptions leading to the specification of $p(\theta) \propto k$ should also lead to $p(\phi) \propto k$, which contradicts the above deduction. Ideally, we would like to state an invariant rule that would not violate results about variable transformation. A possibility consists of considering the procedure advocated by Sir Harold Jeffreys.

3.4.1 Jeffreys non-informative priors

Definition 3.2 (Jeffreys non-informative prior) Consider an observation X with probability (density) function $p(x \mid \theta)$. The Jeffreys non-informative prior has density given by

$$p(\theta) \propto [I(\theta)]^{1/2}, \quad \theta \in \Theta,$$

where $I(\theta)$ is the expected Fisher information measure.

Example 3.5 (Binomial case) Let X be an observation from the $Bin(n,\theta)$ distribution, with probability function

$$p(x \mid \theta) = \binom{n}{x} \theta^x (1-\theta)^{n-x}.$$

Taking the logarithm, it follows that

$$\log p(x \mid \theta) = \ln \left[\frac{n!}{x!(n-x)!} \right] + x \ln \theta + (n-x) \ln(1-\theta).$$

The first and second order derivatives of the log likelihood are

$$\frac{\partial \log p(x \mid \theta)}{\partial \theta} = \frac{x}{\theta} + \frac{(n-x)}{(1-\theta)}.(-1) = \frac{x}{\theta} - \frac{(n-x)}{(1-\theta)}$$

and

$$\frac{\partial^2 \log p(x \mid \theta)}{\partial \theta^2} = \frac{\theta.0 - x.1}{\theta^2} - \frac{(1-\theta).0 - (n-x)(-1)}{(1-\theta)^2} = -\frac{x}{\theta^2} - \frac{(n-x)}{(1-\theta)^2}.$$

Then, the Fisher information will be

$$I(\theta) = E\left(-\frac{\partial^2 \log p(X \mid \theta)}{\partial \theta^2}\right)$$

$$= E\left(\frac{X}{\theta^2} + \frac{(n-X)}{(1-\theta)^2}\right)$$

$$= n\theta^{-1}(1-\theta)^{-1},$$

and the non-informative prior is

$$\pi(\theta) \propto [I(\theta)]^{\frac{1}{2}} = \left[\theta^{-1}(1-\theta)^{-1}\right]^{\frac{1}{2}} = \theta^{-\frac{1}{2}}(1-\theta)^{-\frac{1}{2}}.$$

This is the density of a Beta(1/2, 1/2) distributuion.

The following lemma states the invariance of the non-informative prior proposed by Jeffreys under one to one transformations.

Lemma 3.1 *The Jeffreys prior $p(\theta) \propto [I(\theta)]^{1/2}$ is invariant under one to one transformations, that is, if $\phi = \phi(\theta)$ is a one to one transformation of θ, then Jeffreys prior for ϕ is $p(\phi) \propto [I(\phi)]^{1/2}$.*

Proof Let $\phi = \phi(\theta)$ be a one to one transformation of θ. Taking the derivative of $\log p(X \mid \phi)$ with respect to ϕ, it follows that

$$\frac{\partial \log p(X \mid \phi)}{\partial \phi} = \frac{\partial \log p(X \mid \phi(\theta))}{\partial \theta} \frac{\partial \theta}{\partial \phi},$$

where $\theta = \theta(\phi)$ is the inverse transformation of ϕ. To obtain the Fisher transformation of a parameter, the log likelihood of this parameter needs to be differentiated twice. This gives

$$\frac{\partial^2 \log p(X \mid \phi)}{\partial \phi^2} = \frac{\partial \log p(X \mid \phi(\theta))}{\partial \theta} \frac{\partial^2 \theta}{\partial \phi^2} + \frac{\partial^2 \log p(X \mid \phi(\theta))}{\partial \theta^2}\left(\frac{\partial \theta}{\partial \phi}\right)^2.$$

Multiplying both sides by (-1) and calculating the expected value with respect to $p(x \mid \theta)$, gives

$$I(\phi) = E_{X|\theta}\left[-\frac{\partial \log p(X \mid \theta)}{\partial \theta}\right]\frac{\partial^2 \theta}{\partial \phi^2} + I(\theta)\left(\frac{\partial \theta}{\partial \phi}\right)^2$$

$$= I(\theta)\left(\frac{\partial \theta}{\partial \phi}\right)^2,$$

since $E_{X|\theta}[\partial \log p(X \mid \theta)/\partial \theta] = 0$, as seen in Chapter 2. Therefore, $I^{1/2}(\phi) = I^{1/2}(\theta)|\partial \theta/\partial \phi|$. By the rules of probability, if θ has density proportional to $I^{1/2}(\theta)$, then ϕ has density

$$p(\phi) \propto I^{1/2}(\theta(\phi))\,|\partial \theta/\partial \phi| = I^{1/2}(\phi)$$

and the specification is invariant to one-to-one transformation. ☐

There is only one transformation ψ of θ which satisfies the invariance rule and has constant density. This transformation is easily obtained by making

$$p(\psi) \propto I^{1/2}(\theta) |\partial\theta/\partial\psi| \propto k$$

or

$$|\partial\theta/\partial\psi| \propto I^{-1/2}(\theta) \implies |\partial\psi/\partial\theta| \propto I^{1/2}(\theta) \implies \psi \propto \int^{\theta} I^{1/2}(u)\, du.$$

Example 3.6 *(Poisson case) Let $\boldsymbol{X} = (X_1,\ldots,X_n)$ be a sample of the Pois(θ). The joint density of the observations is*

$$p(\boldsymbol{x} \mid \theta) = \frac{e^{-n\theta}\theta^{\Sigma x_i}}{\prod_i x_i!}.$$

Taking the logarithm, it follows that

$$\log p(\boldsymbol{x} \mid \theta) = -n\theta + \sum_{i=1}^{n} x_i \log\theta - \log\prod_{i=1}^{n} x_i!$$

The first and second order derivatives of the log likelihood are

$$\frac{\partial \log p(\boldsymbol{x} \mid \theta)}{\partial\theta} = -n + \frac{\sum_{i=1}^{n} x_i}{\theta} \quad and \quad \frac{\partial^2 \log p(\boldsymbol{x} \mid \theta)}{\partial\theta^2} = -\frac{\sum x_i}{\theta^2}.$$

Then, the Fisher information will be

$$I(\theta) = E_{\boldsymbol{X}\mid\theta}\left[\frac{\sum X_i}{\theta^2}\right] = \frac{1}{\theta^2}\sum E(X_i) = \frac{n\theta}{\theta^2} = \frac{n}{\theta},$$

and the non-informative prior is $p(\theta) \propto \theta^{-1/2}$. So the posterior density will be

$$p(\theta \mid \boldsymbol{x}) \propto p(\boldsymbol{x} \mid \theta)p(\theta) \propto e^{-n\theta}\theta^{\Sigma_i x_i}\theta^{-1/2} = e^{-n\theta}\theta^{\Sigma_i x_i - 1/2},$$

that is, $\theta \mid \boldsymbol{x} \sim G(\Sigma_i x_i + 1/2, n)$, or alternatively, $2n\theta \mid \boldsymbol{x} \sim \chi^2_{2\Sigma_i x_i + 1}$. The transformation leading to the uniform prior is

$$\phi \propto \int_0^{\theta} u^{-1/2}\, du = 2u^{1/2}\,|_0^{\theta} \propto \theta^{1/2}.$$

The non-informative prior is frequently obtained from the conjugate prior by letting the scale parameters go to zero and keeping the other ones constant. In the above example, it can be noted that the reference prior is the limit of the Gamma distribution (the natural conjugate prior for the Poisson model) with $\theta \sim G(1/2, \epsilon)$, when $\epsilon \longrightarrow 0$, since $\theta^{-1/2}e^{\epsilon\theta} \longrightarrow \theta^{-1/2}$, when $\epsilon \longrightarrow 0$.

The class of non-informative priors proposed by Jeffreys (1961) is invariant but in many cases leads to improper distributions. Despite that, in practice, we are concerned with the posterior distribution, which is often proper, even when the prior distribution is not. In this case, one may not need to give much relevance to the impropriety of the prior distribution. Careful examination must be carried out to make sure the posterior is actually proper to proceed confidently with the analysis.

The class of Jeffreys non-informative priors was extensively used by Box and Tiao (1992). Intuitively, this prior tries to provide as little prior information as possible, relative to the sample information. It is not surprising, therefore, that it should depend on Fisher information measures, in the sense that it depends on the notion of information. There is also a connection between inference performed with Jeffreys non-informative priors and inference with the adjusted profile likelihood $l_{AP}(.)$ presented in Chapter 2. Indeed, the penalty applied to the likelihood is similar to a Jeffreys prior (Firth, 1993).

Definition 3.3 *(Multiparameter Jeffreys non-informative prior) Consider an observation X with probability density function $p(x \mid \boldsymbol{\theta})$, where $\boldsymbol{\theta} = (\theta_1, \ldots, \theta_k)$. In the multiparameter case, the Jeffreys non-informative prior has density proportional to the square root of the expected Fisher information matrix $\boldsymbol{I}(\boldsymbol{\theta})$, that is,*

$$p(\boldsymbol{\theta}) \propto |\boldsymbol{I}(\boldsymbol{\theta})|^{1/2},$$

where the elements of $I(\boldsymbol{\theta})$ are given by

$$I_{ij}(\boldsymbol{\theta}) = E\left[-\frac{\partial \log p(x|\boldsymbol{\theta})}{\partial \theta_i}\frac{\partial \log p(x|\boldsymbol{\theta})}{\partial \theta_j}|\boldsymbol{\theta}\right], \quad i, j = 1, \ldots, k.$$

Example 3.7 *(Normal case) Let $\boldsymbol{X} = (X_1, \ldots, X_n)$ be a sample of the $N(\mu, \sigma^2)$, where $\boldsymbol{\theta} = (\mu, \sigma)$, with both unknown parameters. The joint density of the observations is $p(\boldsymbol{x} \mid \boldsymbol{\theta}) = \prod_{i=1}^{n}(1/\sqrt{2\pi})\sigma^{-1}\exp{-(1/2)\sigma^{-2}(x_i - \mu)^2}$. After taking the logarithm of $p(\boldsymbol{x} \mid \boldsymbol{\theta})$, and calculating the first and second order derivatives, it can be proved that the expected information matrix is given by the diagonal matrix,*

$$I_n(\boldsymbol{\theta}) = n\begin{bmatrix} \sigma^{-2} & 0 \\ 0 & (1/2)\sigma^{-4} \end{bmatrix}.$$

Then, the corresponding non-informative prior for $\boldsymbol{\theta} = (\mu, \sigma)$ is given by

$$p(\boldsymbol{\theta}) \propto \sigma^{-2}.$$

Corollary 3.1 *For the multiparameter case, the Jeffreys prior is invariant under one-to-one transformations.*

It is not always possible to apply the rule to obtain the Jeffreys non-informative prior distributions. Indeed, it is inappropriate to use the Jeffreys when $I(\theta)$ does not exist. Another drawback of Jeffreys priors is that it does not satisfy the Likelihood Principle, as it may produce different results for equal likelihoods.

Example 3.8 *(Bernoulli trials) A famous example illustrating that the non-informative prior depends on the sampling model is provided by Bernoulli trials with success probability θ. If the sample design is such that n fixed Bernoulli trials are made and the number X of success observed, then $X \sim Bin(n, \theta)$. From Example 3.5, the non-informative prior is a Beta(1/2,1/2) distribution, that is a proper distribution.*

Now, suppose that the sampling scheme consists in observing the number Y of replications until s successes are obtained. The observation now is Y with negative Binomial distribution denoted by $Y \sim NB(s, \theta)$ and

$$p(y \mid \theta) = \binom{y-1}{s-1} \theta^s (1-\theta)^{y-s}, \ for \ y = s, s+1, \ldots,$$

which implies that the first and second derivatives with respect to θ are

$$\frac{\partial \log p(y \mid \theta)}{\partial \theta} = \frac{s}{\theta} - \frac{y-s}{1-\theta} \ and \ \frac{\partial^2 \log p(y \mid \theta)}{\partial \theta^2} = -\frac{s}{\theta^2} - \frac{y-s}{(1-\theta)^2}.$$

The expected information is

$$\begin{aligned} I_{BN}(\theta) &= \frac{s}{\theta^2} + \frac{E(Y-s \mid \theta)}{(1-\theta)^2} \\ &= \frac{s}{\theta^2(1-\theta)} \ , \ since \ E(Y \mid \theta) = \frac{s}{\theta} \end{aligned}$$

and the non-informative prior is $p_{BN}(\theta) \propto \theta^{-1}(1-\theta)^{-1/2}$. This prior is the limit of a Beta(ϵ, 1/2) when $\epsilon \to 0$, and then it is an improper distribution.

Suppose now that in 10 experiments, three successes were observed. Therefore, $l(\theta; x = 3) \propto \theta^3 (1-\theta)^7$ and $l(\theta; y = 10) \propto \theta^3 (1-\theta)^7$. The sample information, namely the likelihood about θ, is the same for both models. Although the information is the same, the prior and consequently the posterior distributions are different, with

$$p(\theta \mid x = 3) \propto \theta^{2.5} (1-\theta)^{6.5} \ and \ p(\theta \mid y = 10) \propto \theta^2 (1-\theta)^{6.5},$$

showing, then, that the non-informative prior violates the likelihood principle.

More applications of non-informative priors in a few frequently used models described in Chapter 2 are presented below, starting with the location model.

Example 3.9 *(Location model) If X has the location model, then*

$$\frac{\partial \log p(x \mid \theta)}{\partial \theta} = \frac{\partial \log f(x - \theta)}{\partial \theta} = -\frac{f'(x - \theta)}{f(x - \theta)} \text{ where } f' = \frac{\partial f}{\partial \theta}.$$

Then by the last lemma, $I(\theta) = E_{X|\theta}\left[\left(-\frac{f'(X-\theta)}{f(X-\theta)}\right)^2\right]$. Making the transformation $U = X - \theta$, then $I(\theta) = E_U\left[\left(-\frac{f'(U)}{f(U)}\right)^2\right]$ which does not depend on θ. So, $I(\theta) = k$ and then $p(\theta) \propto k$.

The result above is also true for a parameter vector $\boldsymbol{\theta}$. One way to justify this prior is through model invariance directly. Working with an observation vector $\mathbf{X}$ and location parameter $\boldsymbol{\theta}$ is equivalent to working with observation vector $\mathbf{Y} = \mathbf{X} + \mathbf{c}$ and location parameter $\boldsymbol{\eta} = \boldsymbol{\theta} + \mathbf{c}$ for any given constant $\mathbf{c}$. We can then insist on the same non-informative prior specification for $\boldsymbol{\theta}$ as for $\boldsymbol{\eta}$. It is not difficult to show that the only distribution satisfying this requirement has density $p(\boldsymbol{\theta}) \propto k$.

Example 3.10 *(Scale model) If X has the scale model, then*

$$\begin{aligned}
\frac{\partial \log p(x \mid \sigma)}{\partial \sigma} &= \frac{\partial \log[\sigma^{-1} f(x/\sigma)]}{\partial \sigma} \\
&= \frac{\partial}{\partial \sigma}\left[-\log \sigma + \log f\left(\frac{x}{\sigma}\right)\right] \\
&= -\frac{1}{\sigma} + \frac{\partial \log f(x/\sigma)}{\partial \sigma}\left(-\frac{x}{\sigma^2}\right) \\
&= -\frac{1}{\sigma}\left[1 + \frac{x}{\sigma}\frac{f'(x/\sigma)}{f(x/\sigma)}\right] \text{ where } f' = \frac{\partial f}{\partial \sigma}.
\end{aligned}$$

Therefore, the information measure about σ will be

$$\begin{aligned}
I(\sigma) &= \frac{1}{\sigma^2} E_{X|\sigma}\left[\left(1 + \frac{X}{\sigma}\frac{f'(X/\sigma)}{f(x/\sigma)}\right)^2 \mid \sigma\right] \\
&= \frac{1}{\sigma^2} E\left(1 + U\frac{f'(U)}{f(U)}\right)^2
\end{aligned}$$

after the transformation $U = X/\sigma$. Since the distribution of U does not depend on σ, then $I(\sigma) = k \times \sigma^{-2}$ and $p(\sigma) \propto \sigma^{-1}$ is the non-informative prior distribution.

Once again, model invariance can be invoked directly by assuming equivalence between a model with observation X and scale parameter σ and a model with observation $Y = cX$ and scale parameter $\eta = c\sigma$. Insisting on the same non-informative prior for σ and η leads to $p(\sigma) \propto \sigma^{-1}$.

3.4.2 Reference priors

Bernardo (1979) advocates the construction of a posterior distribution that can be used as a reference posterior distribution, obtained in the case of ignorance a priori. The Bernardo argument is based on the fact that the amount of information which may be expected from an experiment clearly depends on prior knowledge. The amount of information from the data is expected to get smaller as available prior knowledge gets larger. The reference prior is the one which maximizes the missing information about the parameter of interest in an infinitely large experiment.

Definition 3.4 *The amount of unknown information about θ in n replications of the experiment is defined as*

$$
\begin{aligned}
I(\boldsymbol{X}_n, \theta) &= E_{(\boldsymbol{X}_n, \theta)}\left[\log \frac{p(\theta \mid \boldsymbol{x}_n)}{p_n(\theta)}\right] \\
&= E_{\boldsymbol{X}_n}\left[E_{\theta \mid \boldsymbol{X}_n}\left[\log \frac{p(\theta \mid \boldsymbol{x}_n)}{p_n(\theta)}\right]\right] \\
&= \int p(\boldsymbol{x}_n \mid \theta)\int p(\theta \mid \boldsymbol{x}_n)\log\frac{p(\theta \mid \boldsymbol{x}_n)}{p_n(\theta)}\, d\theta d\boldsymbol{x}_n,
\end{aligned}
$$

where $\boldsymbol{X}_n = (X_1, \ldots, X_n)$ and $\boldsymbol{x}_n = (x_1, \ldots, x_n)$.

Definition 3.5 *(Reference prior) The reference prior $p(\theta)$ is defined as the distribution that maximizes the amount of missing information about θ in an infinite number of replications of the experiment. The amount of missing information about θ in an infinite number of replications of the experiment is obtained as the limit of the information based on n replications when $n \to \infty$, $I(\boldsymbol{X}_\infty, \theta) = \lim_{n \to \infty} I(\boldsymbol{X}_n, \theta)$. That is, the reference prior distribution is given by*

$$
p(\theta) = \arg\max_{p_n(\theta)} I(\boldsymbol{X}_\infty, \theta).
$$

According to Bernardo (1979), it is adequate to consider $n \to \infty$, since, when choosing a prior, there is an intention of considering the information which might be obtained from many future experiments in addition to the information obtained from a particular experiment. A drawback however with the reference prior definition above is that the maximization of $I(\boldsymbol{X}_\infty, \theta)$ is often not analytically tractable. An alternative is to use asymptotic theory, to be described in Chapter 5.

Reference priors have several desirable properties including invariance with respect to one-to-one transformations and sampling consistency (the posterior densities from an ensemble of experiments tend to cluster around the true values of the parameters).

In the multivariate case, Bernardo proposed a modification to Jeffreys rule. He suggested a two-stage procedure. The parametric vector is divided into two components: $\boldsymbol{\theta}$ denoting the parameters of interest and $\boldsymbol{\phi}$ the nuisance parameters. First, a (conditional) reference prior distribution $p(\boldsymbol{\phi} \mid \boldsymbol{\theta})$ is obtained. This prior is used to eliminate the parameters $\boldsymbol{\phi}$ from the likelihood and gives a marginal likelihood $p(x \mid \boldsymbol{\theta})$ as we have seen in Section 2.6. Then, this likelihood is used to obtain the (marginal) reference prior $p(\boldsymbol{\theta})$. Finally, the complete reference prior is obtained by the multiplication rule $p(\boldsymbol{\phi}, \boldsymbol{\theta}) = p(\boldsymbol{\phi} \mid \boldsymbol{\theta})p(\boldsymbol{\theta})$. This procedure seems to provide better results than that proposed by Jeffreys, although it depends on an arbitrary partition of the parametric vector, once reference priors are not invariant to the choice of the parameter partition. Another similar procedure, based on information maximization, was proposed by Zellner (1971).

Example 3.11 *(Location-scale model) If X has location-scale model, a reference prior (θ, σ) can be obtained following the procedure proposed by Bernardo with θ being the parameter of interest and σ the nuisance parameter. This partition corresponds to the majority of the cases of interest. Now, if θ is supposed known, we are restricted to a scale model and its prior is $p(\sigma \mid \theta) \propto \sigma^{-1}$. The distribution of $X \mid \theta$ is now obtained as*

$$p(x \mid \theta) = \int p(x \mid \sigma, \theta)\, p(\sigma \mid \theta)\, d\sigma,$$

where we can observe that the dependence on X and θ will continue to be of the form $f(x - \theta)$. Thus, the (marginal) reference prior for θ has constant density. Therefore, in a location-scale model, the reference prior is

$$p(\theta, \sigma) = p(\theta)\, p(\sigma \mid \theta) \propto \frac{1}{\sigma}.$$

This prior is also recommended by Jeffreys (1961) even though it is not the prior implied by straight application of his rule.

There are other difficulties associated with improper prior distributions. They may lead to incoherent inferences in the sense that if analysis conditional on the sample is replaced by analysis conditional on a sufficient statistic, which should not affect inferences, different posterior distributions may be obtained for some parameter transformations (see Exercise 3.25). This marginalization paradox is obviously an unpleasant situation that fortunately does not occur very often.

These criticisms only reinforce the point that non-informative priors must be used with extreme care. Nevertheless, if proper care is taken, they provide a useful benchmark in a number of problems and may be a useful input to the analysis.

3.5   Hierarchical priors

A good strategy to specify the prior distribution, or to better describe an experimental situation, is to divide it in stages inside a hierarchy. The idea of using a hierarchical structure with multiple stages to set a prior distribution was formalized by Lindley and Smith (1972). Following this recipe, the prior specification is made in two phases:

1. structural, for the division on stages, and

2. subjective, for quantitative specification at each stage.

Example 3.12 *Suppose that* $Y_1, \ldots, Y_n$ *are such that* $Y_i \sim N(\theta_i, \sigma^2)$, *with* σ^2 *known. Among many possibilities depending on the situation under study, many choices are available for specification of the prior for* $\boldsymbol{\theta} = (\theta_1, \ldots, \theta_n)$. *The following options can be used*

- *the* θ_i's *are independent, that is,* $p(\boldsymbol{\theta}) = \Pi_i p(\theta_i)$;

- *the* θ_i's *are a sample from a population with* $p(\theta \mid \boldsymbol{\lambda})$ *where* $\boldsymbol{\lambda}$ *contains the parameters describing the population.*

For this last option,

$$p(\boldsymbol{\theta} \mid \boldsymbol{\lambda}) = \prod_{i=1}^{n} p(\theta_i \mid \boldsymbol{\lambda}),$$

which corresponds to the specification of the first stage. To complete the prior setting, it is necessary to specify the second stage: the distribution of $\boldsymbol{\lambda}$, $p(\boldsymbol{\lambda})$.
One can then obtain the marginal prior distribution of $\boldsymbol{\theta}$ *by*

$$p(\boldsymbol{\theta}) = \int p(\boldsymbol{\theta}, \boldsymbol{\lambda}) \, d\boldsymbol{\lambda} = \int p(\boldsymbol{\theta} \mid \boldsymbol{\lambda}) \, p(\boldsymbol{\lambda}) \, d\boldsymbol{\lambda} = \int \prod_{i=1}^{n} p(\theta_i \mid \boldsymbol{\lambda}) \, p(\boldsymbol{\lambda}) \, d\boldsymbol{\lambda}.$$

Note that the θ_i's *are supposedly exchangeable as their subscripts are irrelevant in terms of this prior. Since the distribution of* $\boldsymbol{\lambda}$ *is independent of the first stage, it can be specified as*

1. *concentrated:* $P(\boldsymbol{\lambda} = \boldsymbol{\lambda}_0) = 1$, *for some value* $\boldsymbol{\lambda}_0 \in \Lambda$.

2. *discrete:* $P(\boldsymbol{\lambda} = \boldsymbol{\lambda}_j) = p_j, j = 1, \ldots, k$, *with* $\Sigma_j p_j = 1$. *In this case the distribution of* $\boldsymbol{\theta}$ *will be a finite mixture of the densities* $p(\boldsymbol{\theta} \mid \boldsymbol{\lambda}_j)$ *with weights* $p_j, j = 1, \ldots, k$.

3. *continuous: as above, the distribution of* $\boldsymbol{\theta}$ *will be a mixture of* $p(\boldsymbol{\theta} \mid \boldsymbol{\lambda})$ *but now with continuous weights given by* $p(\boldsymbol{\lambda})$.

If the first stage prior assumes that $\theta_i \sim N(\mu, \tau^2)$, $i = 1, \ldots, n$ *then* $\boldsymbol{\lambda} = (\mu, \tau^2)$. *Assuming that* $P(\tau^2 = \tau_0^2) = 1$ *and* μ *is normally distributed then*

$\boldsymbol{\theta}$ *has multivariate Normal distribution. On the other hand, assuming that* $P(\mu = \mu_0) = 1$ *and* τ^{-2} *has Gamma prior distribution implies that* $\boldsymbol{\theta}$ *has multivariate Student-t distribution.*

This sub-division in stages is a probabilistic strategy that allows easy iden-tification and specification of coherent priors. Nothing prevents these ideas from going further into the hierarchy. For example, the distribution of $\boldsymbol{\lambda}$ can depend on $\boldsymbol{\phi}$. In this case,

$$p(\boldsymbol{\theta}) = \int_\Phi \int_\Lambda p(\boldsymbol{\theta} \mid \boldsymbol{\lambda}) \, p(\boldsymbol{\lambda} \mid \boldsymbol{\phi}) \, p(\boldsymbol{\phi}) \, d\boldsymbol{\lambda} \, d\boldsymbol{\phi}.$$

The parameters $\boldsymbol{\lambda}$ and $\boldsymbol{\phi}$ are called hyperparameters and are introduced to ease the prior specification. Theoretically one can consider as many states as one thinks are necessary to improve prior specification. In practice, it is very hard to interpret the parameters of third or higher stages, so it is common practice to use a non-informative prior for these levels.

Example 3.13 *(The baseball players problem; Efron and Morris, 1975) A problem that is popular in Statistics literature consists on predicting the batting averages of the 18 best baseball players for the rest of the season based on the average obtained during the first 45 trials at bats. The sample of batting averages for 18 baseball players during the 1970 season is presented in Table 3.2, where* y_i *represents the batting average for the* i^{th} *player,* $i = 1, 2, \ldots, 18,$ *after* $n = 45$ *times at bats and* p_i *the true observed batting percentage observed at the end of the season.*

Table 3.2: *Batting Averages for 18 Major League Players for the 1970 Season.*

Player	y_i	p_i	Player	y_i	p_i
1	0.400	0.346	10	0.244	0.230
2	0.378	0.298	11	0.222	0.264
3	0.356	0.276	12	0.222	0.256
4	0.333	0.222	13	0.222	0.303
5	0.311	0.273	14	0.222	0.264
6	0.311	0.270	15	0.222	0.226
7	0.289	0.263	16	0.200	0.285
8	0.267	0.210	17	0.178	0.316
9	0.244	0.269	18	0.156	0.200

One prediction strategy is to predict the end of season batting average as the average in the first 45 trials (that is, $Y_i, i = 1, \ldots, 18$). *To improve those* *predictions, for example, to borrow strength from the other players, some additional assumptions will be needed. First of all it is reasonable to assume that the number of hits of each player in the first 45 at the bats position follows a Binomial distribution $nY_i \sim Bin(n, p_i)$, where $p_i \in (0, 1)$ is the true bats proportion of player i. An arcsine transformation is applied to the original data giving $X_i = n^{1/2} \sin^{-1}(2Y_i - 1), i = 1, 2, \ldots, k$, thus breaking the relationship existing among the mean and variance of the Binomial model. Actually, the variance is stabilized at 1 and the transformed mean is given as $E(x_i) = \theta_i = n^{1/2} arcsin(2p_i - 1)$. These results are easy to show and are therefore left as an exercise. The central limit theorem provides the approximation $X_i \mid \theta_i \sim N(\theta_i, 1)$, for $i = 1, \ldots, k$.*

Additionally, we assume that the observations are conditionally independent given θ_i and that a priori the θ_i's are random draws from a Normal population or, alternatively, the 18 best payers at the season can be thought of as exchangeable. Then, the hierarchical Normal model

$$\begin{aligned} X_i \mid \theta_i &\sim N(\theta_i, 1) \\ \theta_i \mid \mu, \tau^2 &\sim N(\mu, \tau^2), \quad for \quad i = 1, \ldots, k, \end{aligned}$$

follows. From Theorem 2.1, we know that

$$E(\theta_i \mid x_i) = \mu + \omega_i(x_i - \mu),$$

with $\omega_i = (1 + \tau^2)^{-1}$, which depends on μ and τ^2. Actually, those quantities are unknown and an extra hierarchical level will typically need to be included in the prior. One possibility, among many others, is to propose an independent non-informative prior for (μ, τ^2).

The concept of hierarchical modeling will be returned to in Chapters 4 and 8. Chapter 8 provides an introduction to more elaborate models where the full strength of hierarchical prior specification will be better appreciated.

Exercises

§3.1

1. Let θ represent the maximum temperature at your house door in September.

   (a) Determine, subjectively, the 0.25 and 0.5 quantiles of your prior distribution for θ.

   (b) Obtain the Normal density that fits best these quantiles.

(c) Find subjectively (without using the Normal density obtained in (b)) the 0.1 quantile of your prior distribution for θ. Is it consistent with the Normal obtained in (b)? What can you conclude from this fact?

2. Let θ be the probability that a football team from Rio de Janeiro will be the winner of the next Brazilian championship. Supposing that θ does not vary in time, build a prior distribution for θ based on past information.

§3.2

3. Show that the classes of Beta and Normal-Gamma distributions are closed under sampling.

§3.3

4. Show that the Beta family is conjugate with respect to the Binomial, geometric and negative Binomial sampling distributions.

5. For four pairs of a rare specimen of a bird that nested last season, the number of eggs per nest n and the number of eggs hatched Y were observed, providing the data $n = 2, 3, 3$ and 4 and $y = 1, 2, 3$ and 3. For a fifth pair nesting this season in similar conditions, $n_5 = 3$ eggs were observed and are about to hatch. Let θ be the probability that an egg is hatched.

(a) Obtain the likelihood function for θ, based on the observations $\mathbf{y} = (y_1, \dots, y_4)$.

(b) Assess a conjugate prior that in your opinion is adequate and calculate the posterior of $\theta \mid \mathbf{y}$.

(c) State a probabilistic model for Y_5, the number of eggs hatched in the 5th nest and obtain its predictive distribution.

6. Suppose that a random sample $\mathbf{X} = (X_1, \dots, X_n)$ from the $N(\theta, \sigma^2)$ distribution is observed with θ known and that a χ^2 prior is used for σ^{-2}. If the prior coefficient of variation (CV) of σ^{-2} is equal to 0.5, what must the value of n be to ensure that the posterior CV reduces to 0.1?
Note: The coefficient of variation of X is defined by $\sigma/|\mu|$, where $\mu = E(X)$ and $\sigma^2 = V(X)$.

7. Let $X_1, \dots, X_n$ be a random sample of the $N(\theta, \phi^{-1})$ distribution and consider the conjugate prior distribution for θ and ϕ.

(a) Determine the parameters $(\mu_0, c_0, n_0, \sigma_0)$ of the prior distribution, knowing that $E(\theta) = 0$, $Pr(|\theta| < 1.412) = 0.5$, $E(\phi) = 2$ and $E(\phi^2) = 5$.

(b) In a sample of size $n = 10$, $\bar{X} = 1$ and $\sum_{i=1}^{n}(X_i - \bar{X})^2 = 8$ were observed. Determine the posterior distribution of θ and sketch a graph of the prior, posterior and likelihood functions with ϕ fixed.

(c) Obtain $Pr(|Y| > 1 \mid \mathbf{x})$, where Y is a new observation taken from the same population.

8. A random sample $X_1, \ldots, X_n$ is selected from the $N(\theta, \sigma^2)$ distribution, with σ^2 known. The prior distribution for θ is $N(\mu_0, \sigma_0^2)$. What must the sample size be to reduce the variance:

(a) of the posterior distribution of θ to σ_0^2/k $(k > 1)$?

(b) of the predictive distribution of a future observation drawn from the same population to σ_0^2/k $(k > 1)$?

9. Consider the sampling model $\mathbf{X} \sim N(\boldsymbol{\theta}, \sigma^2 \mathbf{I}_p)$, where the p-dimensional mean vector $\boldsymbol{\theta}$ and the scalar σ^2 are known. Show that the distribution family given by $\boldsymbol{\theta} \mid \sigma^2 \sim N(\boldsymbol{\mu}, \sigma^2 \mathbf{C}_0)$ and $n_0 \sigma_0^2/\sigma^2 \sim \chi_{n_0}^2$ is conjugate to the model presented above, generalizing the results obtained in Section 3.3 for univariate Normal distributions.

10. Let $X_1, \ldots, X_n$ be a random sample from the $Pois(\theta)$ distribution.

(a) Determine the conjugate prior parameters for θ assuming that $E(\theta) = 4$ and $CV(\theta) = 0.5$ and determine n such that $V(\theta \mid \mathbf{x}) < 0.01$.

(b) Show that the posterior mean is of the form

$$\gamma_n \bar{x}_n + (1 - \gamma_n)\mu_0,$$

where $\mu_0 = E(\theta)$ and that $\gamma_n \to 1$ when $n \to \infty$.

(c) Repeat the previous item for a sample from a Bernoulli distribution with success probability θ and $\theta \sim Beta(a, b)$ a priori.

11. Let $\mathbf{X} = (X_1, \ldots, X_n)$ be a random sample of the $U(0, \theta)$ distribution.

(a) Show that the Pareto family of distributions, with parameters a and b, and density $p(\theta) = ab^a/\theta^{1+a}$, $\theta > b$, $(a, b > 0)$, is a conjugate family to the Uniform.

(b) Obtain the mode, mean and median of the posterior distribution of θ.

12. Consider the conjugate model Poisson-Gamma with $n = 1$. Obtain the predictive distribution using

(a) the usual integration procedures;

(b) the approach described in Section 3.3.

(c) Calculate also the mean and variance of this distribution.

13. Consider the conditions of Section 3.3.8. Show that

$$c_0(\theta - \mu_0)^2 + n(\theta - \bar{x})^2 = (c_0 + n)(\theta - \mu_1)^2 + \frac{c_0 n}{c_0 + n}(\mu_0 - \bar{x})^2$$

where $\mu_1 = (c_0\mu + n\bar{x})/(c_0 + n)$.

§3.4

14. Consider the observation of a sample $\mathbf{X} = (X_1, \ldots, X_n)$ with probability (density) function $p(x \mid \theta)$.

   (a) Show that

$$E_{\theta|\mathbf{x}}\left[\log \frac{p(\theta \mid \mathbf{x})}{p(\theta)}\right] \geq 0 \quad \forall \mathbf{x}$$

   with equality obtained only when $p(\theta \mid \mathbf{x}) = p(\theta)$.

   (b) Interpret the above result.

15. Let $X_i \sim p(x_i|\theta_i)$ and $p_i(\theta_i)$ the non-informative prior for θ_i, for $i = 1, \ldots, p$. Assuming that the X_i's are independent, show that the non-informative Jeffreys prior for $\boldsymbol{\theta} = (\theta_1, \ldots, \theta_p)$ is given by $\prod_{i=1}^{p} p_i(\theta_i)$.

16. Consider a random sample of size n from the Pareto distribution, as defined in Exercise 3.11, with parameters θ and b.

   (a) Show that there is a sufficient statistic of fixed dimension for θ.

   (b) Obtain the non-informative prior for θ. Is it improper?

   (c) Show that, for the prior in (b), the posterior for θ is $G(n, n \log(z/b))$, where z is the Geometric mean of the observations.

17. Suppose that $X \mid \theta \sim Exp(\theta)$ and that the prior for θ is non-informative.

   (a) Obtain the predictive distribution of X and show that $p(x)$ and $p(x \mid \theta)$ are monotonically decreasing in x.

   (b) Calculate the mode and the median of the sampling distribution and of the predictive distribution.

18. Suppose that $\mathbf{X} = (X_1, X_2, X_3)$ has a trinomial distribution with parameters n and $\boldsymbol{\pi} = (\pi_1, \pi_2, \pi_3)$, where $\pi_1 + \pi_2 + \pi_3 = 1$ and n is known, with density given by

$$f(x_1, x_2 \mid \theta_1, \theta_2) = \frac{n!}{x_1!x_2!(n - x_1 - x_2)!}\pi_1^{x_1}\pi_2^{x_2}(1 - \pi_1 - \pi_2)^{n - x_1 - x_2},$$

   where $x_i = 0, 1, \ldots, n$, $i = 1, 2$, $0 \leq x_1 + x_2 \leq n$. Show that the non-informative Jeffreys prior for $\boldsymbol{\pi}$ is $p(\boldsymbol{\pi}) \propto [\pi_1\pi_2(1 - \pi_1 - \pi_2)]^{-1/2}$.

19. Suppose that the lifetimes of n bulbs are exponentially distributed with mean θ.

   (a) Obtain a non-informative prior for θ and show that it is improper.

(b) Suppose that the times are observed up until r failures have occurred. Obtain the likelihood expression.

(c) Show that, if no bulbs fail before a pre-specified time limit $c > 0$ where observation stops, then the posterior distribution is also improper.

20. Suppose that θ has non-informative prior $p(\theta) \propto k$. Show that $\phi = a\theta + b$, $a \neq 0$, also has prior $p(\phi) \propto k$. Suppose now that θ has non-informative prior $p(\theta) \propto \theta^{-1}$, $\theta > 0$. Show that $\phi = \theta^a$, $a \neq 0$, also has non-informative prior $p(\phi) \propto \phi^{-1}$ and that $\psi = \log \theta$ has non-informative prior $p(\psi) \propto k$.

21. Let $\mathbf{X} = (X_1, X_2, X_3)$ be a random vector with trinomial distribution with parameters n and $(\theta_1, \theta_2, \theta_3)$. The statistician decides to reparametrize the problem defining $\lambda = \theta_1/(\theta_1+\theta_2)$ and $\psi = \theta_1+\theta_2$. (This is a valid procedure as far as the transformation from (θ_1, θ_2) to (λ, ψ) is 1-to-1.)

(a) Write the density of $\mathbf{X}$ as a function of λ and ψ.

(b) Show that $T = X_1+X_2$ is a sufficient statistic for ψ. Interpret this result in terms of the inference about ψ.

(c) Obtain the non-informative prior for ψ based on the result proved in (b).

(d) Interpreting ψ as the success probability in an experiment and supposing that in n realizations of the experiment t successes were observed, what is the probability of a future experiment being a success?

22. Show that Jeffreys prior $p(\boldsymbol{\theta}) \propto |I(\boldsymbol{\theta})|^{1/2}$ is invariant under one-to-one transformations; that is, if $\boldsymbol{\phi} = \boldsymbol{\phi}(\boldsymbol{\theta})$ is a one-to-one transformation of $\boldsymbol{\theta}$, then Jeffreys prior for $\boldsymbol{\phi}$ is $p(\boldsymbol{\phi}) \propto |\det I(\boldsymbol{\phi})]^{1/2}$.

23. Assuming that to work with an observation vector $\mathbf{X}$ and location parameter $\boldsymbol{\theta}$ is equivalent to working with an observation vector $\mathbf{Y} = \mathbf{X} + \mathbf{c}$ and location parameter $\boldsymbol{\eta} = \boldsymbol{\theta} + \mathbf{c}$ for any given constant $\mathbf{c}$ and insisting on the same non-informative prior specification for $\boldsymbol{\theta}$ as for $\boldsymbol{\eta}$, show that the only possible distribution for $\boldsymbol{\theta}$ has density $p(\boldsymbol{\theta}) \propto k$.

24. Repeat the above exercise under the conditions of the scale model to show that the non-informative prior must have density in the form $p(\sigma) \propto \sigma^{-1}$ by

(a) assuming equivalence between the model with observation X and scale parameter σ and the model with observation $Y = X/c$ and scale parameter $\eta = \sigma/c$.

(b) transforming the problem into a location model with observation $Z = \log X$ and location $\xi = \log \sigma$.

25. (Dawid, Stone and Zidek, 1973) Assume that independent observations

$X_1 \sim N(\mu_1, \sigma^2)$, $X_2 \sim N(\mu_2, \sigma^2)$ and $\nu S^2/\sigma^2 \sim \chi^2_\nu$ are made. Define $Z = (X_1 - X_2)/(S\sqrt{2})$ and the discrimination parameter $\delta = (\mu_1 - \mu_2)/(\sigma\sqrt{2})$.

(a) Show that the above observations correspond to data from two independent samples from Normal distributions with common variance, identifying corresponding parameters with those stated above.

(b) Obtain the joint posterior distribution of (μ_1, μ_2, σ^2) based on Jeffreys prior $p(\mu_1, \mu_2, \sigma^2) \propto 1/\sigma^2$.

(c) Show that the marginal posterior distribution of δ depends on the data only through Z.

(d) Obtain the sampling distribution of Z and show it depends on (μ_1, μ_2, σ^2) only through δ and, hence, that Z is a sufficient statistic for δ.

(e) Show that there is no prior distribution for δ that combined with the likelihood from item (d) would lead to the same posterior obtain in item (c).

26. Y has Student-t distribution with ν degrees of freedom, location μ and scale σ^2 if its density is

$$p(y|\nu) = c(\nu, \sigma^2)\left[\nu + \left(\frac{y-\mu}{\sigma}\right)^2\right]^{-(\nu+1)/2}, y \in \Re$$

where $c(\nu, \sigma^2) = \frac{\Gamma((\nu+1)/2)\nu^{\nu/2}}{\Gamma(\nu/2)\sqrt{\pi\sigma^2}}$.

(a) Consider $\mu = 0$ and $\sigma = 1$. Obtain the likelihood function of ν under the above model.

(b) Show that the limit of the likelihood in (a) when $\nu \to \infty$ is a constant $\delta \neq 0$ and find the value of δ.

(c) Show that Jeffreys prior for ν is given by

$$p(\nu) \propto \left\{2\,h(\nu) - \psi^{(2)}\left(\frac{\nu+1}{2}\right) + \psi^{(2)}\left(\frac{\nu}{2}\right) - \frac{2}{\nu}\right\}^{1/2},$$

where $\psi^{(2)}(z) = \frac{d^2}{dz^2}\log(\Gamma(z))$ is the trigamma function and $h(\nu) = \frac{2}{\nu+1} - \frac{\nu+2}{\nu(\nu+3)}$. Hint: Evaluate first $E[(\nu + Y^2)^{-k}] = \int_{-\infty}^{\infty} c(\nu,1)[\nu + y^2]^{-(\nu+2k+1)/2}dy$.

§3.5

27. Assume that the first stage prior specifies that $\theta_i \sim N(\mu, \tau^2)$, $i = 1, \ldots, n$ and define $\boldsymbol{\lambda} = (\mu, \tau^2)$.

(a) Assuming that $p(\tau^2 = \tau_0^2) = 1$ and μ is normally distributed, prove that $\boldsymbol{\theta}$ has Multivariate Normal distribution.

(b) Assuming that $p(\mu = \mu_0) = 1$ and τ^{-2} has Gamma prior distribution, prove that $\boldsymbol{\theta}$ has multivariate Student-t distribution.

28. (DeGroot, 1970, pg.154) Suppose that the prior distribution $p(\theta)$ for θ is built up hierarchically as follows:

(a) If $\xi = i$, then the conditional prior density for θ is $p_i(\theta)$, $i = 1, \ldots, k$.

(b) The marginal distribution of ξ is $p(\xi = i) = c_i$, $i = 1, \ldots, k$.

Suppose also that X, with density $p(x \mid \theta)$, is observed and define Ψ_i as the class, containing p_i of distributions that are conjugate to the sampling distribution of X, $i = 1, \ldots, k$ and Ψ as the class of distributions given by

$$\{ p : p(\theta) = \sum_{i=1}^{k} \beta_i \, p_i(\theta) \text{ and } p_i \in \Psi_i \}.$$

(a) Show that Ψ is conjugate to the distribution of X; that is, there exist constants b_i, $i = 1, \ldots, k$ such that

$$p(\theta \mid x) = \sum_{i=1}^{k} b_i \, p_i(\theta \mid x).$$

(b) Obtain the relationship between the constants b_i's and the prior and posterior probabilities of $\xi = i$, $i = 1, \ldots, k$.

Hint: Define $h_i(x) = \int p(x \mid \theta) p_i(\theta) \, d\theta$ and $p_i(\theta \mid x) = p(x \mid \theta) p_i(\theta) / h_i(x)$, $i = 1, \ldots, k$.

29. The IQ's of a sample of n senior undergraduate students of Statistics at Federal University of Rio de Janeiro (UFRJ), Brazil, are represented, respectively, by θ_i, $i = 1, \ldots, n$ and the common unknown mean for all the final year students at UFRJ is denoted by μ. Suppose that the θ_i's constitute a random sample of the population of IQ's, with unknown mean but with known variance b. A useful test to assess IQ's is applied, providing the independent observations $Y_1, \ldots, Y_n$, where $Y_i | \theta_i \sim N(\theta_i, a)$, with a known.

(a) Build a hierarchical prior for the parameters $\theta_1, \ldots, \theta_n$.

(b) Calculate $p(\mu | y_1, \ldots, y_n)$ and obtain $E(\mu | y_1, \ldots, y_n)$.

(c) Obtain $p(\theta_i \mid \mu, y_1, \ldots, y_n)$ and $E(\theta_i | \mu, y_1, \ldots, y_n)$, for $i = 1, \ldots, n$.

(d) Obtain $E(\theta_i | y_1, \ldots, y_n)$, for $i = 1, \ldots, n$.

30. Suppose that the prior distribution for $(\theta_1, \ldots, \theta_n)$ is such that the θ_i's constitute a random sample from a $N(\mu, \tau^2)$ distribution and $\mu \mid \tau^2 \sim N(\mu_0, \tau^2/c)$. Obtain the prior distribution for $(\theta_1, \ldots, \theta_n)$ and, in particular, calculate the covariance between θ_i and θ_j, $1 \le i, j \le n$ with $i \ne j$ assuming that

(a) τ^2 is known;

(b) τ^2 is unknown with prior distribution $\tau^{-2} \sim G(\alpha, \beta)$.

Chapter 4

Estimation

One of the central problems of statistical inference is discussed in this chapter. The general problem of decision making is briefly described to motivate estimation as a special case. Estimation is then treated from both Bayesian and frequentist points of view.

In Section 4.1, the decision problem is defined, the concepts of loss function and Bayes risk are discussed and the connection between decision theory and estimation is presented. Different loss functions are considered in the definition of different decision problems. In Section 4.2 Bayes estimators ensuing from these losses are presented. The more important classical methods of estimation, namely maximum likelihood, least squares and method of moments, are presented in Section 4.3. Properties of these methods are extensively discussed. Empirical Bayes estimation is presented in Section 4.4, along with estimation procedures that penalize the likelihood. In Section 4.5, methods of comparison of these estimators are defined. The concepts of bias, (classical) risk and consistency of an estimator are introduced. Following a discussion on point estimation, interval estimation is presented in Section 4.6. Finally, the results are applied in Section 4.7 to the estimation of means and variances in Normal models. Results concerning approximate methods of estimation, including asymptotics, are deferred to the next chapter.

4.1 Introduction to decision theory

A decision problem is completely specified by the description of three spaces:
1. Parameter (or states of the nature) space Θ;

2. Space of possible results of an experiment Ω;

3. Space of possible actions $\mathcal{A}$.

Three essential concepts in decision theory are the decision rule, the loss function and the risk (function) of a decision rule, which are presented as follows

Definition 4.1 *Let a denote a particular action taken and θ denote the state of nature, then $L(\theta, a)$ is the loss incurred by taking action a when θ occurs. So the loss function $L(\theta, a)$ is defined for all $(\theta, a) \in \Theta \times \mathcal{A}$ and takes values in R^+.*

The loss function can be interpreted as the punishment that one suffers for taking decision a when the value of the parameter is θ. This function is denoted by $L(a, \theta)$. Table 4.1 shows a common discrete representation to establish the losses related to possible actions and states of the nature.

Table 4.1 *Losses Related to the Actions $a_1, \ldots, a_p$ and to the State of the Nature $\theta_1, \ldots, \theta_n$.*

State the Nature	Actions			
	a_1	a_2	$\ldots$	a_p
θ_1	$L(a_1, \theta_1)$	$L(a_2, \theta_1)$	$\ldots$	$L(a_p, \theta_1)$
$\vdots$	$\vdots$	$\vdots$		$\vdots$
θ_n	$L(a_1, \theta_n)$	$L(a_2, \theta_n)$	$\ldots$	$L(a_p, \theta_n)$

Since $L(a, \theta)$ will never be known with certainty, one natural method to deal with uncertainty involved in making decisions is to consider the expected loss, and then choose an optimal decision with respect to the expected loss.

Definition 4.2 *A decision rule δ is a function defined in Ω with values in $\mathcal{A}$, that is, $\delta : \Omega \to \mathcal{A}$. It represents the action to be taken if $\boldsymbol{X} = \boldsymbol{x}$ is the observed value from a sample.*

Example 4.1 *Suppose that a doctor must decide if a patient (for example, John from previous chapters) with a given disease must undergo surgery or not. The states of nature are as follows: John is sick ($\theta = 1$) or healthy ($\theta = 0$). Let us simplify the problem by assuming that the doctor will prescribe surgery ($\delta = 1$) only if he thinks John is sick. This way, a decision rule δ directly related to the value of θ gets established. Unfortunately, the value of the parameter θ is unknown for the decision maker, the doctor. Table 4.2 is a possible representation of the losses (measured in a hypothetical monetary unit) associated with all combinations of values of the action and states of nature.*

These losses represent the subjective evaluation of the decision maker with respect to the combination of actions and states of the nature. The smallest loss is null, occurring when the patient is healthy and does not undergo surgery. The largest loss occurs when the patient is sick but is not prescribed surgery. This implies a loss in the doctor's reputation and may even lead him to legal problems. Note that all losses are non-negative and do not take into account the

Table 4.2: *Losses Associated with the Doctor Problem.*

θ	δ	
	No Surgery	Surgery
Healthy	0	500
Sick	1000	100

doctor's fee which is constant or at least should be immaterial to the problem considered.

Roughly speaking, from the classical point of view the decision is made based only on the experiment. On the other hand, from the Bayesian point of view, prior information relevant to the problem can be added to the information obtained by the experiment. We first focus on the classical decision theory.

4.1.1   Classical decision theory

The quality of a decision rule $\delta(x)$ may be quantified via its risk, defined as follows.

Definition 4.3 *The risk (function) of a decision rule $\delta(x)$ is defined as*

$$R(\theta, \delta) = E_{X|\theta}[L(\theta, \delta(x))] = \int_{\mathcal{X}} L(\theta, \delta(x))p(x|\theta)dx,$$

where $p(x|\theta)$ is the sampling probability (density) function of the observation X. The risk can be interpreted as the average loss, at a given θ, that will be incurred if the decision rule $\delta(x)$ is considered.

The importance of the risk is the introduction of a measure that enables one to rank different decision rules. From the practical point of view, it is desirable to use a decision rule which has a small risk regardless of the true value of θ. A decision rule δ_1 is a better decision than a decision rule δ_2 if $R(\theta, \delta_1) \leq R(\theta, \delta_2)$, for all $\theta \in \Theta$. A decision rule δ_1 is equivalent to the decision rule δ_2 if $R(\theta, \delta_1) = R(\theta, \delta_2)$, for all $\theta \in \Theta$. Comparisons may be formalized by the definitions below.

Definition 4.4 *A decision rule δ is admissible if there is no decision rule better than δ. That is, a decision rule δ is an admissible decision rule if $R(\theta, \delta) \leq R(\theta, \delta^*)$, for all $\theta \in \Theta$ and all $\delta^* : \Omega \to \mathcal{A}$. A decision rule δ is inadmissible if it is not admissible.*

However, in some cases decision δ_1 may be preferred (have smaller risk) than δ_2 for some values of θ, while for other values δ_2 may be preferred. This situation makes it difficult to rank these decisions. In this context, the minimax rule may be applied to force a choice between decision rules.

Definition 4.5 *A decision rule is minimax if it minimizes the quantity* $\sup_\theta R(\theta, \delta)$ *over all decision rules* $\delta \in D$. *That is, a minimax decision rule* δ^* *is given by*

$$\delta^* = \inf_{\delta \in D} \sup_{\theta \in \Theta} R(\theta, \delta).$$

Example 4.2 *(Example 4.1 continued) Suppose we are interested in the doctor's minimax decision rule for the patient with a given disease who may undergo surgery (δ_1) or not (δ_2). Since* $\max_\theta L(\delta_1, \theta) = \max\{500, 100\} = 100 < \max_\theta L(\delta_2, \theta) = \max\{0, 1000\} = 1000$, δ_1 *is the doctor's minimax decision rule.*

Remember that probability considerations about states of nature are ruled out in the frequentist approach. The minimax rule considers only the worst case scenario, irrespective of how likely it may be, to circumvent this difficulty.

4.1.2 Bayesian decision theory

From the Bayesian point of view, a decision is guided by taking into account the uncertainty about the unknown state of the nature involved in the problem, as a counterpart to the structure considered in the classical point of view.

Consider the posterior density exhibited in Figure 4.1. This density is not completely uncommon and illustrates the difficulties that may be associated in the learning process involved in a statistical procedure. Nevertheless, this density contains all that is available in terms of probabilistic description of our information about a quantity of interest. Any attempt to summarize the information contained in this density must be made with caution. The graph of the posterior density is the best description of the inferential process. Sometimes, however, it is useful to summarize the information further into a few numerical figures for communication purposes.

The risk and Bayes risk are then defined as following.

Definition 4.6 *The risk of a decision rule, denoted by $R(\delta)$, is the expected posterior loss given by*

$$R(\delta) = E_{\theta|x}[L(\delta, \theta)] = \int_\Theta L(\theta, \delta(x)) p(\theta|x) d\theta.$$

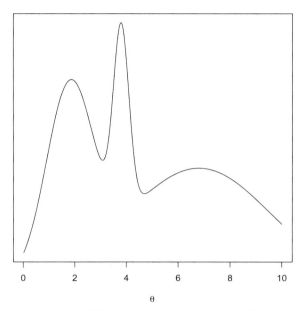

Figure 4.1 *Posterior density of θ with 3 distinct regions: the first containing around 30% of the total probability, the second with 10% and the third with around 60%. The mode of this density is 3.78, the mean is 4.54 and the median is 4.00.*

Definition 4.7 *A decision rule δ^* is optimal if it has minimum risk, namely $R(\delta^*) < R(\delta)$, $\forall \delta$. This rule is called the Bayes rule and its risk is called the Bayes risk.*

Example 4.3 *(Example 4.1 continued) In this case, the unknown is θ and let us assume that its uncertainty is described by its updated distribution, $Pr(\theta = 1) = \pi$ and $Pr(\theta = 0) = 1 - \pi$, for $0 \le \pi \le 1$. This can be a prior distribution or a posterior distribution, obtained after a few tests have been carried out on the patient. Evaluation of the risk of an action δ is straightforward and*

$$R(\delta = 0) = E_\theta[L(\delta = 0, \theta)] = 0(1 - \pi) + 1000\pi = 1000\pi$$

and

$$R(\delta = 1) = E_\theta[L(\delta = 1, \theta)] = 500(1 - \pi) + 100\pi = 500 - 400\pi.$$

As can be seen from Figure 4.2, the two actions have equal risk if $R(\delta = 0) = R(\delta = 1)$, which happens iff $1000\pi = 500 - 400\pi$, or $1400\pi = 500$ or $\pi = 5/14$. For $\pi < 5/14$, the risk associated with $\delta = 0$ is smaller than the

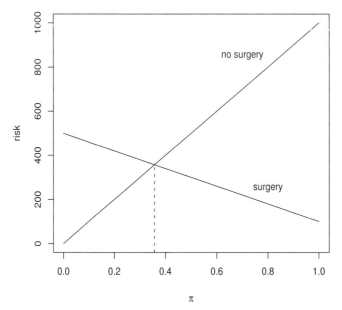

Figure 4.2 *Risks associated with the 2 actions: surgery or not, as functions of the probability of disease* π.

risk associated with $\delta = 1$. In this case, $\delta = 0$ is the Bayes rule and the Bayes risk is 1000π. For $\pi > 5/14$, the problem is reversed, the Bayes rule is $\delta = 1$ and the Bayes risk is $500 - 400\pi$. In summary, the doctor's strategy must be to prescribe John surgery if $\pi > 5/14$.

Example 4.3 shows how sensitive the decision is to the choice of priors. But it seems to be fairer than the minimax rule, which appears to be too conservative. It is important to study also the sensitivity of the decision to the choice of loss function. Estimation is clearly dependent on the specified losses, and variation to their values may lead to different decisions.

4.1.3   Statistical inference as a decision problem

Statistical inference involves making decisions. After the data is observed, say, $X = x$, a decision is made on θ, with the space of possible actions denoted by $\mathcal{A}$. The nature of the space $\mathcal{A}$ of the possible actions will determine whether a given inferential problem is a point estimation, an interval estimation or a hypothesis testing. The first case is the simplest of them all and assumes that

$\mathcal{A} = \Theta$, namely the decisions are characterized by selection of a single value for θ. This case will be considered in the next section.

Interval estimation is the decision problem where $\mathcal{A}$ is the set of all possible subsets of Θ, namely the decisions are characterized by selection of a given interval for θ and will be considered later in this chapter. Finally, hypothesis testing is the decision problem where $\mathcal{A}$ are the values of an indicator function $I_\theta(A)$ where $A \subset \Theta$, that is, the decisions are characterized by determination of whether $\theta \in A$ or not. This type of problem can be generalizes from a binary set of actions to a finite set of actions, leading to the multiple decision problem. For conciseness, we shall concentrate only on the dichotomous case of hypothesis testing, to be addressed in Chapter 6.

In what follows, we consider an important definition for point estimation.

Definition 4.8 *Assume that $\mathcal{A} = \Theta$. An estimator is an optimal decision rule with respect to a given loss function. Its observed value is called an estimate.*

This definition is broad enough to be useful in both classical and Bayesian perspectives, but, from the classical point of view, other optimality criteria will be introduced later. The next section presents further connections between decision theory and Bayesian point estimation.

4.2 Bayesian point estimation

Most loss functions are based on symmetric functions of the form $L(\delta, \theta) = h(\delta - \theta)$, for some function h. These are the most commonly used loss functions.

Lemma 4.1 *Let $L_1(\delta, \theta) = |\delta - \theta|$ be the loss associated with the estimation of θ. The estimator of θ is $\delta_1 = med(\theta)$, the median of the updated distribution of θ.*

The proof of this lemma is cumbersome and will be left as an exercise.

Lemma 4.2 *Let $L_2(\delta, \theta) = (\delta - \theta)^2$ be the loss associated with the estimation of θ by δ. (This loss is usually known as quadratic loss.) The estimator of θ is $\delta_2 = E(\theta)$, the mean of the updated distribution of θ.*

Proof We have to calculate the risk and show that δ_1 minimizes it. So, $R(\delta) = E[(\delta - \theta)^2] = E\{[(\delta - \delta_2) + (\delta_2 - \theta)]^2\}$, where $\delta_2 = E(\theta)$. Therefore,

$$
\begin{aligned}
R(\delta) &= E_\theta[(\delta - \delta_2)^2] + E_\theta[(\delta_2 - \theta)^2] + 2E_\theta[(\delta - \delta_2)(\delta_2 - \theta)] \\
&= (\delta - \delta_2)^2 + E_\theta[(\delta_2 - \theta)^2] + 2(\delta - \delta_2)E_\theta[\delta_2 - \theta] \\
&= (\delta - \delta_2)^2 + E_\theta[(\delta_2 - \theta)^2], \text{ since } \delta_2 = E(\theta) \\
&= (\delta - \delta_2)^2 + V(\theta)
\end{aligned}
$$

and the risk is minimized for $\delta = \delta_2$. In this case, the Bayes risk is $R(\delta_2) = V(\theta)$ and $R(\delta_2) \leq R(\delta)$, $\forall \delta$, with equality iff $\delta = \delta_2$. $\square$

The quadratic loss is sometimes criticized for introducing a penalty that increases very strongly with the estimation error $\delta - \theta$. In many cases, it is desirable to have a loss function that does not overly emphasize large estimation errors. The estimator associated with the absolute loss function considers punishments increasing only linearly with the estimation error.

Another form to reduce the effect of large estimation errors is to consider loss functions that remain constant whenever $|\delta - \theta| > k$ for some k arbitrary. There is some freedom for options of suitable values of k. The most common choice is the limiting value as $k \to 0$. This loss function associates a fixed loss when an error is committed, irrespective of its magnitude. This loss is usually known as the $0 - 1$ loss.

Lemma 4.3 *Let* $L_\infty(\delta, \theta) = \lim_{\varepsilon \to 0} I_{|\theta - \delta|}([\varepsilon, \infty))$. *The estimator of* θ *is* $\delta_\infty = mode(\theta)$, *the mode of the updated distribution of* θ.

Proof (continuous case)

$$E[L_3(\delta, \theta)] = \lim_{\varepsilon \to 0} \left[\int_{-\infty}^{\delta - \varepsilon} 1 \cdot p(\theta) \, d\theta + \int_{\delta - \varepsilon}^{\delta + \varepsilon} 0 \cdot p(\theta) \, d\theta + \int_{\delta + \varepsilon}^{\infty} 1 \cdot p(\theta) d\theta \right]$$

$$= \lim_{\varepsilon \to 0} \left[1 - \int_{\delta - \varepsilon}^{\delta + \varepsilon} p(\theta) \, d\theta \right]$$

$$= 1 - \lim_{\varepsilon \to 0} P(\delta - \varepsilon < \theta < \delta + \varepsilon).$$

But $\lim_{\varepsilon \to 0} P(\delta - \varepsilon < \theta < \delta + \varepsilon)/(2\varepsilon) = p(\delta)$. Thus, $E[L_\infty]$ is minimized when $p(\delta)$ is maximized. Hence, $\delta_\infty = $ mode (θ). $\square$

When the updated distribution is the posterior, the estimator associated with the $0 - 1$ loss is the posterior mode. This is also referred to as the generalized maximum likelihood estimator (GMLE, in short). In the next section, we will see the reason for the name. The GMLE is the easiest to be obtained among the estimators presented so far. In the continuous case, it typically involves finding the solution of the equation $dp(\theta \mid \mathbf{x})/d\theta = 0$. Figure 4.3 illustrates the variation of the loss functions considered here as a function of the estimation error.

Many of these results can be generalized to the multivariate case. Apart from the absolute value that has no clear extension to the multivariate case, the quadratic and $0 - 1$ loss can be respectively extended by

$$L_2(\boldsymbol{\delta}, \boldsymbol{\theta}) = (\boldsymbol{\delta} - \boldsymbol{\theta})'(\boldsymbol{\delta} - \boldsymbol{\theta}) \ \text{ and } \ L_\infty(\boldsymbol{\delta}, \boldsymbol{\theta}) = \lim_{vol(\mathbf{A}) \to 0} I_{\boldsymbol{\delta} - \boldsymbol{\theta}}(\mathbf{A}),$$

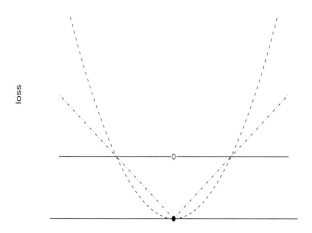

Figure 4.3: *Loss functions: $0 - 1$, ——; absolute, $- - - - -$; quadratic, $\cdot - \cdot - \cdot$.*

where $\mathbf{A}$ is a region containing the origin and $vol(\mathbf{A})$ is the volume of the region $\mathbf{A}$. It is not difficult to show that the Bayes estimators of $\boldsymbol{\theta}$ under loss functions L_2 and L_∞ are respectively given by the joint mean and joint mode of the updated distribution of $\boldsymbol{\theta}$. These concepts are treated with greater depth by Berger (1985), DeGroot (1970) and Ferguson (1967).

Example 4.4 *Let $\boldsymbol{X} = (X_1, \ldots, X_n)$ be a sample from a $N(\theta, \sigma^2)$ distribution and $\phi = \sigma^{-2}$. We have previously seen that in a conjugate analysis, the posterior distribution is $\theta \mid \phi \sim N(\mu_1, (c_1\phi)^{-1})$ and $(n_1\sigma_1^2)\phi \sim \chi^2_{n_1}$.*

To ease derivation of the mode of the joint distribution, it is usual to work with the logarithm of $p(\theta, \phi \mid \boldsymbol{x})$,

$$\log p(\theta, \phi \mid \boldsymbol{x}) = k - \frac{\phi}{2}\left[c_1(\theta - \mu_1)^2 + n_1\sigma_1^2\right] + \left(\frac{n_1 + 1}{2} - 1\right)\log \phi.$$

Differentiation with respect to θ and ϕ leads to

$$\frac{\partial \log p(\theta, \phi \mid \boldsymbol{x})}{\partial \theta} = -\frac{\phi}{2}\left[2c_1(\theta - \mu_1)\right]$$
$$= -\phi c_1(\theta - \mu_1) \Rightarrow \theta = \mu_1 \text{ is a critical point and}$$
$$\frac{\partial \log p(\theta, \phi \mid \boldsymbol{x})}{\partial \phi} = -\frac{c_1(\theta - \mu_1)^2 + n_1\sigma_1^2}{2} + \left(\frac{n_1 + 1}{2} - 1\right)\frac{1}{\phi}.$$

Making $\partial \log p(\mu_1, \hat{\phi} \mid \boldsymbol{x})/\partial \phi = 0$ gives $\hat{\phi} = \sigma_1^{-2}(n_1 - 1)/n_1$. The second order *conditions are satisfied as*

$$\frac{\partial^2 \log p(\theta, \phi \mid x)}{\partial \theta^2}\bigg|_{\theta=\mu_1, \phi=\hat{\phi}} = -c_1\hat{\phi} < 0$$

$$\frac{\partial^2 \log p(\theta, \phi \mid x)}{\partial \phi^2}\bigg|_{\theta=\mu_1, \phi=\hat{\phi}} = -\left(\frac{n_1+1}{2} - 1\right)\frac{1}{\hat{\phi}^2} < 0$$

$$\frac{\partial^2 \log p(\theta, \phi \mid x)}{\partial \theta \partial \phi}\bigg|_{\theta=\mu_1, \phi=\hat{\phi}} = 0.$$

Therefore, $(\mu_1, \hat{\phi})$ is the mode of the joint posterior distribution of (θ, ϕ).

The above calculations do not guarantee that μ_1 is the maximum of the marginal distribution of θ and $\hat{\phi}$ is the maximum of the marginal distribution of ϕ. In this example, it is easy to see that μ_1 is also the marginal mode since the marginal distribution of θ is a Student-t centered at μ_1. This automatically implies that μ_1 is the mean (when it exists) and the median of the marginal posterior distribution of θ.

However, the same is not true with $\hat{\phi}$. It was shown in Section 3.3.8 that $\phi \mid x \sim G(n_1/2, n_1\sigma_1^2/2)$. This distribution has mode

$$\tilde{\phi} = \frac{n_1 - 2}{n_1\sigma_1^2} \neq \hat{\phi}.$$

Note also that the posterior mean of ϕ is σ_1^{-2} and the posterior median cannot be explicitly evaluated.

Another important consequence from probability theory is that the mode and the mean are not invariant under transformations. Let $\sigma^2 = \phi^{-1}$ and denote the mode of σ^2 by $\tilde{\sigma}^2$. $\hat{\phi}^{-1}$ is not the joint nor the marginal mode of σ^2. To evaluate the mode of σ^2, the posterior distribution of σ^2 must be obtained. For the case of the marginal distribution,

$$p(\sigma^2 \mid x) = p(\phi(\sigma^2) \mid x)\left|\frac{d\phi}{d\sigma^2}\right| \quad where \quad \left|\frac{d\phi}{d\sigma^2}\right| = \left|-\frac{1}{\sigma^4}\right| = \frac{1}{\sigma^4} = (\sigma^{-2})^2$$

$$\propto (\sigma^{-2})^{(n_1/2)-1+2}\exp\left(-\frac{n_1\sigma_1^2}{2\sigma^2}\right)$$

and its logarithm is $\log p(\sigma^2 \mid x) = k - \left(\frac{n_1}{2} + 1\right)\log \sigma^2 - \frac{n_1\sigma_1^2}{2\sigma^2}$.

Differentiating it with respect to σ^2 leads to

$$\frac{d\log p(\sigma^2 \mid x)}{d\sigma^2}\bigg|_{\sigma^2=\tilde{\sigma}^2} = -\left(\frac{n_1}{2} + 1\right)\frac{1}{\tilde{\sigma}^2} + \frac{n_1\sigma_1^2}{2\tilde{\sigma}^4} = 0.$$

The solution of the equation is $\tilde{\sigma}^2 = \frac{n_1\sigma_1^2}{n_1+2} \neq \frac{n_1\sigma_1^2}{n_1-2} = \tilde{\phi}^{-1}$.

The second order condition guarantees that $\tilde{\sigma}^2$ is the marginal posterior mode since

$$\frac{d^2 \log p(\tilde{\sigma}^2 \mid x)}{d(\sigma^2)^2} = \left(\frac{n_1}{2} + 1\right)\frac{1}{\tilde{\sigma}^4} - 2\frac{n_1 \sigma_1^2}{2\tilde{\sigma}^6} = -\frac{1}{2}\frac{(n_1 + 2)^3}{(n_1 \sigma_1^2)^2} < 0.$$

4.3 Classical point estimation

In the Bayesian methodology, point estimation is always dealt with by mini-mization of the expected loss. In the classical perspective, many methods have been proposed in an effort to suit the methodology to a variety of problems. In this section, the three most important methods will be cited: the method of maximum likelihood, the method of least squares and the method of moments. We will also briefly present non-parametric estimation. Classical point esti-mation is covered in a clear and concise way in the books by Cox and Hinkley (1974) and Silvey (1970). We recommend both books to the interested reader.

4.3.1 Maximum likelihood

This is currently the most used method of estimation in classical inference. Its use is justified in many instances, and it is useful to note that it is entirely based on the likelihood function. Therefore, it does not violate the likelihood principle. In addition, there is a good body of theory developed in a variety of situations. Its intuitive appeal can be grasped in the very simple example below.

Example 4.5 *Consider a situation where all that is known about an unknown quantity of interest θ is that its values are either 1/4 or 3/4. Assume now that a Bernoulli random variable X is observed and the success probability of X is either 1/6, when $\theta = 1/4$ or 4/5, when $\theta = 3/4$. This probabilistic setup is summarized in Table 4.3.*

Table 4.3: *Table of Probabilities.*

x	θ	
	1/4	3/4
0	5/6	1/5
1	1/6	4/5

Observe that the sum of each column is 1 but the sum of each line is not. Given the value of X, the likelihood function of θ can be constructed as $l(\theta; x) = p(x \mid \theta)$.

If $X = 1$ is observed, $l(1/4; x = 1) = 1/6 < 4/5 = l(3/4; x = 1)$. This means that the model with $\theta = 3/4$ attaches a larger probability to the observed event than the model with $\theta = 1/4$ and therefore seems more plausible or likely. If we had to choose an estimate for θ after observing only that $X = 1$, we would probably opt for the value $3/4$.

If $X = 0$ is observed, the same reasoning would lead to the choice of the value $1/4$ for θ since $l(1/4; x = 0) = 5/6 > 1/5 = l(3/4; x = 0)$.

So, in the above example, we have opted to estimate θ by the value that maximizes the likelihood function for every value of x. This simple and powerful idea is the basis of the method.

Definition 4.9 *Consider the observation of $\boldsymbol{X}$ with joint density $p(\boldsymbol{x}|\boldsymbol{\theta})$. The maximum likelihood estimator (MLE, in short) of $\boldsymbol{\theta}$ is the value of $\boldsymbol{\theta} \in \Theta$ that maximizes $l(\boldsymbol{\theta}; \boldsymbol{X})$. The usual notation for the MLE of $\boldsymbol{\theta}$ is $\hat{\boldsymbol{\theta}}$. Its observed value is called the maximum likelihood estimate.*

In most cases, a p-dimensional $\boldsymbol{\theta}$ varies continuously over a region of R^p. In these cases, irrespective of whether $\boldsymbol{X}$ contains discrete variables or not, the MLE can typically be found by solving the equation $\partial l(\boldsymbol{\theta}; \boldsymbol{X})/\partial \boldsymbol{\theta} = \mathbf{0}$, or equivalently $\partial \log l(\boldsymbol{\theta}; \boldsymbol{X})/\partial \boldsymbol{\theta} = \mathbf{0}$, where $\mathbf{0}$ is a vector of 0's.

We are now in position to compare the GMLE, presented in the previous section, with the MLE. The GMLE generalizes the maximum likelihood estimator just as the posterior density generalizes the likelihood function. Recall that $p(\boldsymbol{\theta} \mid \boldsymbol{X}) \propto l(\boldsymbol{\theta}; \boldsymbol{X})p(\boldsymbol{\theta})$. In the special case, $p(\boldsymbol{\theta}) \propto k$, it follows that $p(\boldsymbol{\theta} \mid \boldsymbol{X}) \propto l(\boldsymbol{\theta}; \boldsymbol{X})$. Therefore, the value of $\boldsymbol{\theta}$ that maximizes the posterior also maximizes the likelihood. So, the MLE is the GMLE in the case $p(\boldsymbol{\theta}) \propto k$.

Despite their similarity, it is important to stress the distinction between classical and Bayesian estimators. The first ones are statistics and therefore have a sampling distribution based on which their properties will be established. Bayesian estimators are based on the posterior distribution which is always conditional on the observed value of the sample, and therefore, their properties are based on the posterior distribution, an entirely different object. Nevertheless, they can be seen as functions of the observed sample and, in this way, compared numerically with classical estimate.

Example 4.6 *Let $\boldsymbol{X} = (X_1, \ldots, X_n)$ be a sample from the $N(\theta, \sigma^2)$ distribution. The likelihood function is*

$$l(\theta, \sigma^2; \boldsymbol{X}) = \left(\frac{1}{\sqrt{2\pi\sigma^2}}\right)^n \exp\left\{-\frac{1}{2\sigma^2} \sum_{i=1}^n (X_i - \theta)^2\right\},$$

with logarithm

$$\log l(\theta, \sigma^2; \boldsymbol{X}) = -\frac{n}{2} \log(2\pi\sigma^2) - \frac{1}{2\sigma^2} \left[\sum_{i=1}^{n} (X_i - \overline{X})^2 + n(\overline{X} - \theta)^2 \right].$$

Differentiating with respect to θ and σ^2 gives

$$\frac{\partial \log l(\theta, \sigma^2; \boldsymbol{X})}{\partial \theta} = \frac{1}{2\sigma^2} (2n)(\overline{X} - \theta)$$

and

$$\frac{\partial \log l(\theta, \sigma^2; \boldsymbol{X})}{\partial \sigma^2} = -\frac{n}{2\sigma^2} + \frac{1}{2(\sigma^2)^2} \left[(n-1)S^2 + n(\overline{X} - \theta)^2 \right],$$

where

$$S^2 = \frac{1}{n-1} \sum_{i=1}^{n} (X_i - \overline{X})^2.$$

Observe that the denominator $(n-1)$ used in the expression of S^2, was modified with respect to the value n used in previous chapters.

Equating derivatives to 0 leads to $\hat{\theta} = \overline{X}$ and $\hat{\sigma}^2 = (n-1)S^2/n$ as critical points.

Differentiating again gives

$$\frac{\partial^2 \log l(\theta, \sigma^2; \boldsymbol{X})}{\partial \theta^2} = -\frac{n}{\sigma^2} < 0, \forall (\theta, \sigma^2) \in R \times R^+,$$

$$\frac{\partial^2 \log l(\theta, \sigma^2; \boldsymbol{X})}{\partial \theta \, \partial \sigma^2} \Big|_{\theta = \hat{\theta}, \sigma^2 = \hat{\sigma}^2} = 0$$

and

$$\frac{\partial^2 \log l(\theta, \sigma^2; \boldsymbol{x})}{\partial (\sigma^2)^2} \Big|_{\theta = \hat{\theta}, \sigma^2 = \hat{\sigma}^2} = \frac{n}{2(\hat{\sigma}^2)^2} - \frac{1}{(\hat{\sigma}^2)^3} [(n-1)S^2 + n(\overline{X} - \hat{\theta})^2]$$

$$= -\frac{n}{2} \frac{1}{(\hat{\sigma}^2)^2} < 0.$$

Therefore, $(\hat{\theta}, \hat{\sigma}^2)$ is the maximum likelihood estimator of (θ, σ^2).

This result can also be established from the similarities between generalized maximum likelihood estimator and MLE and calculations in the previous section. Note that by taking $c_0 \to 0$, $\sigma_0^2 \to 0$ and $n_0 = 2$, the prior becomes constant and the above result becomes a special case of the derivations of the previous section.

There are many interesting properties of the MLE's that can be easily shown:

1. The maximum likelihood estimator is invariant to 1-to-1 transformations (unlike the GMLE). If $\hat{\boldsymbol{\theta}}$ is the maximum likelihood estimator
   of $\boldsymbol{\theta}$ and $\phi = \phi(\boldsymbol{\theta})$ is a 1-to-1 function of $\boldsymbol{\theta}$, then the MLE of ϕ is
   given by $\hat{\phi} = \phi(\hat{\boldsymbol{\theta}})$. To see that, remember that $\hat{\boldsymbol{\theta}}$ maximizes $l(\boldsymbol{\theta})$
   and that the likelihood of ϕ is $l^*(\phi) = l^*(\phi(\boldsymbol{\theta})) = l(\boldsymbol{\theta})$. Therefore,
   if $\hat{\boldsymbol{\theta}}$ maximizes l, it will also maximize $l^*(\phi(\boldsymbol{\theta}))$ by uniqueness of the
   transformation and $\phi(\hat{\boldsymbol{\theta}})$ will maximize l^*. Consequently, $\hat{\phi}$ maximizes
   l^*, and therefore, $\hat{\phi}$ is the MLE of ϕ.

   Example 4.7 *In the case of an $N(\theta, \sigma^2)$ distribution, the maximum
   likelihood estimator of σ^2 is $\hat{\sigma}^2$. Therefore, the maximum likelihood
   estimator of the standard deviation σ is $\hat{\sigma}$ and the maximum likelihood
   estimator of the precision $\phi = \sigma^{-2}$ is $\hat{\phi} = \hat{\sigma}^{-2}$.*

2. As briefly mentioned earlier, the maximum likelihood estimator does
   not depend on the sampling plan. If different experiments $\mathcal{E}_1$ and $\mathcal{E}_2$
   lead to respective likelihood functions $l_1(\boldsymbol{\theta})$ and $l_2(\boldsymbol{\theta})$ and $l_1 = kl_2$ for
   some $k > 0$ that does not depend on $\boldsymbol{\theta}$, their maximum likelihood estimator will be the same. Therefore, the maximum likelihood estimator
   does not violate the likelihood principle.

3. The maximum likelihood estimator may not exist.

   Example 4.8 *Assume that $X_1, \ldots, X_n$ is a sample from the $U(0, \theta)$
   distribution, $\theta > 0$. The likelihood function is*

   $$l(\theta; \boldsymbol{X}) = p(\boldsymbol{X} \mid \theta) = \frac{1}{\theta^n} \prod_{i=1}^{n} I_\theta(X_i, \infty) = \frac{1}{\theta^n} I_\theta(T, \infty),$$

   *where $T = \max_i X_i$. As the likelihood function is a strictly decreasing
   function of θ, its maximum is attained at the lower value of its domain, the interval (T, ∞). As the interval is open, the function does
   not have a maximum and θ does not have an MLE. This technical difficulty is easily remedied by considering closed intervals without loss
   of generality. In this case, the maximum likelihood estimator of θ is
   $T = \max_i X_i$. Note, however, that $Pr(T < \theta) = 1$ and therefore the
   maximum likelihood estimator will underestimate θ with certainty.*

4. The maximum likelihood estimator may not be unique.

   Example 4.9 *Assume that $X_1, \ldots, X_n$ is a sample from the $U(0, \theta +
   1)$ distribution, $\theta \in R$. The likelihood function is*

   $$l(\theta; \boldsymbol{X}) = \prod_{i=1}^{n} I_\theta(X_i - 1, X_i) = I_\theta(T_2 - 1, T_1),$$

where $T_1 = \min_i X_i$ and $T_2 = \max_i X_i$. Therefore, the maximum likelihood estimator of θ will be any value in the interval $(T_2 - 1, T_1)$, if it exists, because the likelihood function is constant over that region.

5. The maximum likelihood estimator and Bayes estimators depend on the sample only through minimal sufficient statistics. By the factorization criterion, $l(\boldsymbol{\theta}; \mathbf{X}) = g(\mathbf{X})f(\mathbf{T}, \boldsymbol{\theta})$, where T is a sufficient statistic. Maximization of $l(\boldsymbol{\theta}; \mathbf{X})$ is therefore equivalent to maximization of $f(\mathbf{T}, \boldsymbol{\theta})$. Since f depends only on the sample through T, the maximum likelihood estimator will have to be a function of T. The same reason applies for the Bayes estimators. As this is valid for every sufficient statistic, it must be valid for the minimal one.

To obtain the (G)MLE, one must obtain the maximum of the likelihood function (posterior density, respectively). This task may be regarded as an optimization problem. This task is considerably simplified by assuming further that $\boldsymbol{\theta}$ varies continuously over the parameter space. The problem can typically be solved through the equation $\partial l(\boldsymbol{\theta}; \mathbf{X})/\partial\boldsymbol{\theta} = \mathbf{0}$ (or $\partial p(\boldsymbol{\theta}|\mathbf{x})/\partial\boldsymbol{\theta} = \mathbf{0}$). In many cases, it is easier to work with the logarithm. Concentrating on the likelihood from now on, let $L(\boldsymbol{\theta}; \mathbf{X}) = \log l(\boldsymbol{\theta}; \mathbf{X})$. The problem can then be rephrased in terms of finding the solution of

$$\mathbf{U}(\mathbf{X}; \boldsymbol{\theta}) = \frac{\partial L(\boldsymbol{\theta}; \mathbf{X})}{\partial\boldsymbol{\theta}} = \mathbf{0},$$

where $\mathbf{U}$ is the score function introduced in Chapter 2. Note that

$$\frac{\partial \log p(\boldsymbol{\theta}|\mathbf{x})}{\partial\boldsymbol{\theta}} = \frac{\partial L(\boldsymbol{\theta}; \mathbf{x})}{\partial\boldsymbol{\theta}} + \frac{\partial \log p(\boldsymbol{\theta})}{\partial\boldsymbol{\theta}} = \mathbf{U}(\mathbf{X}; \boldsymbol{\theta}) + \frac{\partial \log p(\boldsymbol{\theta})}{\partial\boldsymbol{\theta}}$$

and $\partial \log p(\boldsymbol{\theta}|\mathbf{x})/\partial\boldsymbol{\theta}$ can, thus, be referred to as the generalized score function. In some cases, the above equation can be analytically solved and the roots of the (generalized) score function found explicitly. In other applications, typically involving a highly dimensional $\boldsymbol{\theta}$, no analytical solution can be found. Algorithms for solving this problem will be presented in the next chapter.

4.3.2   Method of least squares

Assume now that $\mathbf{Y} = (Y_1, \ldots, Y_n)$ for a random sample such that $E(Y_i \mid \boldsymbol{\theta}) = f_i(\boldsymbol{\theta})$ and $V(Y_i \mid \boldsymbol{\theta}) = \sigma^2$. One can rewrite each Y_i as

$$Y_i = f_i(\boldsymbol{\theta}) + e_i,$$

where $E(e_i) = 0$ and $V(e_i) = \sigma^2, i = 1, \ldots, n$.

One possible criterion for the estimation of $\boldsymbol{\theta}$ is to minimize the observation errors, e_i's, incurred. There are many ways to account globally for the errors. Given the assumption of homoscedasticity (equal error variances), it seems fair to account for all the errors in the same way and with the same weight. Also, it seems reasonable to account for the errors symmetrically to avoid penalizing more positive or negative errors. One possibility is to attempt minimization of the sum of the absolute errors. In fact, this choice is as plausible as the choice of the absolute loss function in the context of Bayesian estimation. However, for historical and mathematical reasons, the criterion preferred in many cases is to consider the squared errors.

Therefore, the estimation criterion can be stated as the minimization of

$$S(\boldsymbol{\theta}) = \sum_{i=1}^{n} e_i^2 = \sum_{i=1}^{n} [Y_i - f_i(\boldsymbol{\theta})]^2.$$

Note that by forming the vector $\mathbf{f}(\boldsymbol{\theta}) = (f_1(\boldsymbol{\theta}), \ldots, f_n(\boldsymbol{\theta}))'$, the quadratic form can be rewritten as $S(\boldsymbol{\theta}) = (\mathbf{Y} - \mathbf{f}(\boldsymbol{\theta}))'(\mathbf{Y} - \mathbf{f}(\boldsymbol{\theta}))$. The value of $\boldsymbol{\theta}$ that minimizes $S(\boldsymbol{\theta})$ is called the (ordinary) least squares (OLS, in short) estimator of $\boldsymbol{\theta}$. Once again, minimization is achieved by solving the equation $\partial S(\boldsymbol{\theta})/\partial \boldsymbol{\theta} = \mathbf{0}$.

Example 4.10 *(Simple linear regression) Assume one knows that one's variable of interest Y is affected by the values of another known quantity X and that this dependence is linear on the mean. One model for this setup is to take $E(Y_i \mid \boldsymbol{\theta}) = \theta_0 + \theta_1 X_i$ where $\boldsymbol{\theta} = (\theta_0, \theta_1)$. The quadratic form is given by*

$$S(\theta_0, \theta_1) = \sum_{i=1}^{n} (Y_i - \theta_0 - \theta_1 X_i)^2.$$

Differentiation gives

$$\frac{\partial S(\boldsymbol{\theta})}{\partial \theta_0} = -2 \sum_{i=1}^{n} (Y_i - \theta_0 - \theta_1 X_i) = -n(\bar{Y} - \theta_0 - \theta_1 \bar{X}),$$

$$\frac{\partial S(\boldsymbol{\theta})}{\partial \theta_1} = -2 \sum_{i=1}^{n} X_i(Y_i - \theta_0 - \theta_1 X_i) = -n(\overline{XY} - \theta_0 \bar{X} - \theta_1 \overline{X^2}),$$

where $\overline{g(X,Y)}$ generically denotes $(1/n) \sum_{i=1}^{n} g(X_i, Y_i)$ for any given function g. It is not difficult to show that the least squares estimator of (θ_0, θ_1) is

$$(\hat{\theta}_0, \hat{\theta}_1) = \left(\bar{Y} - \hat{\theta}_1 \bar{X}, \frac{\overline{XY} - \bar{X}\bar{Y}}{\overline{X^2} - \bar{X}^2} \right).$$

An important theoretical support for the method is the fact that it coincides with the maximum likelihood estimator if the error distribution is

Normal. This method can be extended to the case of error variances that are unequal due to constants, namely $V(e_i) = w_i^{-1}\sigma^2$. It seems reasonable in this case to take into account the different variabilities and weigh more heavily in the sum the more precise observations, those with larger values of w_i. This modification of the criterion leads to the weighted least squares (WLS, in short) estimator obtained by minimization of

$$S(\boldsymbol{\theta}) = \sum_{i=1}^{n} w_i (Y_i - f_i(\boldsymbol{\theta}))^2.$$

The sum can again be written in matrix notation as $[\mathbf{Y} - \mathbf{f}(\boldsymbol{\theta})]'\mathbf{W}[\mathbf{Y} - \mathbf{f}(\boldsymbol{\theta})]$ where $\mathbf{W}$ is the $n \times n$ diagonal matrix with elements $w_1, \ldots, w_n$. Note that in its full generality, the matrix of weights $\mathbf{W}$ does not even need to be diagonal by allowing correlated observation errors.

Ordinary least squares can be obtained as the special case of weighted least squares where $\mathbf{W} = \sigma^2 \mathbf{I}_n$, where $\mathbf{I}_n$ is the $n \times n$ identity matrix.

One useful application of this method is in the case of spurious observations. We would not want our analysis to be influenced by observations that are known to be discordant from the rest of the observations. Reduction of the effect of these variables is achieved by setting smaller values of w_i for them. In the limit, $w_i \to 0$ and the i-th observation has no influence in the estimation. This line of reasoning forms the basis of robustness studies.

In all the cases above, σ^2 is estimated by

$$\frac{1}{k(n)}[\mathbf{Y} - \mathbf{f}(\boldsymbol{\theta}_{LS})]'\mathbf{W}[\mathbf{Y} - \mathbf{f}(\boldsymbol{\theta}_{LS})],$$

where $\boldsymbol{\theta}_{LS}$ is the WLS estimator of $\boldsymbol{\theta}$ and $k(n)$ is a suitable function of the sample size n. Common choices are $k(n) = n$ and $k(n) = n - dim(\boldsymbol{\theta})$, where $dim(\boldsymbol{\theta})$ is the dimension of $\boldsymbol{\theta}$.

4.3.3 Method of moments

The method of moments is one of the oldest method of deriving point estimation. It is based on a minimal set of assumptions, typically moments of the observations. Thus, it is very convenient as it does not need a full description of the data generation process. Actually, we only need a partial specification of the model, as the method of least squares. Although they may not yield the best estimator, they are often feasible to obtain whereas the likelihood analysis may be extremely difficult in some cases.

Assume again $X_1, \ldots, X_n$ are iid from the parametric model $\{p(\cdot|\boldsymbol{\theta}), \boldsymbol{\theta} \in \boldsymbol{\Theta} \subset R^p\}$ and suppose the k-th moment of $p(x|\boldsymbol{\theta})$ exists and is denoted as $\mu_k =$

$E(X^k|\theta)$, for $k = 1, 2, \ldots$. Assume further that $\mu_k = h_k(\boldsymbol{\theta})$, for given functions h_k, for $k = 1, 2, \ldots$. Denote the k-th sample moment by $\hat{\mu}_k = \sum_1^n x_i^k/n = \overline{X^k}$. Pragmatically, the method of moments consists in equating the sample moments to the corresponding population moments, which means estimating μ_k by $\hat{\mu}_k = h_k(\hat{\boldsymbol{\theta}})$ and solving for $\boldsymbol{\theta}$, for some choice of values for k. In practice, if $\boldsymbol{\theta}$ is a p-dimensional parameter, only p equations are required and the most common choice of values for k is $k = 1, \ldots, p$. This strategy is illustrated in the simple examples below.

Example 4.11 *(Pareto Distribution) Assume a random sample $X_1, \ldots, X_n$ from a Pareto distribution with density $p(x \mid \theta) = \theta/x^{\theta+1}$, for $x > 1$, and 0, otherwise. Suppose we are interested in the estimation of θ. The mean $\mu = \theta/(\theta - 1)$ is estimated by $\hat{\mu}_1 = \bar{X}$. Solving for θ as a function of the mean μ gives $\theta = \mu/(\mu - 1)$. Then the method of moments estimate of θ is given by $\tilde{\theta} = \bar{X}/(\bar{X} - 1)$.*

Example 4.12 *Assume a random sample $X_1, \ldots, X_n$ from a Gamma model with parameter $\boldsymbol{\theta} = (\alpha, \beta) \in R^+ \times R^+$. The first and second moments are, respectively, $\mu_1 = h_1(\boldsymbol{\theta}) = \alpha/\beta$ and $\mu_2 = h_2(\boldsymbol{\theta}) = \alpha(\alpha + 1)/\beta$. These moments can be respectively estimated by $\hat{\mu}_1 = \bar{X}$ and $\hat{\mu}_2 = \overline{X^2}$. Applying this idea to $\boldsymbol{\theta}$ gives method of moment estimators $\tilde{\boldsymbol{\theta}} = (\tilde{\alpha}, \tilde{\beta})$ satisfying $\hat{\mu}_1 = \tilde{\alpha}/\tilde{\beta}$ and $\hat{\mu}_2 = \tilde{\alpha}(\tilde{\alpha} + 1)/\tilde{\beta}$. Solving for $\tilde{\alpha}$ and $\tilde{\beta}$ gives*

$$\tilde{\alpha} = \frac{\overline{X^2}}{\bar{X}} - 1 \ and \ \tilde{\beta} = \frac{\tilde{\alpha}}{\bar{X}}.$$

More general settings can be accomodated by the introduction of the concept of moment conditions. A moment condition is defined as a statement involving the data and the parameters: $h : R^n \times R^p \rightarrow R^r, r \geq p$ satisfying $g(\boldsymbol{\theta}) = E[h(X_i, \boldsymbol{\theta})] = 0$, for $i = 1, \ldots, n$. If the expectation is available we can replace the theoretical mean $E[h(X_i, \boldsymbol{\theta})]$ by the sampling mean $\overline{h(X, \boldsymbol{\theta})}$ and solve the above system of equations to find a solution $\tilde{\boldsymbol{\theta}}$ for $\boldsymbol{\theta}$. If there is only one solution satisfying $\overline{h(X, \boldsymbol{\theta})} = 0$, the system is said to be identified; this situation typically occurs when $r = p$. Otherwise, the system is over identified when $r > p$, and underidentified for $r < p$.

Definition 4.10 *The moment estimator is defined as*

$$\tilde{\boldsymbol{\theta}} = \arg_{\boldsymbol{\theta}}\{g_n(\boldsymbol{\theta}) \equiv 0\}, \ where \ g_n(\boldsymbol{\theta}) = \overline{h(X, \boldsymbol{\theta})} = \frac{1}{n}\sum_{i=1}^n h(X_i, \boldsymbol{\theta})$$

is the analogous sample moments condition.

To find an estimator we need at least as many moments conditions as the dimension of the parameter space. The overidentified case $(r > p)$ shall be discussed later in this section. The moment conditions may be obtained after some model assumption, as illustrated in the next examples.

Example 4.13 *(Location model) Let $\mu \in R$ be the location parameter and the observations be $X_i = \mu + \epsilon_i$, where $E(\epsilon_i) = 0$ and $Var(\epsilon_i) = \sigma^2$, for $i = 1, ..., n$, with σ^2 known and thus $p = 1$. Set the moment condition as $h(X, \mu) = X - \mu$ so that $g(\mu) = E[h(X, \mu)] = 0$. In this example $r = p = 1$ and the system is identified. The solution is $\tilde{\mu} = \bar{X}$.*

Example 4.14 *(Location-scale model) Let $\boldsymbol{\theta} = (\mu, \sigma^2) \in R \times R^+$ be the location and scale parameters. Define the moment conditions $h(X_i, \boldsymbol{\theta})) = (X_i, X_i^2) - (\mu, \sigma^2 + \mu^2), \forall i = 1, \ldots, n$, corresponding to $g(\boldsymbol{\theta}) = (0, 0)$. In this case we have $p = r = 2$. Obviously the system is identified since $\tilde{\boldsymbol{\theta}} = (\bar{X}, \overline{X^2} - (\bar{X})^2)$ is its unique solution.*

Typically $h(X_i, \boldsymbol{\theta}) = X_i - \phi_j(\boldsymbol{\theta}), \forall i = 1, \ldots, n, j = 1, \ldots, p$, corresponding to the identified case. The sampling moments conditions can be concisely rewritten as $g_n(\boldsymbol{\theta}) = \hat{\mu} - \boldsymbol{\phi}(\boldsymbol{\theta})$, where $\boldsymbol{\phi} = (\phi_1, \ldots, \phi_p)'$ and $\hat{\mu} = (\hat{\mu}_1, \ldots, \hat{\mu}_p)'$. Therefore, $\tilde{\boldsymbol{\theta}}$ is the solution of $\hat{\mu} = \boldsymbol{\phi}(\tilde{\boldsymbol{\theta}})$ which is uniquely given by $\tilde{\boldsymbol{\theta}} = \boldsymbol{\phi}^{-1}(\hat{\mu})$ iff the inverse of $\boldsymbol{\phi}$ exists. When this is not the case (ϕ is not one to one), the solution is not unique, or we have many alternative solutions.

Example 4.15 *(Degrees of freedom in a Student-t distribution) Let $X_1, \ldots, X_n$ be a random sample from a Student-t distribution with location 0, scale 1 and an unknown number ν of degrees of freedom. It is known that $\mu_2 = E[(X_i)^2|\nu] = \nu/(\nu - 2)$. The moment condition can be $h(x, \nu) = x^2 - (\nu/(\nu - 2))$. Therefore, $g(X, \nu) = E[h(X, \nu)] = 0$ and the corresponding sampling quantity is $g_n(x, \nu) = \frac{1}{n} \sum [X_i - (\nu/(\nu - 2))] = 0$. After some algebra it follows that the method of moment estimator of ν is $\tilde{\nu} = 2\hat{\mu}_2/(\hat{\mu}_2 - 1)$, when $\hat{\mu}_2 > 1$.*

In some applications a number of moments conditions larger than the dimension of the parameter space is quite natural. Let us consider two examples: in the Poisson model, the population mean and variance are both equal to the same parameter λ. What moment condition should someone use? Why should one not combine both conditions? Another example is the point estimator of the number of degrees of freedom in the Student-t distribution. We can simultaneous take into account the variance and kurtosis of the distribution. In both examples $r > p$, so we have more equations than parameters; then there is no unique solution for the sampling equations: $g_n(\boldsymbol{\theta}) = 0$. Since we are not able to obtain a single point in the parameter space which makes the identity valid all we can do is get an approximate solution. So define the distance

$$Q_n(\boldsymbol{\theta}) = g_n(\boldsymbol{\theta})'Wg_n(\boldsymbol{\theta})$$

for an arbitrary weight matrix W with dimension r. The generalized method of moment (GMM, in short) estimator $\tilde{\boldsymbol{\theta}}$ is the solution of

$$\tilde{\boldsymbol{\theta}} = \arg\min Q_n(\boldsymbol{\theta}).$$

Example 4.16 *Consider observations $X_i \sim Pois(\lambda), \lambda > 0$, for $i = 1, ..., n$. Suppose we have $r = 2$ moments conditions, so $r > p = 1$, given by $h(X_i, \lambda) = (X_i - \lambda, (X_i - \lambda)^2 - \lambda)$. Then $g_n = (\bar{X} - \lambda, \hat{\sigma}^2 - \lambda)$, where $\hat{\sigma}^2 = (1/n)\sum_{i=1}^n (X_i - \bar{X})^2$. Using $W = I_2$, we obtain the GMM estimator of λ as $\tilde{\lambda} = (\bar{X} + \hat{\sigma}^2)/2$.*

Example 4.17 *(Example 4.15 continued) Many times we want to choose ν to control simultaneously the variance and the kurtosis of the model. The fourth moment of X is easily obtained as $\mu_4 = E(X_i^4|\nu) = \frac{3\nu^2}{(\nu-2)(\nu-4)}$, for $\nu > 4$. The moment condition is $g(\nu) = E\left[\left(X^2 - \frac{\nu}{\nu-2}, X^4 - \frac{3\nu^2}{(\nu-2)(\nu-4)}\right)|\nu\right] = (0,0)$. Since we have two equation and just one parameters we cannot find a single value for ν. Using $W = I_2$ we can get the solution of $Q(\nu; x) = g_n'Wg_n$. This problem is left as an exercise.*

The GMM estimator includes the ordinary least squares, weighted least squares and the maximum likelihood estimators as special cases. The moment condition for the OLS is $h(X_i, \boldsymbol{\theta}) = X_i - f_i(\boldsymbol{\theta})$, with $f_i(\boldsymbol{\theta})$ defined in Section 4.3.2 and for the MLE, it is natural to set $h(X, \boldsymbol{\theta})$ as the score function, since its expected value is zero.

Some immediate questions appear. How to choose the weight matrix? Is there an optimal weight matrix? What can be said about the asymptotic distribution of the GMM estimator? These questions will not be answered in this book but the interested reader will find answers to them in the book by Greene (2008) on Econometrics, an area where GMM is extensively used.

4.3.4   Empirical distribution function

This is a non-parametric method which involves no knowledge of the distribution function to be estimated. This estimator is useful at least as an initial estimator, thus providing some insight into the form of the distribution function.

Let $\mathbf{X} = (X_1, \ldots, X_n)$ be a random sample from an unknown distribution function F. Recall from the definition that $F(x) = P(X \leq x)$. The empirical distribution function is denoted by $\hat{F}$ and is given by

$$\hat{F}(x) = \frac{\#X_i's \le x}{n}.$$

Observe that just like F, $\hat{F}$ is non-decreasing and contained in the interval $[0, 1]$. It is interesting to note that $\hat{F}(x)$ can be written as $\bar{Z}$ where $Z_i = I_{X_i}(-\infty, x]$, $i = 1, \ldots, n$. The populational quantity equivalent to $\bar{Z}$ is the proportion of population elements that are $\le x$. This is given by $P(X \le x) = F(x)$. So, the empirical distribution function is a form of method of moments estimator of the distribution function.

4.4 Empirical Bayes estimation

Our focus in this subsection is on a simple parametric model distributed according to $p(\mathbf{x}|\boldsymbol{\theta})$, $\boldsymbol{\theta} \in \boldsymbol{\theta}$ with prior distribution $p(\boldsymbol{\theta}|\boldsymbol{\eta})$, where $\boldsymbol{\eta} \in \Xi$ are often called hyperparameters. Thus, the prior distribution for $\boldsymbol{\theta}$ depends on unknown parameters that in turn follow some second-stage prior in a hierarchical fashion, as introduced in Section 3.5. The empirical Bayes (EB, in short) approach assumes that the remaining prior parameters are not known and are not subjectively assessed. Alternatively, the EB approach uses only the observed data to estimate these final stage parameters. Therefore, the approach is not fully Bayesian, since we are using the data to determine the value of these second-stage parameters. The approach is also not entirely frequentist since it relies on a prior specification for $\boldsymbol{\theta}$.

The estimation of $\boldsymbol{\eta}$ may be performed with any of the non-Bayesian techniques presented in the previous section.

The EB method consists of dropping the highest level prior distribution, $p(\boldsymbol{\eta})$ and using instead only the data to provide information about it. Assuming that we known the priori of $\boldsymbol{\theta}$ conditional on $\boldsymbol{\eta}$, estimation of $\boldsymbol{\eta}$ may be obtained from its marginal likelihood (see Section 2.6),

$$l(\boldsymbol{\eta}; \mathbf{x}) = \int p(\mathbf{x}|\boldsymbol{\theta})p(\boldsymbol{\theta}|\boldsymbol{\eta})d\boldsymbol{\theta}.$$

Point estimation of $\boldsymbol{\eta}$ based on the available data can be obtained via any of the methods of the previous section. For example, the method of moments and maximum likelihood estimators are respectively given by

$$\tilde{\boldsymbol{\eta}} = \arg_{\boldsymbol{\eta}}\{g_n(\boldsymbol{\eta}) = 0\} \text{ and } \hat{\boldsymbol{\eta}} = \arg\max_{\boldsymbol{\eta}} l(\boldsymbol{\eta}; \mathbf{x}),$$

where $g_n(\boldsymbol{\eta})$ is the moment condition introduced in Section 4.3.3.

The original likelihood parameters $\boldsymbol{\theta}$ could be estimated in the two stage case from their marginal posterior distribution

$$p(\boldsymbol{\theta}|\mathbf{x}) = \int \frac{p(\mathbf{x}|\boldsymbol{\theta})p(\boldsymbol{\theta}|\boldsymbol{\eta})p(\boldsymbol{\eta})}{p(\mathbf{x})}d\boldsymbol{\eta},$$

where we explicitly assume that $\mathbf{x}$ is conditionally independent of the hyperparameter $\boldsymbol{\eta}$, given $\boldsymbol{\theta}$, and $p(\mathbf{x}) = \int \int p(\mathbf{x}|\boldsymbol{\theta})p(\boldsymbol{\theta}|\boldsymbol{\eta})p(\boldsymbol{\eta})d\boldsymbol{\theta}d\boldsymbol{\eta}$. These marginal distributions are not available due to the lack of a prior distribution for $\boldsymbol{\eta}$. Instead, we can obtain the conditional posterior distribution of $\boldsymbol{\theta}|x, \boldsymbol{\eta}$ as

$$p(\boldsymbol{\theta}|x, \boldsymbol{\eta}) \propto p(x|\boldsymbol{\theta}, \boldsymbol{\eta})p(\boldsymbol{\theta}|\boldsymbol{\eta}).$$

Replacement of $\boldsymbol{\eta}$ by some estimator $\hat{\boldsymbol{\eta}}$ obtained as above would then allow estimation of $\boldsymbol{\theta}$ from $p(\boldsymbol{\theta}|x, \hat{\boldsymbol{\eta}})$. Estimators obtained with such procedures are usually known as plug-in estimators. Note that by using them, the uncertainty associated with estimation of $\boldsymbol{\eta}$ is removed, and one estimates $\boldsymbol{\theta}$ as if one knew that $\boldsymbol{\eta} = \hat{\boldsymbol{\eta}}$ with certainty. Later in the section, we will introduce a seemingly unrelated but extremely similar estimation procedure to estimate $\boldsymbol{\theta}$ from this hierarchical setup.

Example 4.18 Let $X_1, \ldots, X_n$ be a random sampling from a Normal distribution with mean μ and known variance σ^2. The conjugate prior depends on the hyperparameters μ_0 and τ^2, respectively, the prior mean and variance. The marginal likelihood $l(\mu_0, \tau^2; X_1, \ldots, X_n)$ follows easily, and using the method of moments or maximum likelihood, we can obtain the point estimators $\hat{\mu}_0 = \bar{X}$, which coincides with the maximum likelihood estimator of μ, and $\hat{\tau}^2 = \sum(X_i - \bar{X})^2/n - \sigma^2$, which could take negative values. Since $\tau^2 > 0$, it is customary to take the empirical Bayes estimator of τ^2 as $\max\{\hat{\tau}^2, 0\}$.

Example 4.19 Suppose we want a point estimate of $\theta \in (0, 1)$, the chance of success in a Bernoulli experiment. We use a Beta (α, β) distribution as a prior for θ, with the reparametrization given by mean $\mu = \alpha/(\alpha + \beta)$ and precision $\phi = \alpha + \beta$. An alternative route to obtain the marginal likelihood function is to apply the iterated expectation and variance given by

$$\begin{aligned} E(X|\boldsymbol{\eta}) &= E[E(X|\theta)] = \mu \\ V(X|\boldsymbol{\eta}) &= V[E(X|\theta)] + E[V(X|\theta)] = \frac{\mu(1 - \mu)}{\phi + 1}, \end{aligned}$$

where $\boldsymbol{\eta} = (\mu, \phi)$.

Now let $X_1, \ldots, X_n$ be a random sample of the Bernoulli model and define $T = \sum_i X_i$, the sufficient statistic for θ. It is worth pointing out that *calculation of the maximum likelihood estimator is quite cumbersome since it involves the gamma and digamma functions. Instead, it is easy to obtain $E(T|\boldsymbol{\eta}) = n\mu$ and $V(T|\boldsymbol{\eta}) = n\mu(1-\mu)/(\phi+1)$. Therefore, application of the method of moments gives estimators*

$$\tilde{\mu} = \bar{X},$$
$$\tilde{\phi} = \frac{n^2\hat{\mu}_1(1-\hat{\mu}_1)}{S^2} - 1,$$

where $S^2 = \hat{\mu}_2 - \hat{\mu}_1^2$.

Let us now introduce a more general scenario. Suppose that we extend the two stage model introduced earlier by allowing $p(x_i|\theta_i)$, for $i = 1, \ldots, k$, to be conditionally independent observations, where k represents the number of exchangeable groups, and assume that $p(\theta_i|\boldsymbol{\eta}), \forall i = 1, \ldots, n$ are also independent and identically distributed conditional on $\boldsymbol{\eta}$. The marginal likelihood for $\boldsymbol{\eta}$ is obtained as

$$l(\boldsymbol{\eta}; x_1, \ldots, x_k) = \int_{\boldsymbol{\theta}} \prod_{i=1}^{k} p(x_i|\theta_i) \, p(\theta_i|\boldsymbol{\eta}) d\theta_i$$
$$= \prod_{i=1}^{k} \int_{\boldsymbol{\theta}} p(x_i|\theta_i) \, p(\theta_i|\boldsymbol{\eta}) d\theta_i,$$

and can be used to obtain point estimation of $\boldsymbol{\eta}$ based on the maximum likelihood or on the method of moments as exemplified previously.

An estimator of $\boldsymbol{\eta}$ obtained from the above marginal distribution could be plugged into the conditional posterior distribution of θ_i

$$p(\theta_i|x_i, \ldots, x_k, \hat{\boldsymbol{\eta}}) = \frac{p(x_i|\theta_i)p(\theta_i|\hat{\boldsymbol{\eta}})}{p(x_i|\hat{\boldsymbol{\eta}})}.$$

It can be shown that the point estimator $E(\theta_i|x_1, \ldots, x_n, \hat{\boldsymbol{\eta}})$ of θ_i based on the square error loss function is given in many cases by $\omega_i\mu_0 + (1-\omega_1)X_i$, where ω_i are weights based on the relative precision of the prior and the data, and depends on $\hat{\boldsymbol{\eta}}$. Thus, empirical Bayes estimators typically involve shrinkage or borrowing of information. Another aspect of empirical Bayes estimation is that such plug-in estimators do not account for the variability associated with the hyperparameter estimator.

Example 4.20 *Let $X_1, \ldots, X_k$ be independent observations of a Poisson distributions with parameters $\lambda_i > 0$, and suppose the λ_i are iid sampled from a Gamma(α, β) distribution, with $\alpha, \beta > 0$. From the conjugacy, it follows that the posteriori distribution is a Gamma$(\alpha + x_i, \beta + 1)$ distribution, then* the point estimate using square loss function is $E[\theta_i|x_i] = \omega_i\mu + (1 - \omega_i)x_i$ where $\mu = \alpha/\beta$ and $\omega_i = \beta/(1 + \beta)$. The marginal distribution of X_i given $\boldsymbol{\eta} = (\alpha, \beta)$ is Negative Binomial, with probability function $p(x_i|\boldsymbol{\eta}) = \binom{\alpha-1}{x_i-\alpha-1}\left(\frac{\beta}{\beta+1}\right)^{\alpha}\left(\frac{1}{\beta+1}\right)^{x_i}$. It follows that the maximum likelihood estimator of β is given by $\hat{\beta} = \alpha/\bar{X}$.

When both hyperparameters are unknown we can use the method of moments, where estimators must satisfy equations $\hat{\mu}_1 = \tilde{\alpha}/\tilde{\beta}$ and $\hat{\mu}_2 = \tilde{\alpha}(1 + \tilde{\alpha} + \tilde{\beta})/\tilde{\beta}^2$. After some algebraic manipulations, we get estimators $\tilde{\alpha} = \hat{\mu}_1^2/[\hat{\mu}_2 - \hat{\mu}_1(1 + \hat{\mu}_1)]$ and $\tilde{\beta} = \hat{\mu}_1/[\hat{\mu}_2 - \hat{\mu}_1(1 + \hat{\mu}_1)]$. These estimators can be plugged into the above expression of the posterior mean of $\boldsymbol{\theta}$ to obtain empirical Bayes estimators for the θ_i's.

Example 4.21 *Part a: Let us consider the model:*

$$X_i \sim N(\mu_i, \sigma^2), i = 1, \ldots, k$$
$$\mu_i \sim N(\mu_0, \tau^2),$$

where the observation and also the μ_i's in the second stage are conditionally independent normally distributed. This is the popular one-way layout model, to be addressed later in this book. First we will take into account the case where the variance components are known. It is easy to show that $X_i|\mu_0, \tau^2, \sigma^2 \sim N(\mu_0, \sigma^2 + \tau^2)$. The maximum likelihood estimator of μ_0 is $\hat{\mu}_0 = \bar{X}$; therefore, the posterior distributions of μ_i are given by

$$\mu_i|X_i, \sigma \sim N[\hat{\mu}_i, (\sigma^{-2} + \tau^{-2})^{-1}],$$

where $\hat{\mu}_i = \omega X_i + (1 - \omega)\bar{X}$, with $\omega = 1/(1 + \sigma^2/\tau^2)$.

Part b: Assuming that n independent observations are made in each of the k groups, we have an extension of the previous example.

$$X_{i,j} \sim N(\mu_i, \sigma^2), \text{ for } i = 1, \ldots, k, \ j = 1, \ldots n$$
$$\mu_i \sim N(\mu_0, \tau^2), \text{ for } i = 1, \ldots, k.$$

In each group we can make a reduction by sufficiency to obtain observations $\bar{X}_i|\mu_i, \sigma^2 \sim N(\mu_i, \sigma^2/n)$.

The Bayes estimators of the μ_i's under squared error loss are given by $\hat{\mu}_i = \omega \bar{X}_i + (1-\omega)\bar{X}$, where $\bar{X}_i = \sum_{j=1}^{n} X_{ij}/n$, $\bar{X} = \sum_{i=1}^{k}\sum_{j=1}^{n} X_{ij}/nk$ and $\omega = 1/(1 + \sigma^2/n\tau^2)$. If τ^2 is unknown and σ^2 known, it can be shown that $(k-3)/\sum_{I=1}^{k}(\bar{X}_i - \bar{X})^2$ is a suitable estimator of $1/(\sigma^2 + n\tau^2)$, and so we have the empirical Bayes estimators of the μ_is as

$$\hat{\mu}_i = \left(1 - \frac{(k-3)\sigma^2}{\sum_{I=1}^{k}(\bar{X}_i - \bar{X})^2}\right)\bar{X}_i + \left(\frac{(k-3)\sigma^2}{\sum_{I=1}^{k}(\bar{X}_i - \bar{X})^2}\right)\bar{X}.$$

The above estimator is also known as the James and Stein (1961) estimator, shown to have superior properties over the standard maximum likelihood estimator given by the group means (more on that in Section 4.5.2).

Example 4.22 (*Example 3.13 continued*). *For the data in Table 3.2, Efron and Morris (1975) obtained the empirical Bayes estimate of* $(1 + \tau^2)^{-1}$ *equal to 0.791. Therefore, the standard deviation of the second level is estimated by* $\hat{\tau} = 0.514$, *which represents a substantial amount of prior information. Since* $\bar{X} = -3.275$, *the empirical Bayes estimators are obtained as*

$$\hat{\theta}_i^{(EB)} = 0.791\bar{X} + 0.209X_i = 0.209X_i - 2.590.$$

The expression for the conditional posterior distribution for $\boldsymbol{\theta}$, given by $p(\boldsymbol{\theta}|\mathbf{x}, \boldsymbol{\eta}) \propto p(\mathbf{x}|\boldsymbol{\theta}, \boldsymbol{\eta})p(\boldsymbol{\theta}|\boldsymbol{\eta})$, suggests an alternative estimation procedure. Assume that we are willing to use a likelihood-based approach for estimation but want to impose further restriction on the parameter space. These restrictions may be deterministic. For example, in the case of n group means θ_i's, the restriction could be $\sum_{i=1}^{n}(\theta_i - \mu)^2 = c$, for some known constants c and μ. This restriction controls the dispersion of the group means by forcing their squared distances to a common value by a fixed amount. In this case, Lagrange multipliers may be used to incorporate these restrictions in the estimation procedure and restricted maximum likelihood estimators may be obtained.

A more general and suitable form of restrictions may be incorporated in more general terms through penalizations functions. These functions do not impose deterministic relations to be satisfied but rather impose penalties for departures from the desired specifications. These penalties are usually applied in additive form to the logarithm of the likelihood. An example of penalty function in the case of n group means θ_i's is given by $q(\boldsymbol{\theta}) = \sum_{i=1}^{n}(\theta_i - \mu)^2$. When these penalties are additively applied to the log likelihood, a general alternative objective function given by

$$L^*(\boldsymbol{\theta}, \boldsymbol{\eta}; \boldsymbol{X}) = L(\boldsymbol{\theta}; \boldsymbol{X}) + \lambda q(\boldsymbol{\theta})$$

appears. It becomes clear from the above expression that the penalty functions such as $\sum_{i=1}^{n}(\theta_i - \mu)^d$ force (but do not impose) the estimated components of $\boldsymbol{\theta}$ to be closer. The amount of closeness is largely controlled by the multiplier λ. The parameter $\boldsymbol{\eta}$ consists of all other unknowns in the expression of L^* that must be specified and/or estimated. In the above example, it consists of λ and μ.

It is not difficult to realize that functions added to the log likelihood are mathematically equivalent to their exponentials being multiplied to the likelihood function. This is exactly the role played by the prior distribution. Hence, procedures based on penalization on the likelihood bear a strong resemblance to empirical Bayes procedures. These procedures based on penalized likelihood functions are becoming increasingly popular and bridge the gap between Bayesian and frequentist estimation procedures. Surprisingly, they are rarely referred to as empirical Bayes procedures. A variety of option is available for assessing the hyperparameters $\boldsymbol{\eta}$, giving rise to a number of variations of penalized estimation procedures.

4.5 Comparison of estimators

Given that there is no unifying criteria for choice of frequentist estimators in any given problem, it is important that a set of criteria is established to compare them. The main criteria for comparison are bias, (frequentist) risk or mean squared error and consistency. Much effort had been concentrated on this area during the 50's and 60's. These studies led to the characterization of uniformly minimum variance unbiased (UMVU, in short) estimators.

4.5.1 Bias

Definition 4.11 *Let $\boldsymbol{X} = (X_1, \ldots, X_n)$ be a random sample from $p(x \mid \boldsymbol{\theta})$ and $\boldsymbol{\delta} = \boldsymbol{\delta}(\boldsymbol{X})$ an estimator of $\boldsymbol{h}(\boldsymbol{\theta})$, for any given function $\boldsymbol{h}$. $\boldsymbol{\delta}$ is an unbiased estimator of $\boldsymbol{h}(\boldsymbol{\theta})$ if $\boldsymbol{E}[\boldsymbol{\delta} \mid \boldsymbol{\theta}] = \boldsymbol{h}(\boldsymbol{\theta}), \forall \boldsymbol{\theta}$. The estimator $\boldsymbol{\delta}$ is said to be biased otherwise. In this case, the bias is denoted by $\boldsymbol{b}(\boldsymbol{\theta})$ and defined as $\boldsymbol{b}(\boldsymbol{\theta}) = \boldsymbol{E}[\boldsymbol{\delta} \mid \boldsymbol{\theta}] - \boldsymbol{h}(\boldsymbol{\theta})$.*

The frequentist interpretation of the definition is that after repeating sampling of $\boldsymbol{X}$ from $p(x \mid \boldsymbol{\theta})$ many times, averaging the corresponding values of $\boldsymbol{\delta}$ will produce $\boldsymbol{h}(\boldsymbol{\theta})$ as a result. This is a desirable property because one formulates an estimator $\boldsymbol{\delta}$ in an effort to obtain the value of $\boldsymbol{h}(\boldsymbol{\theta})$. The difficulty is that in most cases only a single sample $\boldsymbol{X}$ is observed for time and/or financial restrictions.

Example 4.23 *Assume a random sample $X_1, \ldots, X_n$ with expectation μ and variance σ^2. Define the sample mean and variance as $\bar{X} = (1/n) \sum_{i=1}^{n} X_i$ and $\hat{\sigma}^2 = (1/n) \sum_{i=1}^{n} (X_i - \bar{X})^2$. The sample mean is an unbiased estimator of the population mean, but the sample variance, as defined above, is biased, since $E(\hat{\sigma}^2) = \sigma^2 - E(\bar{X} - \mu)^2 = \sigma^2 - \sigma^2/n = (n-1)\sigma^2/n \neq \sigma^2$.*

Note also that unbiased estimation is always related to a given parametric function; an estimator can be biased with respect to a given function but unbiased with respect to another one.

Example 4.24 *In the above example, if we define the sample variance as* $S^2 = [1/(n-1)]\sum_{i=1}^{n}(X_i - \bar{X})^2$, *its expected value equals the population variance* σ^2. *So,* S^2 *is an unbiased estimator of* σ^2.

None of the three parametric methods of estimation proposed in the previous section can guarantee unbiased estimators. The empirical distribution function however is an unbiased estimator of the distribution function. Moreover, the bias of an estimator can be large.

Example 4.25 *Here is an example where the bias may be huge. Let a unknown quantity* n *of balls numbered from 1 to* n *be placed in an urn. Suppose we are interested in the number of balls inside the urn, and one ball, say* X, *is selected at random. The maximum likelihood estimator for* n *is given by* $\hat{n}_1 = X$, *which is biased since* $E(X) = (n+1)/2$. *Also, we are certain that* n *is at least* X, *but it may be much larger. In the present case, an unbiased estimator for* n *is given by* $\hat{n}_2 = 2X - 1$.

Bias may be corrected however. If we focus on the MLE, which has desirable properties, the bias correction may be based on obtaining its first-order and/or the second-order bias terms and subtracting them from the MLE. Although some authors differentiate between bias correction and bias reduction, these concepts are not differentiated here and are understood as a procedure for removing bias.

In a regular statistical model (Cox and Hinkley, 1974) with one-dimensional parameter θ, the asymptotic expansion for the bias of a maximum likelihood estimator $\hat{\theta}$ is given by

$$B(\theta) = \frac{B_1(\theta)}{n} + \frac{B_2(\theta)}{n^2} + O(n^{-3}),$$

where n is the sample size and B_j, for $j = 1$ and 2, are $O(1)$ functions of θ, which can be explicitly obtained once the model is specified. The basic idea is to remove the first-order term, $B_1(\theta)$, in the asymptotic expansion above, or even to eliminate both $B_1(\theta)/n$ and $B_2(\theta)/n^2$.

Following the same argument, the variance of the maximum likelihood estimator $\hat{\theta}$ may be expanded as

$$D(\theta) = \frac{D_1(\theta)}{n} + \frac{D_2(\theta)}{n^2} + O(n^{-3}).$$

In this context, the following bias corrected estimators were proposed,

$$\tilde{\theta}_1 = \hat{\theta} - \frac{B_1(\hat{\theta})}{n},$$

$$\tilde{\theta}_2 = \hat{\theta} - \frac{B_1(\hat{\theta})}{n} - \frac{B_2(\hat{\theta})}{n^2},$$

and

$$\tilde{\theta}_3 = \hat{\theta} - \frac{B_1(\hat{\theta})}{n} - \frac{B_2^*(\hat{\theta})}{n^2},$$

where $B_2^*(\hat{\theta}) = B_2(\theta) - B_1(\hat{\theta})B_2'(\theta) - B_1''(\theta)D_1(\theta)/2$.

For more details on the calculation of the above bias corrected estimators, interested readers may refer to Cordeiro (1987) and Ferrari, Botter, Cordeiro and Cribari-Neto (1996). The latter showed that these three modified estimators do not present bias to order n^{-1}, while $\tilde{\theta}_3$ does not present bias to order n^{-2} and is therefore preferred.

Example 4.26 *(Exponential Family) Let the random variables $X_1, \ldots, X_n$ form a sample from a distribution with density given in Chapter 2 by*

$$p(x \mid \theta) = a(x) \exp \left\{ U(x)\phi(\theta) + b(\theta) \right\},$$

where θ is a scale parameter, and $a(\cdot)$, $\phi(\cdot)$, $U(\cdot)$ and $b(\cdot)$ are known functions. Suppose $\phi = \phi(\theta)$ and $\xi = \xi(\theta) = \exp\{-b(\theta)\}$ with the first 5 derivatives continuous with respect to θ and $\phi'(\theta) \neq 0$, for all $\theta \in \Theta$.

Then, defining

$$A(\theta) = \xi'(\theta)[\xi(\theta)\phi'(\theta)]^{-1} \ and \ \bar{U} = \frac{1}{n} \sum_{i=1}^{n} U(x_i),$$

the score function $U(\boldsymbol{X}; \theta)$ is given by

$$U(\boldsymbol{X}; \theta) = n\phi'(\theta)[\bar{U} - A(\theta)].$$

The maximum likelihood estimator is then obtained as the solution of

$$A(\hat{\theta}) = \frac{\sum U(X_i)}{n}.$$

The bias corrected estimators $\tilde{\theta}_1$, $\tilde{\theta}_2$ and $\tilde{\theta}_3$ are constructed by considering the expansion terms given by

$$B_1(\theta) = -\frac{A''}{2\phi' A'^2},$$

$$B_2(\theta) = \frac{1}{24\phi'^3 A'^5} \begin{pmatrix} -12A'\phi'' A''^2 - 33\phi' A''^3 + 4A'^2\phi'' A''' + \\ 26\phi' A'A'' A''' - 3\phi' A'^2 A^{iv} \end{pmatrix},$$

$$B_2^*(\theta) = \frac{1}{24\phi'^4 A'^5} \begin{pmatrix} 12A'^2\phi''^2 A'' + 18\phi' A'\phi'' A''^2 + 15\phi'^2 A''^3 - \\ 6\phi' A'^2 A''\phi''' - 8\phi' A'^2\phi'' A''' - 16\phi'^2 A'A'' A''' + \\ 3\phi'^2 A'^2 A^{iv} \end{pmatrix},$$

$$D_1\left(\theta\right) = \frac{1}{\phi' A'} \quad and$$

$$D_2\left(\theta\right) = \frac{1}{2\phi'^3 A'^4}\left(2A'\phi'' A'' + 5\phi' A''^2 - 2\phi' A' A'''\right).$$

Another general method for removing the first-order term in the asymptotic expansion of the bias of the maximum likelihood estimator was proposed by Firth (1993), based on a suitable modification of the score function. The basic idea is that the bias of $\hat{\theta}$ can be reduced by introducing a small bias in the score function. If $\hat{\theta}$ has a positive bias $B(\theta)$ and $J(\theta) = -U'\left(\mathbf{X};\theta\right)$ is the local gradient, the score function may be shifted downward at each point θ by an amount $J(\theta)B(\theta)$, defining a modified score function given by

$$U_{mod}(\mathbf{X};\theta) = U(\mathbf{X};\theta) - J\left(\theta\right) B(\theta).$$

Then a modified unbiased estimator $\tilde{\theta}_4$ is obtained as the solution of $U_{mod}(\mathbf{X};\theta) = 0$. The appropriate modification of $U(\mathbf{X};\theta)$ for $U_{mod}(\mathbf{X};\theta)$ can be understood geometrically via triangulation, as illustrated in Figure 4.4.

Although in general the bias $B(\theta)$ is unknown, it is possible to obtain its first-order term $B_1(\theta)/n$. Firth (1993) showed that $\tilde{\theta}_4$ is a solution of

$$U(\mathbf{X};\theta) - \frac{J(\theta)B_1(\theta)}{n} = 0.$$

This means that the Firth estimator $\tilde{\theta}_4$ is equivalent to $\tilde{\theta}_1$ and $\tilde{\theta}_2$ in terms of asymptotic bias. Firth (1993) pointed out that this bias correction corresponds to Bayesian estimators obtained with the use of Jeffreys prior in canonical exponential family models.

Example 4.27 (*Example 4.26 continued*) *We obtain here the Firth estimator for the parameter θ under the exponential family. Taking $I(\theta) = E[-U'\left(\mathbf{X};\theta\right)] = -n\alpha' A'$ and solving $U_{mod}(\mathbf{X};\theta) = 0$, we obtain*

$$\tilde{\theta}_4 = -A + \gamma/n,$$

where $\gamma = \gamma\left(\theta\right) = A'' / \left(2\alpha' A'\right)$.

In addition to the use of these analytical methods for bias correction, resampling methods described in the next chapter are also widely used.

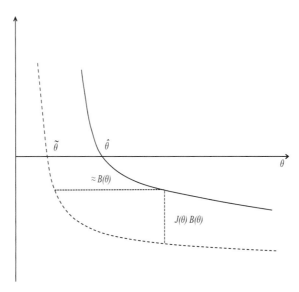

Figure 4.4 *Geometric representation of the score functions: score function, full line; modified score function, dashed line (based on Firth, 1993).*

4.5.2   Risk

Definition 4.12 *Let $X = (X_1, \ldots, X_n)$ be a random sample from $p(x \mid \boldsymbol{\theta})$ and let $\boldsymbol{\delta} = \boldsymbol{\delta}(X)$ be an estimator of $\boldsymbol{h}(\boldsymbol{\theta})$. The frequentist risk of the estimator $\boldsymbol{\delta}$ is defined as $R_{\boldsymbol{\delta}}(\boldsymbol{\theta}) = \boldsymbol{E}_{X|\boldsymbol{\theta}}[L(\boldsymbol{\delta}(X), \boldsymbol{\theta})]$. In the case of a quadratic loss function L, the risk is given by $R_{\boldsymbol{\delta}}(\boldsymbol{\theta}) = \boldsymbol{E}_{X|\boldsymbol{\theta}}[(\boldsymbol{\delta} - \boldsymbol{h}(\boldsymbol{\theta}))'(\boldsymbol{\delta} - \boldsymbol{h}(\boldsymbol{\theta}))]$ and is also called mean squared error (MSE, in short). In the scalar case, the MSE reduces to $E_{X|\theta}[\delta - h(\theta)]^2$.*

Comparing with the Bayes risk, once again we see the presence of an expected loss. The change with respect to the evaluation of the expectation must be stressed. The Bayesian risk considers expectation with respect to the posterior distribution of $\boldsymbol{\theta}|\mathbf{x}$ whereas here expectations are taken here with respect to the sampling distribution of $\mathbf{X}|\boldsymbol{\theta}$. Although the dependence on $\mathbf{x}$ causes no harm to the Bayesian estimators, the dependence on $\boldsymbol{\theta}$ will cause additional problems, to be described below.

In terms of risk, the estimation task consists of finding the estimator of smallest risk. Let $\boldsymbol{\delta}_1 = \boldsymbol{\delta}_1(\mathbf{X})$ and $\boldsymbol{\delta}_2 = \boldsymbol{\delta}_2(\mathbf{X})$ be estimators of $\mathbf{h}(\boldsymbol{\theta})$. Their respective risks are $R_{\boldsymbol{\delta}_1}(\boldsymbol{\theta})$ and $R_{\boldsymbol{\delta}_2}(\boldsymbol{\theta})$, and $\boldsymbol{\delta}_1$ is better than $\boldsymbol{\delta}_2$ if $R_{\boldsymbol{\delta}_1}(\boldsymbol{\theta}) \leq R_{\boldsymbol{\delta}_2}(\boldsymbol{\theta})$, for all $\boldsymbol{\theta}$, with strict inequality for at least one value of $\boldsymbol{\theta}$. An estimator is admissible if there is no better estimator than it. When an estimator is unbiased and its variance is uniformly smaller for all possible values of $\boldsymbol{\theta}$ over all possible estimators, it is referred to as a uniformly minimal variance unbiased estimator.

If the estimator $\boldsymbol{\delta}$ is unbiased, its quadratic risk is given by $\text{tr}[\mathbf{V}(\boldsymbol{\delta}|\boldsymbol{\theta})]$, the trace of its sampling covariance matrix. If it is biased, the quadratic risk is given by $\text{tr}[\mathbf{V}(\boldsymbol{\delta})] + [\mathbf{b}(\boldsymbol{\theta})]'\mathbf{b}(\boldsymbol{\theta})$. In the case of a scalar θ, the quadratic risk of an unbiased estimator is given by its sampling variance $V(\delta|\theta)$ and if δ is biased, its quadratic risk is given by $V(\delta|\theta) + b^2(\theta)$.

Example 4.28 *Let $\boldsymbol{X} = (X_1, \ldots, X_n)$ be a random sample from the $N(\theta, \sigma^2)$ distribution with σ^2 known and $h(\theta) = \theta$. Taking $\delta_1(\boldsymbol{X}) = \overline{X}$ and $\delta_2(\boldsymbol{X}) = X_1$ gives*

$$E[\delta_1(\boldsymbol{X}) \mid \theta] = E(\overline{X} \mid \theta) = \frac{1}{n}\sum_{i=1}^{n} E(X_i \mid \theta) = \frac{n\theta}{n} = \theta,$$

$$E[\delta_2(\boldsymbol{X}) \mid \theta] = E(X_1 \mid \theta) = \theta,$$

and therefore, δ_1 and δ_2 are unbiased estimators of θ. Their quadratic risks will coincide with their sampling variances and will be respectively given by

$$R_{\delta_1}(\theta) = V(\overline{X} \mid \theta) = \frac{\sigma^2}{n} \text{ and } R_{\delta_2}(\theta) = V(X_1) = \sigma^2.$$

Of course, $R(\delta_1) < R(\delta_2)$, if $n > 1$, for all values of θ, and therefore, δ_1 is better than δ_2.

Example 4.29 *(Example 4.22 continued) The quadratic risk of the maximum likelihood estimator is given by $E[\sum(X_i - \theta_i)^2] = 17.56$ and the quadratic risk of the James and Stein (1961) estimator is given by $E[\sum(\hat{\theta}_i^{(EB)} - \theta_i)^2] = 5.01$. Thus, the efficiency of the Stein rule estimator over the MLE, defined by the ratio of quadratic risks is approximately equal to 3.5 for this data. The empirical Bayes estimator is therefore better than the maximum likelihood estimator.*

It is not always possible to find an estimator that completely dominates the other ones in terms of risk. In the ideal situation one would eliminate $\boldsymbol{\theta}$ by suitably weighing the risks over their different values. This is performed naturally in the Bayesian context. Here, however, $\boldsymbol{\theta}$ is fixed and no such natural weighing scheme exists. An alternative is to consider the worst possible risk for each estimator and choose the estimator with smallest worst possible risk. This is the definition of the minimax estimator, previously introduced in this chapter.

Returning to Example 4.28, it is not surprising that δ_1 is better than δ_2. After all, δ_1 seems to be using the sample information better than δ_2. Once again, the key concept here is sufficiency and the result is formalized in the Rao-Blackwell theorem below.

Theorem 4.1 *Let $\boldsymbol{X} = (X_1, \ldots, X_n)$ be a random sample from $p(x|\boldsymbol{\theta})$, $\boldsymbol{\delta} = \boldsymbol{\delta}(\boldsymbol{X})$ an unbiased estimator of $\boldsymbol{h}(\boldsymbol{\theta})$, for some function $\boldsymbol{h}$ and $\boldsymbol{T} = \boldsymbol{T}(\boldsymbol{X})$ a sufficient statistic for $\boldsymbol{\theta}$. Then, $\boldsymbol{\delta}^* = \boldsymbol{\delta}^*(\boldsymbol{X}) = \boldsymbol{E}(\boldsymbol{\delta} \mid \boldsymbol{T})$ is an unbiased estimator of $\boldsymbol{h}(\boldsymbol{\theta})$ with $\boldsymbol{V}(\boldsymbol{\delta}^* \mid \boldsymbol{\theta}) \leq \boldsymbol{V}(\boldsymbol{\delta} \mid \boldsymbol{\theta})$,[1] $\forall \boldsymbol{\theta}$. In the case of a scalar $h(\boldsymbol{\theta})$, the result states that $V(\delta^* \mid \boldsymbol{\theta}) \leq V(\delta \mid \boldsymbol{\theta})$.*

Before proving the result there are a few important comments to be made. The first one is that the theorem states that whenever one finds an unbiased estimator, it can always be improved in terms of risk by conditioning on a sufficient statistic. Also, the conditional expectation used in the definition of $\boldsymbol{\delta}^*$ does not depend on $\boldsymbol{\theta}$ because of the definition of a sufficient statistic.

Proof Initially note that $\boldsymbol{\delta}^*$ is unbiased because

$$\boldsymbol{E}[\boldsymbol{\delta}^*(\boldsymbol{X}) \mid \boldsymbol{\theta}] = \boldsymbol{E}\{\boldsymbol{E}[\boldsymbol{\delta}(\boldsymbol{X}) \mid \boldsymbol{T}(\boldsymbol{X})] \mid \boldsymbol{\theta}\} = \boldsymbol{E}[\boldsymbol{\delta}(\boldsymbol{X}) \mid \boldsymbol{\theta}] = \boldsymbol{h}(\boldsymbol{\theta}).$$

Finally note that

$$
\begin{aligned}
\boldsymbol{V}(\boldsymbol{\delta}^* \mid \boldsymbol{\theta}) &= \boldsymbol{V}[\boldsymbol{E}(\boldsymbol{\delta} \mid \boldsymbol{T}) \mid \boldsymbol{\theta}] - \boldsymbol{E}[\boldsymbol{V}(\boldsymbol{\delta} \mid \boldsymbol{T}) \mid \boldsymbol{\theta}] \\
&= \boldsymbol{V}(\boldsymbol{\delta} \mid \boldsymbol{\theta}) - \boldsymbol{E}[\boldsymbol{V}(\boldsymbol{\delta} \mid \boldsymbol{T}) \mid \boldsymbol{\theta}].
\end{aligned}
$$

As $\boldsymbol{V}(\boldsymbol{\delta} \mid \boldsymbol{T}) \geq 0$, its expectation is also non-negative positive, and therefore, $\boldsymbol{V}(\boldsymbol{\delta} \mid \boldsymbol{\theta}) \geq \boldsymbol{V}(\boldsymbol{\delta}^* \mid \boldsymbol{\theta})$. □

The important message of the theorem is that estimators have their risks reduced if they are functions of sufficient statistics. Risks are indeed reduced by properties of non-negative definite matrices (see Chapter 1). Yet again, maximal improvement in terms of risk is achieved if minimal sufficient statistics are used. A related interesting question is to know if the reduction in risk is the smallest possible. The search for maximal reduction is helped in a sense by the concept of complete families of distributions.

Definition 4.13 *Let $\boldsymbol{X} = (X_1, \ldots, X_n)$ be a random sample from $p(x \mid \boldsymbol{\theta})$, $\boldsymbol{T} = \boldsymbol{T}(\boldsymbol{X})$ any statistic and $\boldsymbol{g}$ any function of $\boldsymbol{T}$. The family of distributions of $\boldsymbol{T}$ is complete if $\forall \boldsymbol{\theta}$,*

$$\boldsymbol{E}(\boldsymbol{g}(\boldsymbol{T})|\boldsymbol{\theta}) = \boldsymbol{0} \Rightarrow \boldsymbol{g}(\boldsymbol{T}) = \boldsymbol{0}, \text{ with probability 1.}$$

---

[1]Recall from Chapter 1 that if $\mathbf{A}$ and $\mathbf{B}$ are squared matrices of the same dimension, $\mathbf{A} \leq \mathbf{B}$ means that the matrix $\mathbf{A} - \mathbf{B}$ is a non-positive definite matrix.

Verification of completeness of families directly from the definition is cumbersome. Fortunately, for exponential families with k parameters, it can be shown that the family of distributions of the k-dimensional statistic $(U_1(\mathbf{X}), \ldots, U_k(\mathbf{X}))$ is complete if the variation space of $(\phi_1(\boldsymbol{\theta}), \ldots, \phi_k(\boldsymbol{\theta}))$ is k-dimensional. The definitions of the U_i's and ϕ_i's were given in Section 2.3. The proof of this result requires elements that are beyond the scope of this book and will therefore be omitted. The interested reader is referred to Lehmann (1986).

Example 4.30 *Let $\mathbf{X} = (X_1, \ldots, X_n)$ be a random sample from the $N(\theta, \sigma^2)$ distribution. Then, $U_1 = \Sigma X_i$, $U_2 = \Sigma X_i^2$, $\phi_1 = \theta/\sigma^2$ and $\phi_2 = -1/2\sigma^2$, and (ϕ_1, ϕ_2) vary over a bidimensional space. Therefore, the family of distributions of (U_1, U_2) is complete. If one assumes that $\theta = \sigma^2$, the space of variation of (ϕ_1, ϕ_2) is reduced to a single dimension and the family of distributions of (U_1, U_2) is no longer complete.*

The concept of completeness is useful to ensure uniqueness of the UMVU estimator. It can be shown that the UMVU estimator is unique in the presence of complete families because if $\boldsymbol{\delta}_1$ and $\boldsymbol{\delta}_2$ are unbiased estimators and functions of the minimal sufficient statistic $\mathbf{T}$ then $\mathbf{E}[(\boldsymbol{\delta}_1^* - \boldsymbol{\delta}_2^*) \mid \boldsymbol{\theta}] = \mathbf{0}$. From completeness, this means that $\boldsymbol{\delta}_1^* - \boldsymbol{\delta}_2^* = \mathbf{0}$, with probability 1. Therefore, $\boldsymbol{\delta}_1^*$ and $\boldsymbol{\delta}_2^*$ must be equal.

Another interesting aspect of risk calculation is the existence of a lower bound for the quadratic risk of unbiased estimators. This result is known as the Cramer-Rao inequality. This inequality, central in the theory of unbiased estimation, is due to Fisher although it was independently stated in its present form by Cramer and Rao in the 40's, as cited in Cox and Hinkley (1974).

Theorem 4.2 *Let $\mathbf{X} = (X_1, \ldots, X_n)$ be a random sample from $p(x \mid \boldsymbol{\theta})$ and $\boldsymbol{\delta}$ an unbiased estimator of $h(\boldsymbol{\theta})$, for some function $\boldsymbol{h}$. Assume further that $\{\boldsymbol{x} : p(\boldsymbol{x} \mid \boldsymbol{\theta}) > 0\}$ does not depend on $\boldsymbol{\theta}$, the differentials $\partial p(\boldsymbol{x} \mid \boldsymbol{\theta})/\partial \boldsymbol{\theta}$ and $\partial h(\boldsymbol{\theta})/\partial \boldsymbol{\theta}$ exist, $\mathbf{E}(\boldsymbol{\delta} \mid \boldsymbol{\theta})$ is differentiable inside the integral sign and that Fisher information $\mathbf{I}(\boldsymbol{\theta})$ is finite. Then*

$$\boldsymbol{V}(\boldsymbol{\delta}|\boldsymbol{\theta}) \geq \frac{\partial \boldsymbol{h}(\boldsymbol{\theta})}{\partial \boldsymbol{\theta}}[\boldsymbol{I}(\boldsymbol{\theta})]^{-1}\left(\frac{\partial \boldsymbol{h}(\boldsymbol{\theta})}{\partial \boldsymbol{\theta}}\right)'.$$

In the case of a scalar θ, the inequality reduces to

$$V[\delta \mid \theta] \geq \frac{[dh(\theta)/d\theta]^2}{I(\theta)}.$$

Proof (scalar case) $E(\delta \mid \theta) = \int \delta(\mathbf{x}) p(\mathbf{x} \mid \theta)\, dx = h(\theta)$ because δ is unbiased. Differentiating both sides with respect to θ gives

$$
\begin{aligned}
\frac{dh(\theta)}{d\theta} &= \frac{\partial}{\partial\theta} \int \delta(\mathbf{x}) p(\mathbf{x} \mid \theta)\, dx \\
&= \int \delta(\mathbf{x}) \frac{\partial p(\mathbf{x} \mid \theta)}{\partial\theta}\, dx, \text{ interchange of signs valid by hypothesis} \\
&= \int \delta(\mathbf{x}) \frac{1}{p(\mathbf{x} \mid \theta)} \frac{\partial p(\mathbf{x} \mid \theta)}{\partial\theta} p(\mathbf{x} \mid \theta)\, dx \\
&= E\left[\left(\delta(\mathbf{X}) \frac{\partial \log p(\mathbf{X} \mid \theta)}{\partial\theta} \right) \mid \theta \right] \\
&= E[\delta(\mathbf{X})\, U(\mathbf{X}; \theta) \mid \theta].
\end{aligned}
$$

As previously seen in Section 2.4, $E[U(\mathbf{X}; \theta)] = 0$ and

$$
\begin{aligned}
\frac{dh(\theta)}{d\theta} &= E\left\{ [\delta(\mathbf{X}) - h(\theta)]\, U(\mathbf{X}; \theta) \mid \theta \right\} \\
&= Cov\left\{ [\delta(\mathbf{X}),\, U(\mathbf{X}; \theta)] \mid \theta \right\}.
\end{aligned}
$$

Since the absolute value of the correlation between two random variables is never larger than 1, the squared covariance will never be larger than the product of the two variances. Hence,

$$
[dh(\theta)/d\theta]^2 \leq V[\delta(\mathbf{X}) \mid \theta]\, V[U(\mathbf{X}; \theta) \mid \theta].
$$

But

$$
V[U(\mathbf{X}; \theta) \mid \theta] = E[U^2(\mathbf{X}; \theta) \mid \theta] = I(\theta),
$$

completing the proof.   □

The proof of the theorem in the multiparameter case is left as an exercise.

Observe that the unbiased estimator attains the lower bound when it has maximal correlation with the score function. In other words, when there are functions $\mathbf{c}$ and $\mathbf{d}$ of $\boldsymbol{\theta}$ such that

$$
\boldsymbol{\delta}(\mathbf{X}) = \mathbf{c}(\boldsymbol{\theta})\, \mathbf{U}(\mathbf{X}; \boldsymbol{\theta}) + \mathbf{d}(\boldsymbol{\theta}),
$$

with probability 1. Taking expectation of both sides with respect to $\mathbf{X}|\boldsymbol{\theta}$ gives that $\boldsymbol{\delta}(\mathbf{X})$ is an unbiased estimator of $\mathbf{d}(\boldsymbol{\theta})$ and therefore $\mathbf{d} = \mathbf{h}$.

Also, when the maximum likelihood estimator is unbiased, it attains the Cramer-Rao lower bound. This can be seen by solving the above equation for $\boldsymbol{\theta}$. This leads to

$$
\mathbf{U}(\mathbf{X}; \boldsymbol{\theta}) = \frac{\partial \log p(\mathbf{X} \mid \boldsymbol{\theta})}{\partial\boldsymbol{\theta}} = \frac{\boldsymbol{\delta}(\mathbf{X}) - \mathbf{d}(\boldsymbol{\theta})}{\mathbf{c}(\boldsymbol{\theta})}.
$$

Equating to 0 implies that $\delta(\mathbf{X})$ is the maximum likelihood estimator of $\mathbf{d}(\boldsymbol{\theta})$. But we have already seen that $\mathbf{d} = \mathbf{h}$ and, by hypothesis, δ is unbiased for $\mathbf{h}(\boldsymbol{\theta})$. Hence, it attains the Cramer-Rao lower bound.

Definition 4.14 *The estimator δ of $\mathbf{h}(\boldsymbol{\theta})$ is said to be efficient if it is unbiased and its variance attains the Cramer-Rao lower bound, $\forall\boldsymbol{\theta}$. The efficiency of an unbiased scalar estimator is given by the ratio between the Cramer-Rao bound and its variance.*

Note that there is no guarantee that UMVU estimators will attain the Cramer-Rao bound, and they may have their efficiency smaller than 1. However, the converse is true with efficient estimators being necessarily UMVU estimators.

Example 4.31 *Let $\mathbf{X} = (X_1, \ldots, X_n)$ be a random sample from the $Pois(\theta)$ distribution. Then*

$$\log p(\mathbf{X} \mid \theta) = -n\theta + \sum_{i=1}^{n} X_i \log \theta - \sum_{i=1}^{n} \log X_i!$$

and therefore $U(\mathbf{X}; \theta) = -n + \Sigma X_i / \theta$. As estimators cannot possibly depend on the parameter they are supposed to estimate, define $c(\theta) = \theta/n$ and $d(\theta) = \theta$. This way, it is immediate to show that $\bar{X}$ is an efficient estimator of its mean θ. In fact, any linear function of $\bar{X}$ is an efficient estimator of the respective linear function of θ. More than that, these are the unique efficient estimators that can be found in the presence of a random sample from the $Pois(\theta)$ distribution.

4.5.3 Consistency

It is to be expected that the information contained in the sample increases with an increase in the sample size. This is certainly true at least for the Fisher measures of information. One would then expect that reasonable estimators will tend to get closer and closer to their estimands. This subsection discusses theoretical properties of the estimators as the sample size gets larger and larger. The relevant question is how close are the estimator and its estimand. Related questions of interest: Is the bias getting smaller when sample size increases? Is the variance getting smaller as well? These questions will be deferred to the next chapter.

Definition 4.15 *Let $\mathbf{X}_n = (X_1, \ldots, X_n)$ be a random sample of size n from $p(x|\boldsymbol{\theta})$ and $\delta_n(\mathbf{X})$ an estimator of $\mathbf{h}(\boldsymbol{\theta})$ based on a sample of size n. As the* *sample size n varies, a sequence of estimators for $h(\boldsymbol{\theta})$ is obtained. This sequence is said to be (weakly) consistent for $h(\boldsymbol{\theta})$ if $\boldsymbol{\delta}_n(\boldsymbol{X}) \to h(\boldsymbol{\theta})$, in probability, when $n \to \infty$.*

In practice, the definition is shortened by saying that the estimator, instead of a sequence of estimators, is or is not consistent. The definition means that $\forall \epsilon > 0$, $P(|\boldsymbol{\delta}_n(\mathbf{X}) - \mathbf{h}(\boldsymbol{\theta})| > \epsilon) \to 0$, when $n \to \infty$. This result is usually denoted by $plim\, \boldsymbol{\delta}_n(\mathbf{X}) = \mathbf{h}(\boldsymbol{\theta})$. As an example, $\hat{F}_n$ (the empirical distribution function) is a consistent estimator for F.

When Bayes estimators are considered as functions of the sample $\mathbf{X}_n$ instead of its observed value $\mathbf{x}_n$, they can be studied for their sampling properties just like any other estimator. In particular, it can be shown that Bayes estimators are invariably biased. Also, it may be reasoned that as the sample size increases, the influence of any non-degenerate prior becomes smaller and Bayes estimators will become closer to the MLE. So, intuitively, one can expect them to inherit all the properties of the MLE, irrespective of the loss function used.

Example 4.32 *Let $\boldsymbol{X}_n = (X_1, \ldots, X_n)$ be a random sample from the $Ber(\theta)$ distribution, with $\theta > 0$. We know that the maximum likelihood estimator of θ is $\hat{\theta}_n = \bar{X}_n$ and that, by the laws of large numbers, $\bar{X}_n \to \theta$, in probability and almost surely. Therefore, $\bar{X}_n$ is a consistent estimator of θ.*

In the case of a conjugate prior $Beta(\alpha, \beta)$ and quadratic loss function, the Bayes estimator is $\delta_n^*(\boldsymbol{x}_n) = (\alpha + n\bar{x}_n)/(\alpha + \beta + n)$ which converges to $\bar{x}_n$ when $n \to \infty$. Therefore, $|\delta_n^*(\boldsymbol{X}_n) - \bar{X}_n| \to 0$ in probability, and as $\bar{X}_n \to \theta$, almost surely, δ_n^* is also a consistent estimator of θ.

It follows readily from Tchebychev's inequality that

$$Pr(|\boldsymbol{\delta}_n(\mathbf{X}) - \mathbf{h}(\boldsymbol{\theta})| > \epsilon) < \frac{E\{[\boldsymbol{\delta}_n(\mathbf{X}) - \mathbf{h}(\boldsymbol{\theta})]'[\boldsymbol{\delta}_n(\mathbf{X}) - \mathbf{h}(\boldsymbol{\theta})] \mid \theta\}}{\epsilon^2}, \forall \epsilon > 0$$

but $E\{[\boldsymbol{\delta}_n(\mathbf{X}) - \mathbf{h}(\boldsymbol{\theta})]'[\boldsymbol{\delta}_n(\mathbf{X}) - \mathbf{h}(\boldsymbol{\theta})] \mid \theta\} = R_{\mathbf{T}_n(\mathbf{X})}(\boldsymbol{\theta})$. So, if a sequence of estimators has quadratic risk tending to 0, the estimator is consistent.

One can also define strong consistency of a sequence of estimators if the convergence for the parameter is almost sure instead of in probability. The theory of probability ensures us that almost sure convergence implies convergence in probability, and therefore, strong consistency implies weak consistency. It can be shown that the maximum likelihood estimator is strongly consistent under the same regularity conditions of the Cramer-Rao inequality. Nevertheless, the concept of weak consistency retains the essence of what is needed and will be used hereafter to define consistency.

4.6   Interval estimation

4.6.1   Bayesian approach

Returning to the Bayesian point of view, the most adequate form to express available information about unknown parameters is through the posterior distribution. Despite its coherent specification through expected loss functions, point estimation presents some inconvenient features. The main restriction is that it simplifies the multitude of information from a distribution into a single figure. It is important at least to have some information about how precise the specification of this figure is. One possibility is to associate point estimators with a measure of the uncertainty about them. So, for the mean, one can use the variance or the coefficient of variation. For the mode, the observed information given by the curvature at the mode is usually adequate. Finally, for the median, the interquartile distance could be used.

In this section, another line of work is sought. The aim is to provide a compromise between the complete posterior distribution and a single figure extracted from it. This compromise is reached by providing a range of values extracted from the posterior distribution. Typically, one attaches a probability to this region, and when the probability is large, one gets a good idea of the probable or likely values of the unknown of interest. Ideally, one would like to report a region of values of $\boldsymbol{\theta}$ that is as small as possible but that contains as much probability as possible. The size of the interval informs about the dispersion of the values of $\boldsymbol{\theta}$.

Definition 4.16 *Let $\boldsymbol{\theta}$ be an unknown quantity defined in Θ. A region $C \subset \Theta$ is a $100(1 - \alpha)\%$ credibility or Bayesian confidence region for $\boldsymbol{\theta}$ if $Pr(\boldsymbol{\theta} \in C|\boldsymbol{x}) \geq 1 - \alpha$. In this case, $1 - \alpha$ is called the credibility or confidence level. In the scalar case, the region C is usually given by an interval, $[c_1, c_2]$, say, hence the name.*

It should be clear from the above definition that the intervals are defined by simple probability evaluation over the posterior distribution of $\boldsymbol{\theta}$. Many Bayesian authors reject the use of the word *confidence* for Bayesian intervals. As will be seen shortly, *confidence* has a very precise meaning in the definition of frequentist intervals that differs substantially from the meaning given here. These authors consider important to dissociate the concepts. Hereafter, we will refer to confidence intervals whenever we refer to the method in general or in the classical context and will use the word *credibility* in reference to Bayesian intervals.

Note that $\mathbf{C}$ is never an interval in a multidimensional case. Even in the uniparameter case, there is nothing in the definition enforcing the region $\mathbf{C}$

to be an interval. Therefore, there is a slightly misleading use of the word *interval* and we will use *region* instead, whenever appropriate.

The above probability is evaluated over the updated distribution of $\boldsymbol{\theta}$ which will be taken from now on as the posterior. In general, one would want both α and $\mathbf{C}$ to be as small as possible. This in turn implies that the posterior distribution is as concentrated as possible. The requirement of a larger posterior probability than the confidence level is essentially technical. It is mainly due to the use in discrete distributions where it is not always possible to find a region that exactly satisfies the probability required by a given level. By and large, the inequality can be taken as an equality, thus implying that the region $\mathbf{C}$ will be as small as possible.

Note also that credibility intervals are invariant under 1-to-1 transformations of the parameter. So, if $\mathbf{C}$ is a $100(1-\alpha)\%$ credibility region for $\boldsymbol{\theta}$ and $\phi = \phi(\boldsymbol{\theta})$ is a 1-to-1 transformation of $\boldsymbol{\theta}$, then $\phi(\mathbf{C})$, the image of $\mathbf{C}$ under ϕ, is a $100(1-\alpha)\%$ credibility region for ϕ. This useful property is also shared by frequentist confidence intervals.

Example 4.33 *Let $\boldsymbol{X} = (X_1, \ldots, X_n)$ be a random sample from the $N(\theta, \sigma^2)$ distribution with σ^2 known. The non-informative prior for θ is $p(\theta) \propto k$ and the likelihood is*

$$l(\theta; \boldsymbol{x}) \propto \exp\left\{-\frac{n}{2\sigma^2}(\theta - \bar{x})^2\right\},$$

providing the posterior $p(\theta \mid \boldsymbol{x}) \propto l(\theta; \boldsymbol{x})p(\theta) \propto l(\theta; \boldsymbol{x})$ and therefore $\theta \mid \boldsymbol{x} \sim N(\bar{x}, \frac{\sigma^2}{n})$ or equivalently $\sqrt{n}(\theta - \bar{x})/\sigma \mid \boldsymbol{x} \sim N(0, 1)$. From there, many $100(1-\alpha)\%$ confidence intervals may be constructed for θ with the use of the standard Normal distribution function Φ. Defining $\Phi(x) = P(X \leq x)$ if $X \sim N(0, 1)$, then z_c is such that $\Phi(z_c) = 1 - c$, $0 < c < 1$ and intervals can be constructed from

1. $1 - \alpha = Pr(\sqrt{n}(\theta - \bar{x})/\sigma \leq z_\alpha \mid \boldsymbol{x})$ *which implies that $\theta \leq z_\alpha \sigma/\sqrt{n} + \bar{x}$ with posterior probability $1 - \alpha$. Hence, the interval $C_1 = (-\infty, \bar{x} + z_\alpha \sigma/\sqrt{n}]$ is a $100(1-\alpha)\%$ credibility interval for θ. The length of C_1 however is infinity which is not very useful for our summarization purposes. As previously mentioned, one would like to have C as small as possible. The problem with this interval is that it includes (infinitely) many values that have very negligible probability around them.*

2. *Let z_β and z_γ be numbers such that $1 - \alpha = P(-z_\beta \leq \sqrt{n}(\theta - \bar{x})/\sigma \leq z_\gamma \mid \boldsymbol{x})$. Using the symmetry of the Normal distribution,*

$$\Phi(-z_\beta) = P(X \leq -z_\beta) = P(X \geq z_\beta) = 1 - P(X < z_\beta) = \beta$$

and the probability of the above interval is given by $\Phi(z_\gamma) - \Phi(-z_\beta) =$

$1 - (\gamma + \beta)$, and therefore $\gamma + \beta = \alpha$. With this assumption,

$$
\begin{aligned}
1 - \alpha &= Pr\left(z_\beta \leq \sqrt{n}\frac{(\theta - \overline{x})}{\sigma} \leq z_\gamma\right) \\
&= Pr\left(-\frac{\sigma}{\sqrt{n}}z_\beta + \overline{x} \leq \theta \leq z_\gamma\frac{\sigma}{\sqrt{n}} + \overline{x}\right).
\end{aligned}
$$

The interval $C = [c_1, c_2]$ where

$$
c_1 = \overline{x} - \frac{\sigma}{\sqrt{n}}z_\beta \text{ and } c_2 = \overline{x} + \frac{\sigma}{\sqrt{n}}z_\gamma
$$

is a $100(1-\alpha)\%$ credibility interval for θ. Note that it has length $(z_\gamma + z_\beta)\sigma/\sqrt{n}$. There still remains the issue associated with minimization of the length of the interval subject to $\gamma + \beta = \alpha$.

Note that if $\phi = \phi(\theta)$ is a monotonically increasing transformation of θ, then $[\phi(c_1), \phi(c_2)]$ is also a $100(1-\alpha)\%$ Bayesian confidence interval for ϕ. If $\phi = \phi(\theta)$ is a monotonically decreasing transformation of θ, then $[\phi(c_2), \phi(c_1)]$ is a $100(1-\alpha)\%$ Bayesian confidence interval for ϕ.

Consider without loss of generality that $z_\gamma \leq z_{\alpha/2} \leq z_\beta$ and define $a = z_{\alpha/2} - z_\gamma \geq 0$, $b = z_\beta - z_{\alpha/2} \geq 0$ and A and B as the areas under the $N(0,1)$ density between $z_{\alpha/2}$ and z_γ and between z_β and $z_{\alpha/2}$, respectively. The length of the confidence interval becomes $2z_{\alpha/2} + b - a$ and $A = B$. It is clear from Figure 4.5 that the density over the first interval is strictly larger than under the second interval. Therefore, $b \geq a$ and $b - a \geq 0$. The shortest possible interval is then obtained by taking $b = a = z_{\alpha/2}$. Therefore, the symmetric interval is the shortest one and every value of θ inside it has larger density than any point lying outside the interval.

This simple example provides the key to finding intervals of shortest length. It indicates that length of the interval is inversely proportional to density height. Shortest intervals are then provided by inclusion of points of higher density. This idea is mathematically expressed in the definition below and represented graphically in Figure 4.6.

Definition 4.17 A $100(1 - \alpha)\%$ credibility interval of highest posterior density (HPD, in short) for $\boldsymbol{\theta}$ is the $100(1 - \alpha)\%$ credibility interval $\boldsymbol{C}$ given by $\boldsymbol{C} = \{\boldsymbol{\theta} \in \boldsymbol{\Theta} : p(\boldsymbol{\theta} \mid \boldsymbol{x}) \geq k(\alpha)\}$ where $k(\alpha)$ is the largest constant such that $P(\boldsymbol{\theta} \in \boldsymbol{C} \mid \boldsymbol{x}) \geq 1 - \alpha$.

Example 4.34 (Berger, 1985) Let $\boldsymbol{X} = (X_1, \ldots, X_n)$ be a random sample from the $Cauchy(\theta, 1)$ distribution and θ has non-informative prior $p(\theta) \propto k$.

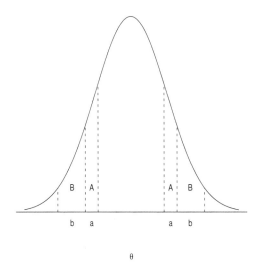

Figure 4.5: *Density of the standard Normal distribution*

The posterior density of θ is

$$p(\theta \mid \boldsymbol{x}) \propto \prod_{i=1}^{n} \frac{1}{1 + (x_i - \theta)^2}.$$

Assume now that the observed sample was $\boldsymbol{x} = (4.0, 5.5, 7.5, 4.5, 3.0)$, with sampling average $\bar{x} = 4.9$. Then, the 95% HPD credibility interval for θ can be numerically obtained as $[3.10, 6.06]$. Had we assumed an $N(\theta, 1)$ sampling distribution, the 95% HPD interval would be $[4.02, 5.86]$ which is more affected by the suspect value 7.5. In both cases, the intervals are easily obtained with computational help. It will be seen in the sequel that the exercise is far from trivial for the Cauchy case in the frequentist approach. The main reasons are absence of a univariate sufficient statistic for θ and the absence of asymptotic results to allow for approximations.

Note however that despite their appeal, HPD regions are not invariant under 1-to-1 transformations. The main reason is the existence of the Jacobian required when obtaining the density of any parametric transformation. Because of the invariance, transformations of HPD regions remain valid credibility regions with the same credibility level. All they lose is the HPD property.

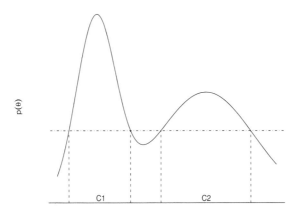

Figure 4.6: *The HPD interval for the density above is given by $C_1 \cup C_2$.*

4.6.2   Classical approach

In the case of classical confidence intervals, only sampling distributions can be used since parameters are unknown but fixed. Therefore, they are not liable to the probabilistic description they get under the Bayesian treatment. That is why the concept of confidence instead of probability becomes relevant in the context of interval estimation. Before describing the general formulation, it is useful to see it applied in an example.

Example 4.35 *Let $\boldsymbol{X} = (X_1, \ldots, X_n)$ be a random sample from the $N(\theta, \sigma^2)$ distribution, with σ^2 known. To draw classical inference one should ideally base calculations on a minimal sufficient statistic for θ. In this case, we have seen that $\overline{X}$ is such a statistic and*

$$\overline{X} \sim N\left(\theta, \frac{\sigma^2}{n}\right) \quad \text{and} \quad U = \frac{\overline{X} - \theta}{\sigma/\sqrt{n}} \sim N(0, 1).$$

Observe that U is a function of the sample and of the parameter θ, the parameter of interest and its distribution does not depend on θ. It can be said that

$$P\left(-z_{\alpha/2} \leq U \leq z_{\alpha/2}\right) = 1 - \alpha,$$

and isolating θ yields

$$P\left(\overline{X} - z_{\alpha/2} \frac{\sigma}{\sqrt{n}} \leq \theta \leq \overline{X} + z_{\alpha/2} \frac{\sigma}{\sqrt{n}}\right) = 1 - \alpha.$$

So, even though an interval for θ was obtained, it cannot be understood as a probability interval for θ as in Bayesian procedures. It can only be interpreted in the sampling framework by saying that if the same experiment were to be repeated many times, in approximately $100(1-\alpha)\%$ of them, its random limits $\overline{X} - z_{\alpha/2}\sigma/\sqrt{n}$ and $\overline{X} + z_{\alpha/2}\sigma/\sqrt{n}$ would include the value of θ. Also, this assertion is useless from a practical perspective since it is based on unobserved samples. What can be done is to replace the observed value of $\overline{X}$ in the expression and state that one can have $100(1-\alpha)\%$ confidence instead of probability that the so formed numerical interval contains θ.

The general procedure to obtain confidence intervals in the frequentist framework is based on a generalization of the steps of the above example to any statistical problem. These are

1. A quantity $\mathbf{U} = \mathbf{G}(\mathbf{X}, \boldsymbol{\theta}) \in \mathcal{U}$ with distribution that does not depend on $\boldsymbol{\theta} \in \boldsymbol{\Theta}$ must be found. Ideally, this statistic must depend on $\mathbf{X}$ through minimal sufficient statistics and have a known distribution. Both requirements were met in the example since U depended on the sample through $\bar{X}$ and had a standard Normal distribution.

2. With knowledge of the distribution of $\mathbf{U}$, find a region $\mathbf{A} \subset \mathcal{U}$ such that $Pr(\mathbf{U} \in \mathbf{A}) = 1 - \alpha$. When θ is scalar, then a scalar U can be found in many cases with $Pr(a_1 \leq U \leq a_2) = 1 - \alpha$ and, in this case, $A = [a_1, a_2]$.

3. The confidence region $\mathbf{C} \subset \boldsymbol{\Theta}$ is obtained by isolating $\boldsymbol{\theta}$ in the above expression and replacing sample values.

A useful complementary reading about construction of confidence intervals is Silvey (1970).

Definition 4.18 *Let $\boldsymbol{\theta}$ be an unknown quantity defined in $\boldsymbol{\Theta}$, $\mathbf{U}$ be a function $\mathbf{U} = \mathbf{G}(\mathbf{X}, \boldsymbol{\theta})$ with values in $\mathcal{U}$ and $\mathbf{A}$ be a region in $\mathcal{U}$ such that $Pr(\mathbf{U} \in \mathbf{A}) \geq 1 - \alpha$. A region $\mathbf{C} \subset \boldsymbol{\Theta}$ is a $100(1-\alpha)\%$ confidence region for $\boldsymbol{\theta}$ if*

$$\mathbf{C} = \{\boldsymbol{\theta} : \mathbf{G}(\boldsymbol{x}, \boldsymbol{\theta}) \in \mathbf{A}\}.$$

In this case, $1-\alpha$ is called the confidence level. In the scalar case, the inversion in terms of θ usually leads to an interval, $C = [c_1, c_2]$, say, hence the name.

Once again, the use of the word *interval* for the general case is an abuse of language. The inequality in the definition is taken as an equality, whenever possible.

The quantity $\mathbf{U}$ is usually called a pivot or a pivotal quantity and finding one such quantity is fundamental. The choice of $\mathbf{U}$ is crucial to the success of the method. It is not at all obvious that reasonable options are available in any given problem. The effort towards the use of minimal sufficiency is in the direction of shortening intervals as much as possible.

Example 4.36 *Let $X_1, \ldots, X_n$ be a random sample from the uniform distribution over interval $[0, \theta]$. Suppose we are interested in interval estimation of θ. Although $Y = \max(X_1, \ldots, X_n)$ is a minimal sufficient statistics for θ, it is not a pivotal quantity since its distribution depends on θ, that is, $p_Y(y) = ny^{n-1}/\theta^n$, for $0 \leq y \leq \theta$. The quantity $U = G(\mathbf{X}, \theta) = Y/\theta$ however is a pivot since its distribution given by $p_U(u) = nu^{n-1}$, for $0 \leq u \leq 1$, does not depend on θ.*

Then, we can find c_1 and c_2 in $p_U(u)$ such that

$$\int_{c_1}^{c_2} p_U(u)du = 1 - \alpha,$$

which is satisfied for infinite pairs (c_1, c_2). One way to overcome this difficulty is to assume that

$$\int_0^{c_1} p_U(u)du = \frac{\alpha}{2} \quad and \quad \int_{c_2}^1 p_U(u)du = \frac{\alpha}{2},$$

leading to the so-called equal tail probablity intervals.

Solving the above equations, we obtain $c_1 = (\alpha/2)^{1/n}$ and $c_2 = [1 - (\alpha/2)]^{1/n}$ such that

$$P\left(c_1 \leq \frac{Y}{\theta} \leq c_2\right) = P\left(\frac{Y}{c_2} \leq \theta \leq \frac{Y}{c_1}\right) = 1 - \alpha.$$

Then, a $100(1-\alpha)\%$ classical confidence interval for θ based on the pivot U is given by

$$\left(\frac{y}{[1 - (\alpha/2)]^{1/n}} \leq \theta \leq \frac{y}{(\alpha/2)^{1/n}}\right),$$

where y is the observed value of Y. Note however that this is not the confidence interval of minimal length (see Exercise 4.30).

All randomness present here is due to the sample $\mathbf{X}$ leading to a probability interval for $\mathbf{U}$ and not for $\boldsymbol{\theta}$. The procedure involves an elaboration that is absent from the Bayesian definition. It provides an interval to which a numerical value, the confidence of the interval, is associated. For that reason, it is often interpreted misleadingly as a probability interval, as in the Bayesian framework. Care must be exercised to ensure a correct interpretation of the intervals. The existent symmetry in many canonical situations leads to intervals that coincide numerically when obtained by a frequentist or a non-informative Bayesian approach. This happened in the above example but should not be used to unduly equate the two approaches.

4.6.3   Bonferroni inequalities

Returning to the Bayesian approach, assume now that $Pr(\theta_i \in C_i) \geq 1 - \alpha_i$, $i = 1, \ldots, r$ and let $\mathbf{C} = C_1 \times \ldots \times C_r$. If the θ_i's are independent a posteriori then

$$Pr(\boldsymbol{\theta} \in \mathbf{C}) = \prod_{i=1}^{r} Pr(\theta_i \in C_i) \geq \prod_{i=1}^{r}(1 - \alpha_i)$$

and $\mathbf{C}$ is a $100(1 - \alpha)\%$ credibility region for $\boldsymbol{\theta}$ if $\prod_i(1 - \alpha_i) \geq 1 - \alpha$. If the θ_i's are not independent then

$$Pr(\boldsymbol{\theta} \in \mathbf{C}) \geq 1 - \sum_{i=1}^{r} Pr(\theta_i \notin C_i).$$

This result is also known as the Bonferroni inequality. As $Pr(\theta_i \notin C_i) \leq \alpha_i$, if one takes $\sum_i \alpha_i = \alpha$, taking for example $\alpha_i = \alpha/r$, then $\mathbf{C}$ is a $100(1 - \alpha)\%$ credibility region for $\boldsymbol{\theta}$.

This development is equally applicable to the construction of confidence intervals. In particular, approximations such as those from Bonferroni inequality are more used in classical inference where the search for the pivotal quantity $\mathbf{U}$ does not always lead to independent components. This problem typically does not occur in the Bayesian approach where most problems lie in the computation of the credibility level.

4.7   Estimation in the Normal model

This section deals with applications of point and interval estimation of means and variances to problems of one and two Normal populations. The Bayesian perspective with both non-informative and proper conjugate priors and frequentist perspective are presented and compared. We will particularly emphasize the similarity between the results with the frequentist and the non-informative Bayesian points of view. The similarity is only numerical since we have seen that the derivations are completely different.

4.7.1   One sample case

Assume initially a single sample $\mathbf{X} = (X_1, \ldots, X_n)$ from the $N(\theta, \sigma^2)$ distribution with $\phi = \sigma^{-2}$. If ϕ is known and the prior distribution is $\theta \sim N(\mu_0, \tau_0^2)$, we have already obtained that $\theta | \mathbf{x} \sim N(\mu_1, \tau_1^2)$ where

$$\mu_1 = \frac{n\sigma^{-2}\overline{x} + \tau_0^{-2}\mu_0}{n\sigma^{-2} + \tau_0^{-2}} \quad \text{and} \quad \tau_1^2 = \frac{1}{n\sigma^{-2} + \tau_0^{-2}}.$$

So, posterior mean, median and mode coincide and the posterior precision and curvature of the log posterior are given by τ_1^{-2}. A common abuse of notation is to use $J_{\mathbf{x}}$ to denote this curvature. The reason for doing this will be more fully discussed in the next chapter.

Credibility intervals can be obtained by noticing that

$$\frac{\theta - \mu_1}{\tau_1} \mid \mathbf{x} \sim N(0, 1)$$

and therefore

$$
\begin{aligned}
1 - \alpha &= P(-z_{\alpha/2} < (\theta - \mu_1)/\tau_1 < z_{\alpha/2} \mid \mathbf{x}) \\
&= P(\mu_1 - z_{\alpha/2}\tau_1 < \theta < \mu_1 + z_{\alpha/2}\tau_1 \mid \mathbf{x})
\end{aligned}
$$

and, due to the symmetry of the Normal, $(\mu_1 - z_{\alpha/2}\tau_1, \mu_1 + z_{\alpha/2}\tau_1)$ is the $100(1 - \alpha)\%$ HPD interval for θ.

A non-informative prior can be obtained by letting $\tau_0^2 \to \infty$. In this case, $\tau_1^{-2} \to n\sigma^{-2}$ and $\mu_1 \to \bar{x}$. Posterior mean, median and mode coincide with the moment and maximum likelihood estimators $\overline{X}$. It is easy to check that $\overline{X}$ is also an unbiased, efficient and (strongly) consistent estimator for θ. Also, the $100(1 - \alpha)\%$ HPD credibility interval for θ coincides with the confidence interval obtained in the previous section.

Assuming now that θ is known and the prior for σ^2 is $n_0\sigma_0^2\phi \sim \chi_{n_0}^2$ leads to the posterior for $(n_0\sigma_0^2 + ns_0^2)\phi \mid \mathbf{x} \sim \chi_{n+n_0}^2$ where

$$s_0^2 = \frac{1}{n} \sum_{i=1}^{n} (x_i - \theta)^2.$$

The following quantities can be obtained:

$$
\begin{aligned}
E(\phi \mid \mathbf{x}) &= \frac{n_0 + n}{n_0\sigma_0^2 + ns_0^2}, \\
\{E(\phi \mid \mathbf{x})\}^{-1} &= \frac{n_0\sigma_0^2 + ns_0^2}{n_0 + n} = \frac{n_0}{n_0 + n}\sigma_0^2 + \frac{n}{n_0 + n}s_0^2,
\end{aligned}
$$

which is a weighted average between the prior estimate σ_0^2 and the maximum likelihood estimate s_0^2 with weights $n_0/(n_0 + n)$ and $n/(n_0 + n)$, respectively. Also,

$$E(\sigma^2 \mid \mathbf{x}) = E(\phi^{-1} \mid \mathbf{x}) = \frac{n_0\sigma_0^2 + ns_0^2}{n + n_0 - 2},$$

which coincides with the inverse of the posterior mode (but not of the mean) of ϕ. The main dispersion measures are

$$V(\phi \mid \mathbf{x}) = \frac{2(n + n_0)}{(n_0\sigma_0^2 + ns_0^2)^2} \quad \text{and} \quad J(\text{mode}) = \frac{(n_0\sigma_0^2 + ns_0^2)^2}{2(n + n_0 - 2)}.$$

Note that once again the expression of $J(\text{mode})$ is very similar to the posterior precision $V^{-1}(\phi \mid \mathbf{x})$.

Confidence intervals can be obtained with the percentiles of the χ^2 that are available in many tables and statistical software. Defining $\chi^2_{\alpha,\nu}$ and $\overline{\chi}^2_{\alpha,\nu}$ as the $100\alpha\%$ and $100(1-\alpha)\%$ respective percentiles of the χ^2 distribution with ν degrees of freedom gives

$$
\begin{aligned}
1 - \alpha &= P\left(\chi^2_{\alpha/2,n_1} < (n_0\sigma_0^2 + ns_0^2)\phi < \overline{\chi}^2_{\alpha/2,n_1} \mid \mathbf{x}\right) \text{ where } n_1 = n_0 + n \\
&= P\left(\frac{\chi^2_{\alpha/2,n_1}}{n_0\sigma_0^2 + ns_0^2} < \phi < \frac{\overline{\chi}^2_{\alpha/2,n_1}}{n_0\sigma_0^2 + ns_0^2} \mid \mathbf{x}\right).
\end{aligned}
$$

This gives rise to a $100(1-\alpha)\%$ credibility interval for ϕ. Given the asymmetry of the χ^2 distribution, this is not an HPD interval. As $\sigma^2 = 1/\phi$,

$$
\left(\frac{n_0\sigma_0^2 + ns_0^2}{\overline{\chi}^2_{\alpha/2,n_1}}, \frac{n_0\sigma_0^2 + ns_0^2}{\chi^2_{\alpha/2,n_1}}\right)
$$

is a $100(1-\alpha)\%$ credibility interval for σ^2.

The non-informative prior can be obtained by letting $n_0 \to 0$. In this case, the posterior is $ns_0^2\phi \mid \mathbf{x} \sim \chi^2_n$ and $\{E[\phi \mid \mathbf{x}]\}^{-1} = s_0^2$, which coincides with the maximum likelihood estimate of $\phi^{-1} = \sigma^2$. The maximum likelihood estimator is S_0^2 with sampling distribution $nS_0^2/\sigma^2|\sigma^2 \sim \chi^2_n$. Therefore, it is an unbiased estimator. Since

$$
\frac{\partial \log p(\mathbf{X} \mid \sigma^2)}{\partial \sigma^2} = \frac{n}{2\sigma^4}(S_0^2 - \sigma^2),
$$

S_0^2 is also an efficient estimator of σ^2. Its variance is $2\sigma^4/n$ and tends to 0 as $n \to \infty$ which means that the estimator is also consistent. The $100(1-\alpha)\%$ credibility interval for σ^2 becomes

$$
\left(\frac{ns_0^2}{\overline{\chi}^2_{\alpha/2,n}}, \frac{ns_0^2}{\chi^2_{\alpha/2,n}}\right).
$$

The pivotal quantity used for the construction of the interval is nS_0^2/σ^2 with a χ^2_n sampling distribution. The classical confidence interval can be easily obtained and shown to coincide with the above interval, obtained for the non-informative prior.

If θ and σ^2 are both unknown quantities, using the conjugate prior $\theta \mid \phi \sim N(\mu_0, (c_0\phi)^{-1})$ and $n_0\sigma_0^2\phi \sim \chi^2_{n_0}$ gives the marginal posterior distributions $\theta \mid \mathbf{x} \sim t_{n_1}(\mu_1, \sigma_1^2/c_1)$ and $n_1\sigma_1^2\phi \mid \mathbf{x} \sim \chi^2_{n_1}$, where $n_1\sigma_1^2 = n_0\sigma_0^2 + (n-1)s^2 +$

$c_0 n(\mu_0 - \bar{x})^2/(c_0 + n)$ and $s^2 = \Sigma(x_i - \bar{x})^2/(n-1)$. Once again, posterior mean, mode and median of θ coincide and are given by μ_1. Also,

$$V(\theta \mid \mathbf{x}) = \frac{n_1}{n_1 - 2} \frac{\sigma_1^2}{c_1} \text{ and } J(\mu_1) = \frac{n_1 + 1}{n_1} \frac{c_1}{\sigma_1^2}.$$

Denoting the $100(1-\alpha)\%$ percentile of the $t_\nu(0,1)$ distribution by $t_{\alpha,\nu}$, gives by symmetry of the Student-t that

$$1 - \alpha = P\left(-t_{\alpha/2,n_1} < \sqrt{c_1} \frac{\theta - \mu_1}{\sigma_1} < t_{\alpha/2,n_1}\right)$$

$$= P\left(\mu_1 - t_{\alpha/2,n_1} \frac{\sigma_1}{\sqrt{c_1}} < \theta < \mu_1 + t_{\alpha/2,n_1} \frac{\sigma_1}{\sqrt{c_1}}\right)$$

and the above interval is an HPD interval.

For σ^2, by analogy with the results obtained for known θ, $E(\phi \mid \mathbf{x}) = \sigma_1^{-2}$ and

$$E(\sigma^2 \mid \mathbf{x}) = E(\phi^{-1} \mid \mathbf{x}) = \frac{n_1 \sigma_1^2}{n_1 - 2},$$

which coincides with the inverse of the posterior mode (but not of the mean) of ϕ. The main dispersion measures are

$$V(\phi \mid \mathbf{x}) = \frac{2n_1}{(n_1 \sigma_1^2)^2} \text{ and } J(\text{mode}) = \frac{(n_1 \sigma_1^2)^2}{2(n_1 - 2)}.$$

Once again, the expression of $J(\text{mode})$ is very similar to the posterior precision $V^{-1}(\phi \mid \mathbf{x})$.

Confidence intervals can again be obtained with the χ^2 percentiles leading to

$$1 - \alpha = P\left(\underline{\chi}_{\alpha/2,n_1}^2 < n_1 \sigma_1^2 \phi < \overline{\chi}_{\alpha/2,n_1}^2\right)$$

$$= P\left(\frac{\underline{\chi}_{\alpha/2,n_1}^2}{n_1 \sigma_1^2} < \phi < \frac{\overline{\chi}_{\alpha/2,n_1}^2}{n_1 \sigma_1^2}\right).$$

This gives a $100(1-\alpha)\%$ confidence interval for ϕ (that is also not an HPD interval). As $\sigma^2 = 1/\phi$,

$$\left(\frac{n_1 \sigma_1^2}{\overline{\chi}_{\alpha/2,n_1}^2}, \frac{n_1 \sigma_1^2}{\underline{\chi}_{\alpha/2,n_1}^2}\right)$$

is a $100(1-\alpha)\%$ confidence interval for σ^2.

The non-informative prior in this case is $p(\theta, \phi) \propto \phi^{-1}$. This gives marginal posterior distributions $\theta \mid \mathbf{x} \sim t_{n-1}(\bar{x}, s^2/n)$ and $(n-1)s^2 \phi \mid \mathbf{x} \sim \chi_{n-1}^2$. Again, posterior mean, mode and median of θ coincide with $\bar{x}$, which is the maximum likelihood and moment estimate.

The classical (moments and maximum likelihood) estimator $\bar{X}$ for θ is unbiased, efficient and consistent. The dispersion measures for θ are

$$V(\theta \mid \mathbf{x}) = \frac{n-1}{n-3} \frac{s^2}{n} \text{ and } J(\bar{x}) = \frac{n^2}{(n-1)s^2}.$$

Since $\sqrt{n}(\theta - \bar{x})/s \mid \mathbf{x} \sim t_{n-1}(0, 1)$, the HPD $100(1 - \alpha)\%$ confidence interval for θ is analogously obtained as

$$\left(\bar{x} - t_{\alpha/2, n-1} \frac{s}{\sqrt{n}} , \ \bar{x} + t_{\alpha/2, n-1} \frac{s}{\sqrt{n}} \right).$$

Classical confidence intervals for θ cannot be obtained with the same pivotal quantity used earlier because it depends on the unknown σ. A new pivotal quantity depending only on $\mathbf{X}$ and θ and with a distribution that is known and does not depend on any of the unknown parameters must be found. Fortunately, this is possible in the Normal case with the following results.

Theorem 4.3 Let $\mathbf{X} = (X_1, \ldots, X_n)$ be a random sample from the $N(\theta, \sigma^2)$ distribution and let $\bar{X}$ and S^2 be the sample mean and variance, respectively. Then, conditional on θ and σ^2, $\bar{X}$ and S^2 are independent with respective sampling distributions

$$\sqrt{n} \frac{\bar{X} - \theta}{\sigma} \sim N(0, 1) \text{ and } \frac{(n-1)S^2}{\sigma^2} \sim \chi^2_{n-1}.$$

Proof Define $\mathbf{Z} = (Z_1, \ldots, Z_n)$ where $Z_i = (X_i - \theta)/\sigma$. Then, the Z_i's are iid $N(0, 1)$, $\bar{Z} = (\bar{X} - \theta)/\sigma$ and, in matrix notation, $\mathbf{Z} \sim N(\mathbf{0}, \mathbf{I}_n)$, where $\mathbf{I}_n$ is the $n \times n$ identity matrix. Let $\mathbf{A}$ be an orthogonal matrix with first row given by $\sqrt{n}\mathbf{1}'_n$ where $\mathbf{1}_n$ is an n-dimensional vector of 1's. There are many methods in linear algebra available for completing orthogonally the other $n-1$ rows of $\mathbf{A}$. From the invariance of the multivariate Normal distribution under linear transformations (Exercise 1.6), it follows that

$$\mathbf{Y} = \mathbf{A} \ \mathbf{Z} \sim N(\mathbf{A0}, \mathbf{AI}_n\mathbf{A}') = N(\mathbf{0}, \mathbf{AA}') = N(\mathbf{0}, \mathbf{I}_n).$$

Therefore, the Y_i's are iid $N(0, 1)$ variables,

$$Y_1 = \sqrt{n}\mathbf{1}'_n\mathbf{Z} = \sqrt{n} \ \bar{Z} \text{ and } \mathbf{Y}'\mathbf{Y} = \sum_{i=1}^{n} Y_i^2 \sim \chi^2_n.$$

Also, from the independence of the Y_i's, $\sum_{i=2}^{n} Y_i^2 \sim \chi_{n-1}^2$. So,

$$
\begin{aligned}
(n-1)\frac{S^2}{\sigma^2} &= \sum_{i=1}^{n} \frac{(X_i - \bar{X})^2}{\sigma^2} \\
&= \sum_{i=1}^{n} (Z_i - \bar{Z})^2 \\
&= \sum_{i=1}^{n} Z_i^2 - \frac{\bar{Z}^2}{n} \\
&= \sum_{i=1}^{n} Z_i^2 - Y_1^2 \\
&= \mathbf{Z'Z} - Y_1^2.
\end{aligned}
$$

But

$$
\sum_{i=1}^{n} Y_i^2 = \mathbf{Y'Y} = (\mathbf{A\,Z})'\mathbf{A\,Z} = \mathbf{Z'A'AZ} = \mathbf{Z'Z} = \sum_{i=1}^{n} Z_i^2.
$$

Therefore, $(n-1)S^2/\sigma^2 = \sum_{i=2}^{n} Y_i^2 \sim \chi_{n-1}^2$ and is independent of Y_1^2 and, consequently, of Y_1 and $\bar{X}$, completing the proof. □

Lemma 4.4 *If $T \sim N(0,1)$ and $W \sim \chi_\nu^2$ and T and W are independent then $T/\sqrt{W/\nu} \sim t_\nu(0,1)$.*

Proof The proof of this lemma is an adaptation of results previously shown and is left as an exercise. □

Corollary 4.1 *Let $\mathbf{X} = (X_1, \ldots, X_n)$ be a random sample from the $N(\theta, \sigma^2)$ distribution and let $\bar{X}$ and S^2 be the sample mean and variance, respectively. Then, conditional on θ and σ^2, $\bar{X}$ has sampling distribution*

$$
\sqrt{n}\,\frac{(\bar{X} - \theta)}{S} \sim t_{n-1}(0,1).
$$

Proof This is a straightforward application of the last lemma with $T = \sqrt{n}(\bar{X} - \theta)/\sigma$, $W = (n-1)S^2/\sigma^2$ and $\nu = n-1$. Then, $T/\sqrt{W/\nu} = \sqrt{n}(\bar{X} - \theta)/S$, completing the proof. □

The above results indicate how to obtain the pivotal quantities required for construction of confidence intervals for θ and σ^2. In the case of θ, σ is replaced by its estimator S leading to the new pivotal quantity $\sqrt{n}(\bar{X} - \theta)/S$, whose sampling distribution is $t_{n-1}(0,1)$. Note the similarity with the Bayesian standardization over the marginal posterior of θ. It is easy to verify that the classical interval will coincide with the non-informative Bayesian one. Even if S

could estimate σ without error, this substitution implies an increase in the uncertainty bounds since $t_{\beta,n} > z_\beta$ for small β.

For σ^2, we have that $\{E(\phi \mid \mathbf{x})\}^{-1} = s^2$, the mode of ϕ is $(n-3)/[(n-1)S^2] = \hat{\phi}$ and the dispersion measures are

$$V(\phi \mid \mathbf{x}) = \frac{2}{(n-1)s^4} \quad \text{and} \quad J(\hat{\phi}) = \frac{(n-1)^2 s^4}{2(n-3)}.$$

The maximum likelihood estimator of σ^2 is $\hat{\sigma}^2 = (n-1)S^2/n$; that is biased and is usually replaced by the unbiased estimator S^2 with sampling distribution $(n-1)S^2/\sigma^2 \sim \chi^2_{n-1}$ and $V(S^2) = 2\sigma^4/(n-1)$. Since S^2 is unbiased and $V(S^2) \to 0$ as $n \to \infty$, S^2 is a consistent estimator of σ^2. The difference between S^2 and $\hat{\sigma}^2$ becomes negligible as n increases which means that $\hat{\sigma}^2$ is also consistent.

Non-informative Bayesian and classical intervals for σ^2 once again coincide and are given by

$$\left(\frac{(n-1)s^2}{\overline{\chi}^2_{\alpha/2,n-1}}, \frac{(n-1)s^2}{\underline{\chi}^2_{\alpha/2,n-1}} \right).$$

When making inference about one of the parameters, the other one becomes a nuisance parameter. In the Bayesian approach it is eliminated by integration. In the frequentist approach it is eliminated by appropriate choices of pivotal quantities. In the Normal case, we were able to find suitable quantities given by $\sqrt{n}(\bar{X} - \theta)/S$ and $(n-1)S^2/\sigma^2$. These are based on minimal sufficient statistics and do not depend on the respective nuisance parameters. This fortunate coincidence does not necessarily occur in all statistical problems. In those cases, alternative approaches based on some form of approximation must be used.

4.7.2 Two samples case

From now on, until the end of the section, we will concentrate on two Normal samples where $\mathbf{X}_1 = (X_{11}, \ldots, X_{1n_1})$ is a random sample from the $N(\theta_1, \sigma_1^2)$ distribution and $\mathbf{X}_2 = (X_{21}, \ldots, X_{2n_2})$ is a random sample from the $N(\theta_2, \sigma_2^2)$ distribution. In addition, the two samples will be assumed to be independent.

If σ_1^2 and σ_2^2 are known, the likelihood is

$$
\begin{aligned}
p(\mathbf{x}_1, \mathbf{x}_2 \mid \theta_1, \theta_2) &= p(\mathbf{x}_1 \mid \theta_1)p(\mathbf{x}_2 \mid \theta_2) \\
&\propto \exp\left\{ -\frac{n_1}{2\sigma_1^2}(\theta_1 - \bar{x}_1)^2 \right\} \exp\left\{ -\frac{n_2}{2\sigma_2^2}(\theta_2 - \bar{x}_2)^2 \right\},
\end{aligned}
$$

which factors out into separate likelihoods for θ_1 and θ_2. So, if θ_1 and θ_2

are prior independent, they will remain posterior independent. One class of conjugate prior is given by independent $\theta_i \sim N(\mu_i, \tau_i^2)$ distributions for $i = 1, 2$. Another class is given by bivariate Normal distributions. It includes the previous class by allowing also non-null prior correlation between θ_1 and θ_2. The first class will be used here for simplicity.

Combining the adopted prior with the likelihood leads to the independent posteriors $\theta_i \mid \mathbf{x}_i \sim N(\mu_i^*, \tau_i^{*2})$ where

$$\mu_i^* = \frac{n_i \sigma_i^{-2} \bar{x}_i + \tau_i^{-2} \mu_i}{n_i \sigma_i^{-2} + \tau_i^{-2}} \text{ and } \tau_i^{*2} = \frac{1}{n_i \sigma_i^{-2} + \tau_i^{-2}}, i = 1, 2.$$

The analysis is exactly like two separate conjugate analyses and all the results for one sample follow. The same comments are true for non-informative priors and for classical inference. The non-informative prior for θ_1 and θ_2 is $p(\theta_1, \theta_2) \propto k$ and

$$\theta_i \mid \mathbf{x}_i \sim N\left(\bar{x}_i, \frac{\sigma_i^2}{n_i}\right), i = 1, 2 \text{ independent.}$$

The equivalent sampling result is $\bar{X}_i \mid \theta_i \sim N(\theta_i, \sigma_i^2/n_i)$, $i = 1, 2$ independent, from where point and interval estimation can be processed in the same way.

An interesting problem absent in the single sample case is comparison of means. This can be done by estimation of $\beta = \theta_1 - \theta_2$. The posterior distribution of β is obtained from the properties of the Normal distribution as $N(\mu_1^* - \mu_2^*, \tau_1^{*2} + \tau_2^{*2})$. This posterior reduces in the non-informative case to the $N(\hat{\beta}, \sigma_1^2/n_1 + \sigma_2^2/n_2)$ distribution where $\hat{\beta} = \bar{x}_1 - \bar{x}_2$. In the case of classical inference, estimation is based on $\bar{X}_1 - \bar{X}_2$ with sampling distribution $N(\beta, \sigma_1^2/n_1 + \sigma_2^2/n_2)$. Observe that all above distributions are symmetric, which eases the calculation of estimators, HPD credibility intervals and minimum length confidence intervals.

Assume now that σ_1^2 and σ_2^2 are unknown but equal with $\phi = \sigma_1^{-2} = \sigma_2^{-2}$. Then a conjugate prior can be constructed in the following way: $\theta_i \mid \phi \sim N(\mu_i, (c_i\phi)^{-1})$, $i = 1, 2$ are conditionally independent and $n_0\sigma_0^2\phi \sim \chi_{n_0}^2$. The prior density of $(\theta_1, \theta_2, \phi)$ is

$$
\begin{aligned}
p(\theta_1, \theta_2, \phi) &= p(\theta_1, \theta_2 \mid \phi)\, p(\phi) \\
&= p(\theta_1 \mid \phi) p(\theta_2 \mid \phi)\, p(\phi) \\
&\propto \phi^{1/2} \exp\{-\frac{\phi}{2} c_1 (\theta_1 - \mu_1)^2\} \phi^{1/2} \exp\{-\frac{\phi}{2} c_2 (\theta_2 - \mu_2)^2\} \\
&\quad \times \phi^{(n_0/2)-1} \exp\{-\frac{\phi}{2} n_0\sigma_0^2\} \\
&\propto \phi^{n_0/2} \exp\{-\frac{\phi}{2}[n_0\sigma_0^2 + c_1(\theta_1 - \mu_1)^2 + c_2(\theta_2 - \mu_2)^2]\}.
\end{aligned}
$$

In particular, the prior distribution of $\beta \mid \phi$ is $N(\mu_1 - \mu_2, \phi^{-1}(c_1^{-1} + c_2^{-1}))$. Therefore, using the results from Section 3.4, one can obtain its marginal prior $\beta \sim t_{n_0}(\mu_1 - \mu_2, \sigma_0^2(c_1^{-1} + c_2^{-1}))$. The likelihood is

$$p(\mathbf{x}_1, \mathbf{x}_2 \mid \theta_1, \theta_2, \phi) = p(\mathbf{x}_1 \mid \theta_1, \phi)\, p(\mathbf{x}_2 \mid \theta_2, \phi)$$

$$\propto \prod_{i=1}^{2} \phi^{n_i/2} \exp\left\{-\frac{\phi}{2}\left[(n_i - 1)s_i^2 + n_i(\theta_i - \bar{x}_i)^2\right]\right\},$$

where

$$s_i^2 = \frac{1}{n_i - 1}\sum_{j=1}^{n_i}(x_{ij} - \bar{x}_i)^2, \quad i = 1, 2.$$

Combining the likelihood with the prior gives the posterior density of $(\theta_1, \theta_2, \phi)$,

$$\phi^{(n_0 + n_1 + n_2)/2} \exp\left\{-\frac{\phi}{2}\left[n_0\sigma_0^2 + \nu s^2 + \sum_{i=1}^{2}\frac{c_i n_i}{c_i + n_i}(\mu_i - \bar{x}_i)^2 + \frac{(c_i + n_i)}{(\theta_i - \mu_i^*)^{-2}}\right]\right\},$$

where $\mu_i^* = (c_i\mu_i + n_i\bar{x}_i)/(c_i + n_i)$, $i = 1, 2$, $\nu = n_1 + n_2 - 2$ and $\nu s^2 = (n_1 - 1)s_1^2 + (n_2 - 1)s_2^2$. Note that the posterior and prior densities have the same kernel. Hence, the following results can be established, by analogy:

$$\theta_i \mid \phi, \mathbf{x} \sim N\left(\mu_i^*, \frac{1}{c_i^*\phi}\right) \text{ independent and } \theta_i \mid \mathbf{x} \sim t_{n_0^*}\left(\mu_i^*, \frac{\sigma_0^{*2}}{c_i^*}\right),$$

for $i = 1, 2$, and $n_0^*\sigma_0^{*2}\phi \mid \mathbf{x} \sim \chi_{n_0^*}^2$, where $c_i^* = c_i + n_i$, $n_0^* = n_0 + n_1 + n_2$ and

$$n_0^*\sigma_0^{*2} = n_0\sigma_0^2 + \nu s^2 + \sum_{i=1}^{2}\frac{c_i n_i}{c_i + n_i}(\mu_i - \bar{x}_i)^2.$$

In terms of β, one can obtain the conditional posterior distribution $\beta \mid \phi, \mathbf{x} \sim N(\mu_1^* - \mu_2^*, \phi^{-1}(c_1^{*-1} + c_2^{*-1}))$ and marginal posterior distribution $\beta \mid \mathbf{x} \sim t_{n_0^*}(\mu_1^* - \mu_2^*, \sigma_0^{*2}(c_1^{*-1} + c_2^{*-1}))$. Once again, the posterior is symmetric and posterior mean, mode and median of β coincide, and HPD intervals for β can be obtained using percentiles of Student-t distribution. For ϕ, we have that $\{E(\phi \mid \mathbf{x})\}^{-1} = \sigma_0^{*2}$ and credibility intervals can be constructed using percentiles of the χ^2 distribution.

The non-informative prior distribution can be obtained by noting that this is a location-scale model with location parameters θ_1 and θ_2 and scale parameter ϕ. Therefore, the non-informative prior distribution is given by $p(\theta_1, \theta_2, \phi) \propto \phi^{-1}$. This can be seen as a limiting case of the conjugate prior above when $c_1, c_2, \sigma_0^2 \to 0$ and $n_0 = -2$. Replacing these values in the expression of the conjugate posterior gives $c_i^* = n_i$, $\mu_i^* = \bar{x}_i$, $n_0^* = \nu$ and $n_0^*\sigma_0^{*2} = \nu s^2$.

Therefore, posterior mean, mode and median of β are given by $\hat{\beta}$, and the $100(1-\alpha)\%$ HPD interval for β has limits $\hat{\beta} \pm t_{\alpha/2,\nu}\, s\sqrt{n_1^{-1} + n_2^{-1}}$. A possible estimate for σ is given by s, that is, a weighted average of the s_i^2 obtained within each sample with weights given by $n_i - 1$, $i = 1, 2$. The $100(1-\alpha)\%$ credibility interval for σ^2 obtained as previously is given by

$$\left(\frac{\nu s^2}{\overline{\chi}_{\alpha/2,\nu}^2}, \frac{\nu s^2}{\underline{\chi}_{\alpha/2,\nu}^2} \right).$$

In the case of classical inference, $\hat{\beta}$ and $\hat{\sigma}^2 = \nu S^2/(n_1 + n_2)$ are the maximum likelihood estimators of β and σ^2, respectively. It is not difficult to show that $\hat{\beta}$ is an unbiased, efficient and consistent estimator for β and $\hat{\sigma}^2$ is consistent but biased estimator for σ^2. The estimator $\hat{\sigma}^2$ is usually replaced by S^2 that is an unbiased, efficient and consistent for σ^2. The relevant sampling distributions are

$$\frac{\hat{\beta} - \beta}{S\sqrt{n_1^{-1} + n_2^{-1}}} \sim t_\nu(0,1) \text{ and } \frac{\nu S^2}{\sigma^2} \sim \chi_\nu^2.$$

Note that these variables provide pivotal quantities for the construction of confidence intervals for β and σ^2, respectively. The resulting confidence intervals coincide numerically with those provided by the non-informative prior.

If $\sigma_1^2 = \phi_1^{-1}$ and $\sigma_2^2 = \phi_2^{-1}$ are unknown and unequal, the likelihood factors according to

$$p(\mathbf{x} \mid \theta_1, \theta_2, \sigma_1^2, \sigma_2^2) = p(\mathbf{x}_1 \mid \theta_1, \sigma_1^2)\, p(\mathbf{x}_2 \mid \theta_2, \sigma_2^2).$$

Adopting independent Normal-Gamma conjugate priors with parameters $(\mu_i, c_i, \nu_i, s_{0i}^2)$, $i = 1, 2$ for each of the samples leads to the independent posterior distributions

$$\theta_i \mid \mathbf{x} \sim t_{\nu_i^*}\left(\mu_i^*, \frac{s_{0i}^{*\,2}}{c_i^*} \right), \ i = 1, 2 \text{ and } \nu_i^* s_{0i}^{*\,2} \phi_i \mid \mathbf{x} \sim \chi_{\nu_i^*}^2, \ i = 1, 2,$$

where the relevant quantities are obtained by the usual one sample operations associated with the conjugacy of the Normal-Namma by the Normal observational model.

If interest lies in the comparison of the means, the posterior distribution of β must be obtained. First, let τ and ω be such that

$$\tau = \frac{\beta - (\mu_1^* - \mu_2^*)}{\sqrt{s_{01}^{*\,2}/c_1^* + s_{02}^{*\,2}/c_2^*}} \text{ and } \tan \omega = \frac{s_{01}^*/\sqrt{c_1^*}}{s_{02}^*/\sqrt{c_2^*}}.$$

It follows that

$$\sin \omega = \frac{s_{01}^*/\sqrt{c_1^*}}{\sqrt{s_{01}^*{}^2/c_1^* + s_{02}^*{}^2/c_2^*}} \quad \text{and} \quad \cos \omega = \frac{s_{02}^*/\sqrt{c_2^*}}{\sqrt{s_{01}^*{}^2/c_1^* + s_{02}^*{}^2/c_2^*}}$$

and therefore

$$\tau = \frac{\theta_1 - \mu_1^*}{s_{01}^*/\sqrt{c_1^*}} \sin \omega - \frac{\theta_2 - \mu_2^*}{s_{02}^*/\sqrt{c_2^*}} \cos \omega,$$

where the fractions on the right-hand sides of the equations have independent standard Student-t distributions with respective ν_1^* and ν_2^* degrees of freedom. A random quantity under these conditions is said to have a Behrens-Fisher distribution with parameters ν_1^*, ν_2^* and ω. This distribution is similar to the Student-t distribution and has been tabulated, enabling easy construction of confidence intervals.

In the case of a non-informative prior $p(\theta_1, \theta_2, \sigma_1^2, \sigma_2^2) \propto \sigma_1^{-2}\sigma_2^{-2}$. This can be seen as a limiting case of the conjugate prior above when $c_i, s_{0i}^2 \to 0$ and $\nu_i = -1$, $i = 1, 2$. Replacing these values in the expression of the conjugate posterior gives $c_i^* = n_i$, $\mu_i^* = \bar{x}_i$, $\nu_i^* = n_i - 1$ and $s_{0i}^* = s_i$, where s_i^2 is the usual unbiased estimate of σ_i^2, $i = 1, 2$. Estimators and confidence intervals can be obtained accordingly. The problem is more difficult to treat under the classical perspective. No pivotal quantity for β can be easily found with known distribution although approximations based on the Behrens-Fisher distribution have been proposed.

Another situation of interest is the comparison of variances. Since variances measure the scale of a distribution and are always positive, it makes more sense to compare them through their ratio instead of their difference as we did for the means. Therefore, we will focus on the posterior distribution of $\psi = \sigma_2^2/\sigma_1^2 = \phi_1/\phi_2$. We have just obtained that ϕ_1 and ϕ_2 are independent a posteriori with joint density

$$p(\phi_1, \phi_2 \mid \mathbf{x}) \propto \prod_{i=1}^{2} \phi_i^{\nu_i^*/2-1} \exp\left\{-\frac{\nu_i^* s_{0i}^*{}^2}{2} \phi_i\right\}.$$

The easiest form to obtain the posterior distribution of ψ is to complete the transformation to ensure a bijection and use results available for densities of 1-to-1 transformations of random quantities. Let $\psi_2 = \phi_2$ complete the transformation. It follows that the inverse relation is $\phi_1 = \psi\phi_2 = \psi\psi_2$. The Jacobian of the transformation is

$$J = \left| \frac{\partial(\phi_1, \phi_2)}{\partial(\psi, \psi_2)} \right| = \begin{vmatrix} \psi_2 & \psi \\ 0 & 1 \end{vmatrix} = \psi_2 > 0$$

and the posterior density of (ψ, ψ_2) is

$$p(\psi, \psi_2 \mid \mathbf{x}) \propto \psi^{\nu_1^*/2-1} \psi_2^{(\nu_1^*+\nu_2^*)/2-1} \exp\left\{-\frac{\psi_2}{2}\left[\nu_2^* s_{02}^{*}{}^2 + \nu_1^* s_{01}^{*}{}^2\psi\right]\right\}.$$

Finally, the marginal density of ψ is

$$p(\psi \mid \mathbf{x}) = \int p(\psi, \psi_2 \mid \mathbf{x})\, d\psi_2$$

$$\propto \psi^{\nu_1^*/2-1} \int \psi_2^{(\nu_1^*+\nu_2^*)/2-1} \exp\left\{-\frac{\psi_2}{2}\left[\nu_2^* s_{02}^{*}{}^2 + \nu_1^* s_{01}^{*}{}^2\psi\right]\right\} d\psi_2$$

$$= \psi^{\nu_1^*/2-1} \frac{\Gamma\left((\nu_1^*+\nu_2^*)/2\right)}{[(\nu_2^* s_{02}^{*}{}^2 + \nu_1^* s_{01}^{*}{}^2\psi)/2]^{(\nu_1^*+\nu_2^*)/2}}$$

$$\propto \psi^{\nu_1^*/2-1}\left(\nu_2^* + \nu_1^* \frac{s_{01}^{*}{}^2}{s_{02}^{*}{}^2}\psi\right)^{-(\nu_1^*+\nu_2^*)/2}.$$

It can then be shown that

$$\frac{s_{01}^{*}{}^2}{s_{02}^{*}{}^2}\psi \mid \mathbf{x} \sim F(\nu_1^*, \nu_2^*).$$

Even though the distribution function of the F distribution cannot be obtained analytically, it is tabulated in many books and by computer software. Its percentiles can be used in the construction of confidence intervals. The main properties of the F distribution are given in the list of distributions. An interesting property for probability evaluation with the F distribution can be derived from the fact that if $X \sim F(\nu_2, \nu_1)$ then $X^{-1} \sim F(\nu_1, \nu_2)$ by simple inversion in the ratio of independent χ^2 distributions. Therefore, denoting the α and $1-\alpha$ quantiles of the $F(\nu_1, \nu_2)$ distribution, respectively, by $\underline{F}_\alpha(\nu_1, \nu_2)$ and $\overline{F}_\alpha(\nu_1, \nu_2)$ gives that $\underline{F}_\alpha(\nu_1, \nu_2) = \overline{F}_\alpha^{-1}(\nu_2, \nu_1)$.

Point estimators are given by

$$E\left[\psi \mid \mathbf{x}\right] = \frac{s_{02}^{*}{}^2}{s_{01}^{*}{}^2}\frac{\nu_2^*}{\nu_2^* - 2} \quad \text{and mode of } \psi = \frac{s_{02}^{*}{}^2}{s_{01}^{*}{}^2}\frac{\nu_2^*(\nu_1^* - 2)}{\nu_1^*(\nu_2^* + 2)}.$$

Both estimators converge to $(s_{02}^*/s_{01}^*)^2$ when $\nu_2^* \to \infty$.

In the case of a non-informative prior the $s_{0i}^{*}{}^2$'s are given by s_i^2, the unbiased estimates of the populational variances, and the ν_i^*'s are given by $n_i - 1$, $i = 1, 2$. Therefore, for large samples the estimators will be given by the ratio of the variance estimates.

In classical inference, we have already obtained the independent sampling distributions of $(n_i-1)S_i^2\phi_i \sim \chi^2_{n_i-1}$ $i = 1, 2$. Therefore, from the properties of the F distribution, $(S_1^2/S_2^2)\psi \sim F(n_1-1, n_2-1)$. Once again a pivotal quantity was found for inference about ψ and the result is numerically equivalent to the non-informative Bayesian result. For example, the unbiased estimator of $\psi^{-1} = \sigma_1^2/\sigma_2^2$ is

$$\frac{n_2 - 3}{n_2 - 1}\frac{S_1^2}{S_2^2}.$$

Confidence intervals are obtained in either case using quantiles of the F distribution. These are not minimum length intervals due to the asymmetry of the F distribution. Using the results listed about the F distribution, a $100(1-\alpha)\%$ credibility interval for ψ is obtained from

$$
\begin{aligned}
1 - \alpha &= P\left(\overline{F}_{\alpha/2}^{-1}(\nu_2^*, \nu_1^*) < \frac{s_{02}^{*\,2}}{s_{01}^{*\,2}}\psi < \overline{F}_{\alpha/2}(\nu_1^*, \nu_2^*)|\mathbf{x}_1, \mathbf{x}_2\right) \\
&= P\left(\frac{s_{01}^{*\,2}}{s_{02}^{*\,2}}\overline{F}_{\alpha/2}^{-1}(\nu_2^*, \nu_1^*) < \psi < \frac{s_{01}^{*\,2}}{s_{02}^{*\,2}}\overline{F}_{\alpha/2}(\nu_1^*, \nu_2^*)|\mathbf{x}_1, \mathbf{x}_2\right)
\end{aligned}
$$

and, therefore, $\left[(s_{01}^{*\,2}/s_{02}^{*\,2})\overline{F}_{\alpha/2}^{-1}(\nu_2^*, \nu_1^*), (s_{01}^{*\,2}/s_{02}^{*\,2})\overline{F}_{\alpha/2}(\nu_1^*, \nu_2^*)\right]$ is a $100(1-\alpha)\%$ credibility interval for ψ. In the case of non-informative priors, the modifications previously mentioned apply, and the resulting interval is

$$\left[\frac{s_2^2}{s_1^2}\overline{F}_{\alpha/2}^{-1}(n_2 - 1, n_1 - 1), \frac{s_2^2}{s_1^2}\overline{F}_{\alpha/2}(n_1 - 1, n_2 - 1)\right].$$

Finally, the confidence inference derived for classical inference coincides with the above interval obtained with a non-informative prior.

Exercises

§4.1

1. Prove that the estimator of θ associated with the absolute loss is the posterior median of the updated distribution of θ.

2. Show that if L_1 and L_2 are two proportional loss functions, that is, $L_1(\delta, \theta) = kL_2(\delta, \theta)$, then the Bayes estimators associated with these losses coincide.

3. Suppose that X with distribution $N(\theta, \sigma^2)$ with σ^2 known is observed and it is known that $\theta \in (a, b)$ $(a < b)$.

   (a) Obtain the non-informative prior for θ.
   (b) Obtain the complete expression of the resulting posterior distribution.
   (c) Obtain the posterior mean and mode.

§4.2

4. Let $X_1, \ldots, X_n$ be a random sample from the Bernoulli distribution with unknown parameter θ having unit uniform prior distribution. One wishes to estimate θ using the loss function $L(d, \theta) = (\theta - d)^2 / [\theta(1 - \theta)]$.

(a) Calculate the Bayes estimator and obtain its risk.

(b) Determine the predictive distribution for the $(n + 1)$th observation and determine its mean and variance.

(c) Generalize items (a) and (b) to k possible values for each X_i with respective probabilities θ_j, $j = 1, \ldots, k$ and obtain the Bayes estimator of θ_j, $j = 1, \ldots, k$.

5. Suppose that the loss function used to estimate θ through δ

i) equals the distance between δ and θ if δ is smaller than θ, and

ii) triples the distance between δ and θ if δ is larger than θ.

(a) Obtain the mathematical expression of the loss function.

(b) Show that the estimator of θ is the first quartile of the updated distribution of θ.

(c) Generalize the result in (b) for when the loss in (ii) is p times the distance between δ and θ.

6. Suppose that $X \sim Bin(n, \theta)$ and the conjugate prior $\theta \sim Beta(a, b)$ is used.

(a) What is the value of X that minimizes the variance of the posterior distribution of θ?

(b) What is the value of X that maximizes it? Interpret the results.

(c) Repeat items (a) and (b) for the case of a Negative Binomial distribution for X.

7. Assume that $X \sim U[\theta - 1, \theta + 1]$ is observed and assume a prior $p(\theta) \propto \theta^{-1}$, $\theta > 0$.

(a) Prove that $p(\theta | x) = c\theta^{-1}$, $\theta \in (x - 1, x + 1)$, where $c^{-1} = \log[(x+1)/(x-1)]$, $x > 1$.

(b) Calculate the mean, mode and median of the posterior distribution.

8. The income tax policy of a given country establishes that an individual pays tax q iff his/her income is larger than k. Assume that the income distribution for these individuals follows a $Pa(k, \theta)$ distribution (see Exercise 2.32);

(a) Show that $\log(X/k) \sim Exp(\theta)$.

(b) Assuming that little is known a priori about θ, a sample of these individuals is observed and their incomes registered. Show that the posterior distribution of θ is $G(n, n \log(\bar{G}/k))$ where n is the sample size and $\bar{G}$ is the geometric average of the observations.

(c) Assume a change in policy is under study aiming at taxing the rich people more. The threshold would be raised to $l > k$ and tax raised to $r > q$. Show that the expected revenue would increase only if

$$r > q \left(1 + \frac{\log(l/k)}{n \log(\bar{G}/k)}\right)^n.$$

Hint: Calculate first the expected revenue given θ.

§4.3

9. Consider a simple linear regression where $E(Y_i \mid \boldsymbol{\theta}) = \theta_0 + \theta_1 X_i$ with $\boldsymbol{\theta} = (\theta_0, \theta_1)$, $i = 1, \ldots, n$. Obtain the sum of squares $S(\boldsymbol{\theta})$ and show that its minimization leads to the ordinary least squares estimator

$$(\hat{\theta}_0, \hat{\theta}_1) = \left(\bar{Y} - \hat{\theta}_1 \bar{X}, \frac{\overline{XY} - \bar{X}\bar{Y}}{\overline{X^2} - \bar{X}^2}\right) \quad \text{where } \overline{g(X, Y)} = \frac{1}{n} \sum_{i=1}^n g(X_i, Y_i).$$

10. Show that the maximum likelihood estimator of a parameter is a minimal sufficient statistic for this parameter.

11. Suppose that the waiting time in a bank queue has $Exp(\theta)$ distribution with $\theta > 0$. A sample of n customers is observed over a period of T minutes.

(a) Suppose the individual waiting times were discarded and only the number X of clients was recorded. Determine the maximum likelihood estimator of θ based on X.

(b) Determine the maximum likelihood and Bayes estimators for θ, assuming that in a sample of $n = 20$ customers the average serving time was 3.8 minutes and all 20 clients were served.

(c) Suppose that, in addition to the observations reported in (b), an additional observation was made but all that is known is that it lasted for more than 5 minutes. Obtain the maximum likelihood and Bayes estimators of θ in this case.

12. Suppose one wishes to test three types of bulbs: normal life, long life and extra long life. The lifetimes of the bulbs have exponential distribution with means θ, 2θ and 3θ, respectively. Assume the test consists in observing a bulb of each type selected randomly.

(a) Determine the maximum likelihood estimator of θ.

(b) Determine the method of moments estimator of θ.

(c) Let $\psi = 1/\theta$ and assume the prior $\psi \sim G(\alpha, \beta)$. Determine the posterior distribution of θ.

(d) Determine the Bayes estimators of θ using $0 - 1$ and quadratic loss functions.

13. Let $X_1, \ldots, X_n$ be a random sample from the following discrete distribution: $P(X_1 = 1) = \frac{2(1-\theta)}{2-\theta}$ and $P(X_1 = 2) = 1 - P(X_1 = 1)$. Obtain a moment estimator of θ.

14. Suppose that $X_1, \ldots, X_n$ are iid from the Pareto distribution $Pa(a, \theta)$, for $a > 0, \theta > 2$. Determine moment conditions and obtain method of moment estimators for (a, θ).

15. Obtain the point estimator of the degree of freedom in the Student-t case using the moment conditions on the variance and kurtosis.

16. Let $(X_1, Y_1), \ldots, (X_n, Y_n)$ be bidimensional vectors forming a sample from the bivariate Normal with mean vector μ, variances $\sigma_i^2, i = 1, 2$ and correlation coefficient ρ. Determine the maximum likelihood estimator of all model parameters.

§4.4

17. Consider the conditions of Example 4.21. Show that $(k-3)/\sum_{I=1}^{k}(\bar{X}_i - \bar{X})^2$ is an unbiased estimator of $1/(\sigma^2 + n\tau^2)$.

18. Consider again the conditions of Example 4.21. Obtain the empirical Bayes estimator when σ^2 is unknown and τ^2 is known.

§4.5

19. Prove the Cramer-Rao inequality for the multiparameter case.
    Hint: Consider the joint covariance matrix of δ and the score function and explore the non-negative definiteness of this matrix.

20. Consider a random sample $X = (X_1, \ldots, X_n)$ from an unknown distribution function F and let $\hat{F}$ denote the empirical distribution function. Show that

(a) $\hat{F}$ is non-decreasing and contained in the interval $[0, 1]$.

(b) $\hat{F}(x)$ can be written as $\bar{Z}$ where $Z_i = I_{X_i}(-\infty, x], i = 1, \ldots, n$.

(c) Show that $\hat{F}$ is an unbiased estimator of F.

(d) Show that $\hat{F}$ is a consistent estimator of F.

Hint: Obtain the sampling distribution of $\hat{F}(x)$, $\forall x$.

21. Let $X_1, X_2, \ldots, X_n$ be a random sample from the $Pois(\theta)$ distribution and $Y = \sum_{i=1}^{n} X_i$.

   (a) Determine c such that $\exp(-cY)$ is an unbiased estimator of $\exp(-\theta)$.

   (b) Obtain a bound for the variance of the estimator obtained in (a).

   (c) Discuss the efficiency of the estimator obtained in (a).

22. Let $X_1, \ldots, X_n$ be a random sample from the uniform distribution over interval $[0, \theta]$.

   (a) Obtain $\hat{\theta}_n$, the maximum likelihood estimator of θ, and show it is a biased but consistent estimator of θ.

   (b) Using the result from (a), obtain an unbiased estimator of θ.

   (c) Calculate the quadratic risks of the estimators in (a) and (b).

   (d) Find an estimator of θ with smaller risk than those obtained in (a) and (b).

23. It is common in many areas of science (e.g., Genetics) to obtain samples from the Binomial distribution with the impossibility of 0 values for the observations.

   (a) Show that the sampling distribution is given by

$$f(x \mid \theta) = \binom{n}{x} \cdot \frac{\theta^x (1 - \theta)^{n-x}}{1 - (1 - \theta)^n}, \qquad x = 1, \ldots, n.$$

   (b) Obtain the maximum likelihood estimator of θ, assuming that $n=2$.

   (c) Verify whether the above estimator is unbiased.

24. Let $X_1, \ldots, X_n$ be a random sample from the uniform distribution on the interval $[a - b, a + b]$, $a \in R$ and $b > 0$.

   (a) Verify if a and b are location and/or scale parameters.

   (b) Obtain the maximum likelihood estimator of a and b.

   (c) Assume now that $b = 1$ and define

$$T_1 = \overline{X} = \frac{\sum_{i=1}^{n} X_i}{n} \quad \text{and} \quad T_2 = \frac{1}{2} \left(\max_{1 \le i \le n} X_i + \min_{1 \le i \le n} X_i \right).$$

   (d) Compare T_1 and T_2 in terms of bias, risk and consistency, specifying a choice among them and justifying this choice.

25. Let $X_1, \ldots, X_n$ be iid with probability function $f(x \mid \theta) = \theta(1 - \theta)^x$, $x = 0, 1, 2, \ldots$, $\theta \in (0, 1)$ and define $U_i = I_{X_i}(\{0\})$.

(a) Show that $\bar{U}$ is an unbiased estimator of θ and calculate its variance.

(b) Show that the expected Fisher information for θ is $n/[\theta^2(1-\theta)]$ and find the efficiency of $\bar{U}$.

26. Let X_1 and X_2 be iid with $Exp(\theta)$ distribution, and let $T = X_1 + X_2$.

(a) Show that $U = \exp(-X_1)$ is an unbiased estimator of $\psi = \theta/(\theta+1)$.

(b) Show that $(1 - e^{-T})/T$ is UMVU estimator of ψ.

§4.6

27. Show that credibility intervals are invariant under 1-to-1 transformations of the parameter. That is, if C is a $100(1-\alpha)\%$ credibility region for θ and $\phi = \phi(\theta)$ is a 1-to-1 transformation of θ, then $\phi(C)$, the image of C under ϕ, is a $100(1-\alpha)\%$ credibility region for ϕ. Show that HPD intervals are not invariant under 1-to-1 transformations. Show also that confidence intervals are invariant under 1-to-1 transformations.

28. Let $\boldsymbol{\theta} = (\theta_1, \ldots, \theta_r)$. Show that if the θ_i's are not independent then

$$Pr(\boldsymbol{\theta} \in \boldsymbol{C}) \geq 1 - \sum_{i=1}^{r} Pr(\theta_i \notin C_i).$$

Show that if $Pr(\theta_i \in C_i) \geq \alpha_i = \alpha/r$, then $\boldsymbol{C}$ is a $100(1-\alpha)\%$ confidence region for $\boldsymbol{\theta}$. Obtain the equivalent result from a classical perspective. Hint: Define pivotal quantities $U_i = G_i(X, \theta_i)$, $i = 1, \ldots, r$.

29. Let $X \sim Bin(n, \theta)$ and assume the prior $\theta \sim U[0, 1]$. Suppose that the observed value was $X = n$.

(a) Show that the $100(1-\alpha)\%$ HPD interval for θ has form $[a, 1]$, $a < 1$.

(b) Let $\psi = \theta/(1-\theta)$. Show from (a) that $Pr[a/(1-a) \leq \psi|\mathbf{x}] = 1 - \alpha$ and therefore $[a/(1-a), \infty)$ is a $100(1-\alpha)\%$ credibility interval for ψ.

(c) Obtain the posterior distribution of ψ and discuss the form of a $100(1-\alpha)\%$ HPD interval for ψ.

(d) In particular, is the interval obtained in (b) an HPD?

30. Let $X_1, \ldots, X_n$ be a random sample from the uniform distribution over interval $[0, \theta]$.

(a) Obtain a pivot based on the sampling distribution of $\hat{\theta}_n$, the maximum likelihood estimator of θ.

(b) Obtain a $100(1-\alpha)\%$ classical confidence interval for θ based on the pivot used in (a).

(c) Which conditions must be satisfied for a minimum length interval?

(d) Assuming the non-informative prior $p(\theta) \propto \theta^{-1}$, find the $100(1 - \alpha)\%$ HPD interval for θ.

31. (Berger, 1985, pg. 134) Let X be such that $p(x \mid \theta) = \exp[-(x - \theta)]$, $x \geq \theta$ and assume the prior $p(\theta) \propto (1 + \theta^2)^{-1}$, $\theta \geq 0$.

(a) Prove that the posterior distribution of θ given x is monotonically increasing and that the mode of θ is x.

(b) Show that the $100(1 - \alpha)\%$ HPD interval for θ must have the form $[c(\alpha), x]$, where $c(\alpha)$ is such that $P[c(\alpha) \leq \theta \leq x \mid x] = 1 - \alpha$.

(c) Obtain the posterior density of $\eta = \exp(\theta)$ and prove it is a monotonically increasing function of η.

(d) Show that the $100(1 - \alpha)\%$ HPD interval for η must have the form $[1, d(\alpha)]$, where $d(\alpha)$ is such that $P(1 \leq \eta \leq d(\alpha) \mid x) = 1 - \alpha$.

(e) Show that the credibility interval in (d) implies a $100(1 - \alpha)\%$ lowest posterior density interval for θ.

32. Let $\mathbf{X} = (X_1, \ldots, X_n)$ be a random sample from the $Pois(\theta)$. The prior distribution for θ is judged to be based on information equivalent to that obtained after observing a sample of a lifetimes with $Exp(\theta)$ distribution with observed mean lifetime b, where $a, b > 0$.

(a) Show that the prior distribution of θ is $G(a + 1, ab)$.

(b) Obtain the posterior distribution of $\theta \mid \mathbf{x}$.

(c) Suppose that instead of observing the sample, only $T = \sum_{i=1}^n X_i$ was observed. Obtain the distribution of $\theta \mid T = t$.
Hint: If $\mathbf{X}$ consists of iid $Pois(\theta)$ variables then $T \sim Pois(n\theta)$.

(d) Compare the distributions obtained in (b) and (c) when $\sum_{i=1}^n x_i = t$, justifying the result obtained.

(e) Obtain a $100(1 - \alpha)\%$ credibility interval for θ using the percentiles of the χ^2 distribution.
Hint: It is not possible to obtain a pivot for this situation and exact confidence interval cannot be constructed.

33. Let $X_i = \theta t_i + \epsilon_i$, $i = 1, \ldots, n$ where ϵ_i are iid $N(0, \sigma^2)$, σ^2 known and the non-informative prior for θ.

(a) Obtain the posterior distribution of θ.

(b) Obtain the $100(1 - \alpha)\%$ HPD interval for θ.

(c) Show that the maximum likelihood estimator of θ is the UMVU estimator of θ.

(d) Based on the sampling distribution of the maximum likelihood estimator of θ, construct a $100(1-\alpha)\%$ confidence interval for θ.

(e) If $0 \le t_i \le 1$, $\forall i$, what values of the t_i's must be chosen to obtain the confidence intervals in (b) and (d) of shortest possible length?

§4.7

34. Show that if $T \sim N(0,1)$ and $W \sim \chi_\nu^2$ are independent, then $T/\sqrt{W/\nu} \sim t_\nu(0,1)$.

35. Consider the situation with two Normal samples where $X_1 = (X_{11},\ldots,X_{1n_1})$ is a random sample from the $N(\theta_1,\sigma_1^2)$ distribution and $X_2 = (X_{21},\ldots,X_{2n_2})$ is a random sample from the $N(\theta_2,\sigma_2^2)$ distribution. In addition, the two samples will be assumed to be independent.

(a) If σ_1^2 and σ_2^2 are known, show that the class of bivariate Normal distributions for θ_1 and θ_2 is conjugate with the above observation model. Also, obtain the posterior correlation between θ_1 and θ_2 and compare it with the prior correlation.

(b) If $\sigma_1^2 = \sigma_2^2 = \sigma^2$ is unknown, show that

$$\frac{\hat{\beta} - \beta}{S\sqrt{n_1^{-1} + n_2^{-1}}} \sim t_\nu(0,1) \text{ and } \frac{\nu S^2}{\sigma^2} \sim \chi_\nu^2.$$

(c) Show that the resulting confidence intervals for β and σ^2 obtained with the pivotal quantities of the previous item coincide numerically with the credibility intervals obtained with a non-informative prior.

(d) If σ_1^2 and σ_2^2 are unequal and unknown, show that

$$\frac{s_{01}^{*\,2}}{s_{02}^{*\,2}} \psi \mid x \sim F(\nu_1^*,\nu_2^*).$$

36. Let $\mathbf{X} = (X_1,\ldots,X_n)$ be a random sample from the $N(0,\sigma^2)$ distribution and assume that the Normal-χ^2 conjugate prior is used for (θ,ϕ) with $n_0 = 5$ where $\phi = \sigma^{-2}$. What should be the sample size to guarantee that $Pr((\theta - \mu_1)^2 \le 4V_1|\mathbf{x}) \ge 0.95$, where $\mu_1 = E(\theta \mid \mathbf{x})$ and $V_1 = V(\theta \mid \mathbf{x})$?

37. Let X_1 and X_2 be two observations from the $N(0,\sigma^2)$ distribution and assume a non-informative prior for (θ,σ^2). If Y_1 is the smallest of the observation and Y_2 the largest one, show that a posteriori,

$$Pr(y_1 \le \theta \le y_2|y_1,y_2) = 0.5.$$

Hint: Write the posterior distribution of θ as a function of y_1 and y_2.

38. Let $\mathbf{X} = (X_1, \ldots, X_n)$ be a random sample from the $N(\theta_1, \sigma^2)$ distribution and $\mathbf{Y} = (Y_1, \ldots, Y_n)$ be a random sample from the $N(\theta_2, k\sigma^2)$ distribution, k known.

   (a) Assuming a non-informative prior for $(\theta_1, \theta_2, \sigma^2)$, obtain the posterior distribution of $\theta_1 - \theta_2$ and σ^2.

   (b) Construct a $100(1 - \alpha)\%$ HPD interval for $\theta_1 - \theta_2$.

39. Let $X_1, \ldots, X_n$ be independent observations from the $N(0, \sigma^2/k_i)$, where the k_i's are known positive constants, for $i = 1, \ldots, n$.

   (a) Obtain a family of natural conjugate distributions to the observational model above.
   Hint: Define $\bar{X} = \sum_{i=1}^{n} k_i X_i / \sum_{i=1}^{n} k_i$.

   (b) Construct a $100(1 - \alpha)\%$ HPD interval for θ.

Chapter 5

Approximating methods

As we have seen in Chapter 4, the classical approach to Statistics requires the sampling distribution of the estimators to be useful from both a practical and theoretical points of view. In the Bayesian approach, all the information needed is described by the posterior distribution. In both approaches, the evaluation of probabilities or expected values and optimization of some criterion function are often demanded. Evaluation of these quantities must involve analytical or computationally intensive approximation techniques when the exact solution fails. Typically, the accuracy of the analytic procedures depends on the sample size and the accuracy of simulation-based techniques depends on the number of simulations.

In this chapter the central problem of inference is stated in Section 5.1 and efficient numerical solutions are discussed. Many approximating techniques used in Statistics are described in this chapter. Some optimization methods are presented in Section 5.2 and analytical techniques are discussed in Sections 5.3 and 5.4. An introduction to numerical integration is presented in Section 5.5, and methods based on simulation are described in Section 5.6, including Monte Carlo, Monte Carlo with importance sampling, classical and Bayesian bootstrap and an introduction to Markov chain Monte Carlo. A good account of some of the techniques presented in this chapter can be found in the books by Davison and Hinkley (1997), Gamerman and Lopes (2006), Tanner (1996) and Thisted (1976).

5.1 The general problem of inference

In the development of statistical inference, we are often involved with the optimization of some criterion function or with the evaluation of integrals or expected values. In some cases, these problems are not tractable analytically. The classical methods are often based on the maximization of an objective function, such as the likelihood function or the squared error loss. Based on decision theory, the minimization of the expected loss function provides Bayes estimators, where expectation is calculated with respect to the posterior distribution.

Generally speaking we are often faced with one of the two basic problems:

1. maximization of a function $q(\boldsymbol{\theta})$ that can be either a posterior density or a likelihood function;

2. evaluation of an integral

$$E[\mathbf{g}(\boldsymbol{\theta})\,|\,\mathbf{x}] = \int \mathbf{g}(\boldsymbol{\theta})p(\boldsymbol{\theta}\,|\,\mathbf{x})d\boldsymbol{\theta} \text{ or } E[\mathbf{g}(\mathbf{X})\,|\,\boldsymbol{\theta}] = \int \mathbf{g}(\mathbf{x})p(\mathbf{x}\,|\,\boldsymbol{\theta})d\mathbf{x},$$

where $\mathbf{x}$ represents the observed data, $\mathbf{g}(\cdot)$ is an integrable function and $\boldsymbol{\theta}$ is a p-dimensional vector.

The first problem is related to the definition of the (generalized) maximum likelihood estimator and the second includes many alternatives, which are exemplified below:

1. One of the basic problems in Bayesian inference is to find the value of the normalizing constant k of the posterior density. It is obtained as $k^{-1} = \int l(\boldsymbol{\theta};\mathbf{x})p(\boldsymbol{\theta})d\boldsymbol{\theta}$.

2. If one wishes to ascertain the bias of an estimator $\hat{\boldsymbol{\theta}} = \mathbf{g}(\mathbf{X})$ of $\boldsymbol{\theta}$, then one must evaluate the sampling expectation of $g(\mathbf{X})$.

3. In order to find the Bayes estimator with respect to the loss function $L(\boldsymbol{\delta},\boldsymbol{\theta})$, one must evaluate $\int \mathbf{g}(\boldsymbol{\theta})p(\boldsymbol{\theta}\mid\mathbf{x})d\boldsymbol{\theta}$ with $\mathbf{g}(\cdot) = L(\boldsymbol{\delta},\cdot)$.

4. Evaluation of the confidence of a region C involves calculation of $P(\mathbf{G}(\mathbf{X},\boldsymbol{\theta}) \in C) = \int \mathbf{g}(\mathbf{x})p(\mathbf{x}\mid\boldsymbol{\theta})d\mathbf{x}$ where $\mathbf{g}(\mathbf{x}) = I_\mathbf{x}(\{\mathbf{x} : \mathbf{G}(\mathbf{x},\boldsymbol{\theta}) \in C\})$.

5. The predictive density of a future sample $\mathbf{Y}$ with density $p(\mathbf{y}\mid\boldsymbol{\theta})$, independent of $\mathbf{X}$ conditionally on $\boldsymbol{\theta}$ is given by $\int \mathbf{g}(\boldsymbol{\theta})p(\boldsymbol{\theta}\mid\mathbf{x})d\boldsymbol{\theta}$ where $g(\boldsymbol{\theta}) = p(\mathbf{y}\mid\boldsymbol{\theta})$.

In the next section some useful optimization methods will be presented. Their importance is to yield classical and Bayesian point estimators and also confidence and credibility intervals.

5.2 Optimization techniques

The optimization of some criterion function is present in many theoretical developments of the statistical theory, as seen in Chapter 4, from both the classical and the Bayesian perspectives. For example, if we wish to obtain least square estimators, generalized maximum likelihood estimators or the minimum expected loss, then some sort of optimization will be required. Even to calculate the approximate value of some integrals via Laplace methods, maximum values of some functions are needed, as will be seen in Section 5.4. Numerical optimization techniques will be reviewed in this section since in many relevant problems the optimum cannot be obtained analytically. The main goal in this section is to present and illustrate the use of Newton-Raphson techniques and the Fisher scoring methods to locate the maximum of the likelihood function or of the posterior distribution. Many statistical books include chapters on statistical computing with discussion of optimization techniques as for example Kalbfleisch (1975) and Garthwaite, Jollife and Jones (2002).

An algorithm to find the zeros of a twice differentiable function $\mathbf{g} : R^p \to R$ ($p \geq 1$) is easily obtained from the Taylor expansion of $\mathbf{g}$ around an arbitrary point $\mathbf{x}^{(0)} \in R^p$

$$\mathbf{g}(\mathbf{x}) = \mathbf{g}(\mathbf{x}^{(0)}) + (\mathbf{x} - \mathbf{x}^{(0)})' \frac{\partial \mathbf{g}(\mathbf{x}^{(0)})}{\partial \mathbf{x}} + \dots .$$

Neglecting higher order terms in $\mathbf{x} - \mathbf{x}^{(0)}$ for suitably close values of $\mathbf{x}$ and $\mathbf{x}^{(0)}$ gives

$$\mathbf{g}(\mathbf{x}) \simeq \mathbf{g}(\mathbf{x}^{(0)}) + (\mathbf{x} - \mathbf{x}^{(0)}) \frac{\partial \mathbf{g}(\mathbf{x}^{(0)})}{\partial \mathbf{x}} .$$

If $\mathbf{x}^*$ is a zero of $\mathbf{g}$, then solving the above equation for $\mathbf{x}^*$ gives $\mathbf{x}^* \simeq \mathbf{x}^{(1)} = \mathbf{x}^{(0)} - [\partial \mathbf{g}(\mathbf{x}^{(0)})/\partial \mathbf{x}]^{-1} \mathbf{g}(\mathbf{x}^{(0)})$.

It follows that, starting with an initial value $\mathbf{x}^{(0)}$ and using the relation stated earlier, the algorithm provides us with a new value $\mathbf{x}^{(1)}$ closer to the root of the above equation. This new point is the intersection of the tangent line, the linear approximation of $\mathbf{g}$ at $\mathbf{x}_0$, with the $\mathbf{x}$ axis. The procedure is then repeated with $\mathbf{x}^{(1)}$ replacing $\mathbf{x}^{(0)}$. This will lead to an even better approximation for $\mathbf{x}^*$ denoted by $\mathbf{x}^{(2)}$. Repeating the process successively gives the recursive relation

$$\mathbf{x}^{(j)} = \mathbf{x}^{(j-1)} - \left[\frac{\partial \mathbf{g}(\mathbf{x}^{(j-1)})}{\partial \mathbf{x}} \right]^{-1} \mathbf{g}(\mathbf{x}^{(j-1)}).$$

The procedure is graphically illustrated in Figure 5.1.

This is the well-known Newton-Raphson algorithm and it must be repeated until some convergence criteria is achieved. Typical criteria are $|\mathbf{x}^{(j)} - \mathbf{x}^{(j-1)}| < \delta$ and $|\mathbf{g}(\mathbf{x}^{(j)})| < \epsilon$, where δ and ϵ are preset precisions determined arbitrarily. Since it is easier to evaluate proximity at the $\mathbf{x}$ level rather than at the $\mathbf{g}$ level, the former is sometimes preferred.

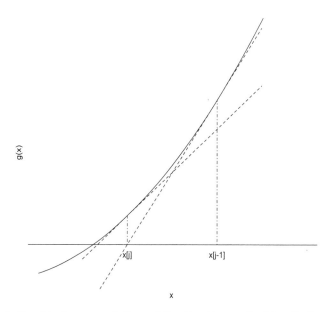

Figure 5.1 *Graphical representation of the iterative method for finding the roots of an equation in the scalar case.*

5.2.1 *Solution of the likelihood equation*

Let $\mathbf{U}(\mathbf{X}; \boldsymbol{\theta}) = \partial \log p(\boldsymbol{x}|\boldsymbol{\theta})/\partial\boldsymbol{\theta}$ be the score function. The maximum likelihood estimator is the solution of the likelihood function

$$\mathbf{U}(\mathbf{X}; \boldsymbol{\theta}) = \mathbf{0}.$$

Remember that $\mathbf{J}(\boldsymbol{\theta}) = -\partial\mathbf{U}(\mathbf{X}; \boldsymbol{\theta})/\partial\boldsymbol{\theta}$ is the observed information matrix. Then, replacing relevant terms in the expression of the Newton-Raphson iteration gives

$$\boldsymbol{\theta}^{(j)} = \boldsymbol{\theta}^{(j-1)} + [\mathbf{J}(\theta^{(j-1)})]^{-1}\mathbf{U}(\boldsymbol{\theta}^{(j-1)}).$$

There is a sense, to be made more precise in the next section, in which the observed information $\mathbf{J}$ approaches the expected information $\mathbf{I}$. This idea can be introduced as a modification to the Newton-Raphson algorithm by replacement of the factors involving $\mathbf{J}$ by other ones involving $\mathbf{I}$. The jth step of the iteration becomes

$$\boldsymbol{\theta}^{(j)} = \boldsymbol{\theta}^{(j-1)} + [\mathbf{I}(\theta^{(j-1)})]^{-1}\mathbf{U}(\boldsymbol{\theta}^{(j-1)}).$$

This revised algorithm is usually known as Fisher scoring and enjoys many of the nice properties of the Newton-Raphson algorithm. Under mild regularity conditions found in many applications, it converges to the maximum likelihood estimator. Its strength lies in the fact that considerable reduction in computation may be achieved by replacement of $\mathbf{J}$ by $\mathbf{I}$ with subsequent elimination of a few terms.

Example 5.1 *In this example an indirect use of the Fisher scoring algorithm is presented. First the profile likelihood is obtained, reducing the dimension of the parameter space, and then the algorithm is implemented. Let $X_1, \ldots, X_n$ be iid random quantities from the Weibull distribution, denoted by $Wei(\alpha, \beta)$ $(\alpha, \beta > 0)$, with*

$$p(x|\alpha, \beta) = \beta \alpha x^{\alpha-1} \exp(-\beta x^{\alpha}), \quad \alpha > 0, \quad \beta > 0.$$

This distribution is sometimes parameterized in terms of α and $\theta = 1/\beta^{\alpha}$. A special case of this model occurs when $\alpha = 1$, corresponding to the exponential model. The log-likelihood function is given by

$$L(\alpha, \beta) = n \log \beta + n \log \alpha + (\alpha - 1) \sum_{i=1}^{n} \log X_i - \beta \sum X_i^{\alpha}.$$

The profile log-likelihood function can be written as

$$L(\alpha, \hat{\beta}(\alpha)) = n \log n - n \log \left(\sum_{i=1}^{n} X_i^{\alpha} \right) + (\alpha - 1) \sum_{i=1}^{n} \log X_i + n \log \alpha - n,$$

after substituting β by its maximum likelihood estimator $\hat{\beta}(\alpha) = n/\sum_{i=1}^{n} X_i^{\alpha}$. Differentiating with respect to α and making some algebraic simplifications gives the score function and the observed information matrix as

$$U(\mathbf{X}; \alpha) = \frac{n}{\alpha} - \frac{n\alpha \sum_{i=1}^{n} X_i^{\alpha} \log X_i}{\sum_{i=1}^{n} X_i^{\alpha}} + \sum_{i=1}^{n} \log X_i,$$

$$J(\alpha) = \frac{n}{\alpha^2} + n \frac{\sum_{i=1}^{n} X_i^{\alpha} \sum_{i=1}^{n} X_i^{\alpha} (\log X_i)^2 - (\sum_{i=1}^{n} X_i^{\alpha} \log X_i)^2}{(\sum_{i=1}^{n} X_i^{\alpha})^2}.$$

Applying the Newton-Raphson with initial value $\alpha^{(0)}$ gives, at the jth step of the iteration,

$$\alpha^{(j)} = \alpha^{(j-1)} + J^{-1}(\alpha^{(j-1)}) U(\mathbf{X}; \alpha^{(j-1)}).$$

As soon as convergence in α is reached, β can be estimated using the relationship between α and its conditional maximum likelihood estimator given by

$$\hat{\beta}(\alpha) = \frac{n}{\sum_{i=1}^{n} X_i^{\alpha}}.$$

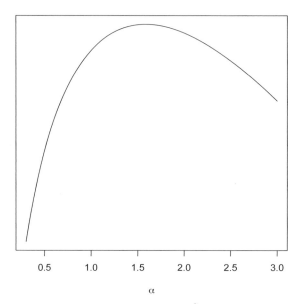

Figure 5.2: *Profile log-likelihood function $L(\alpha, \hat{\beta}(\alpha))$ for different values of α.*

Using $n = 25$ observations artificially generated from a Weibull distribution with $\alpha = 1.5$ and $\theta = 2.0$, initializing the procedure with $\alpha^{(0)} = 0.5$, we obtained via Newton-Raphson the following sequence of estimates: $0.889, 1.324, 1.555, 1.588$, with convergence achieved after $j = 4$ iterations. Then, the maximum likelihood estimates are finally given by $\hat{\alpha} = 1.59$ and $\hat{\theta} = 1.97$. The profile log-likelihood function $L(\alpha, \hat{\beta}(\alpha))$ is graphically illustrated in Figure 5.2.

Example 5.2 *Let $Y_i|x_i$ follow a Bernoulli distribution with success probability $\mu_i \in (0,1)$, with logit $\mu_i = \log[\mu_i/(1-\mu_i)] = \alpha + \beta x_i$, for $i = 1, \ldots, n$. This model is a special case of a linear model, which will be described in Chapter 8.*

The log-likelihood function of $\boldsymbol{\mu} = (\mu_1, \ldots, \mu_n)$ is given by

$$L(\boldsymbol{\mu}) = \sum_{i=1}^{n} y_i \log(\mu_i)(1 - y_i) \log(1 - \mu_i).$$

Letting $\boldsymbol{\theta} = (\alpha, \beta)$ gives, after a few algebraic manipulations, the score function $\boldsymbol{U}(\boldsymbol{\theta})$, obtained via chain rule, $\partial L(\boldsymbol{\theta})/\partial \alpha = [\partial L(\boldsymbol{\mu})/\partial \mu_i][\partial \mu_i/\partial \alpha]$

and $\partial L(\boldsymbol{\theta})/\partial\beta = [\partial L(\boldsymbol{\mu})/\partial\mu_i][\partial\mu_i/\partial\beta]$, with elements

$$\frac{\partial L(\boldsymbol{\theta})}{\partial\alpha} = \sum_{i=1}^{n}(y_i - \mu_i) \quad and \quad \frac{\partial L(\boldsymbol{\theta})}{\partial\beta} = \sum_{i=1}^{n} x_i(y_i - \mu_i).$$

The observed information matrix $\boldsymbol{J}(\boldsymbol{\theta})$ has elements

$$-\frac{\partial^2 L(\boldsymbol{\theta})}{\partial\alpha^2} = \sum_{i=1}^{n}\mu_i(1 - \mu_i),$$

$$-\frac{\partial^2 L(\boldsymbol{\theta})}{\partial\beta^2} = \sum_{i=1}^{n} x_i^2 \mu_i(1 - \mu_i) \quad and$$

$$-\frac{\partial^2 L(\boldsymbol{\theta})}{\partial\alpha\partial\beta} = \sum_{i=1}^{n} x_i\mu_i(1 - \mu_i).$$

Then, the Fisher scoring update equation is readily obtained as referenced above, that is, $\boldsymbol{\theta}^{(j)} = \boldsymbol{\theta}^{(j-1)} + [\boldsymbol{J}(\theta^{(j-1)})]^{-1}\boldsymbol{U}(\boldsymbol{\theta}^{(j-1)})$. The maximum likelihood estimators of the μ_i's are readily obtained as $\hat{\mu}_i = exp(\hat{\alpha} + \hat{\beta}x_i)/[1 + exp(\hat{\alpha} + \hat{\beta}x_i)]$, for $i = 1, \ldots, n$.

5.2.2 The EM Algorithm

This is a general iterative method useful to obtain the (generalized) MLE when we are faced with an incomplete data set or when it is simpler to maximize the likelihood of a more general problem involving unobserved quantities. Dempster, Laird and Rubin (1977) provided the first unified account of the algorithm. A discussion in the context of data imputation is presented in Little and Rubin (1988). The algorithm is iterative and at each iteration it alternates the operations of expectation (**E**) and maximization (**M**).

Let $\boldsymbol{X} \in R^n$ be the n-dimensional vector of observed quantities and $\boldsymbol{Z} \in R^m$ an m-dimensional vector of unobserved quantities. The complete data is denoted by $\boldsymbol{Y} = (\boldsymbol{X}, \boldsymbol{Z}) \in R^{n+m}$ and its density function is $p(\boldsymbol{y}|\boldsymbol{\theta}) = p(\boldsymbol{x}, \boldsymbol{z}|\boldsymbol{\theta})$, $\boldsymbol{\theta} \in \boldsymbol{\Theta}$. On the other hand, let the conditional density of the unobserved data given the observed one be $p(\boldsymbol{z}|\boldsymbol{x}, \boldsymbol{\theta})$, which also depends on $\boldsymbol{\theta}$. The logarithm of the marginal likelihood

$$L(\boldsymbol{\theta}; \boldsymbol{x}) = \log\left(\int p(\boldsymbol{x}, \boldsymbol{z}; \boldsymbol{\theta})d\boldsymbol{z}\right)$$

is usually directly maximized to obtain the maximum likelihood estimator of $\boldsymbol{\theta}$. To avoid the usually high-dimensional integral involved in the marginalization of $p(\boldsymbol{x}, \boldsymbol{z}|\boldsymbol{\theta})$, the following relationship can be used

$$L(\boldsymbol{\theta}; \boldsymbol{x}) = \log\left(\frac{p(\boldsymbol{x}, \boldsymbol{z}|\boldsymbol{\theta})}{p(\boldsymbol{z}|\boldsymbol{x}, \boldsymbol{\theta})}\right) = \log p(\boldsymbol{x}, \boldsymbol{z}|\boldsymbol{\theta}) - \log p(\boldsymbol{z}|\boldsymbol{x}, \boldsymbol{\theta}).$$

Since $\boldsymbol{Z}$ is unobserved, it is necessary to eliminate it earlier maximizing $L(\boldsymbol{\theta}; \mathbf{x})$. One way to do that is to take expected values with respect to the conditional density $p(\boldsymbol{z}|\boldsymbol{x}, \boldsymbol{\theta})$ in the above equation. Noting that $E_{\boldsymbol{Z}|\boldsymbol{X}, \boldsymbol{\theta}}[L(\boldsymbol{\theta}; \boldsymbol{X})] = L(\boldsymbol{\theta}; \boldsymbol{X})$, gives

$$L(\boldsymbol{\theta}; \boldsymbol{x}) = Q(\boldsymbol{\theta}; \boldsymbol{\theta}^{(0)}) - H(\boldsymbol{\theta}; \boldsymbol{\theta}^{(0)}),$$

where

$$
\begin{aligned}
Q(\boldsymbol{\theta}; \boldsymbol{\theta}^{(0)}) &= E_{\boldsymbol{Z}|\boldsymbol{X}, \boldsymbol{\theta}^{(0)}}[\log p(\boldsymbol{X}, \boldsymbol{Z}|\boldsymbol{\theta})], \\
H(\boldsymbol{\theta}; \boldsymbol{\theta}^{(0)}) &= E_{\boldsymbol{Z}|\boldsymbol{X}, \boldsymbol{\theta}^{(0)}}[\log p(\boldsymbol{Z}|\boldsymbol{X}, \boldsymbol{\theta})]
\end{aligned}
$$

and $\boldsymbol{\theta}^{(0)}$ is any given starting value for $\boldsymbol{\theta}$. The expectation involved in the definition of Q is based on the likelihood of the complete data set $\mathbf{Y}$. This is usually straightforward because the data augmentation is performed to simplify the problem, as will be shown in the examples below.

If $\boldsymbol{\theta}^{(j)}$ denotes the value of $\boldsymbol{\theta}$ in the jth iteration then the EM algorithm is defined through the following two steps:

1. **E** (expectation): evaluation of

$$Q(\boldsymbol{\theta}; \boldsymbol{\theta}^{(j-1)}) = E_{\boldsymbol{Z}|\boldsymbol{X}, \boldsymbol{\theta}^{(j-1)}}[L(\boldsymbol{\theta}; \boldsymbol{Y})];$$

2. **M** (maximization): evaluation of $\boldsymbol{\theta}^{(j)}$, the value of $\boldsymbol{\theta}$ that maximizes $Q(\boldsymbol{\theta}; \boldsymbol{\theta}^{(j-1)})$.

The estimation procedure is iterative and alternates the **E** and **M** operations at each iteration. Dempster, Laird and Rubin (1977) showed that the sequence $\boldsymbol{\theta}^{(j)}$, $j \geq 1$, generated by the EM algorithm satisfies $L(\boldsymbol{\theta}^{(j)}; \boldsymbol{X}) \leq L(\boldsymbol{\theta}^{(j+1)}; \boldsymbol{X})$ and is monotonically increasing in the likelihood $l(\boldsymbol{\theta}; \boldsymbol{X})$. Therefore, the sequence $\boldsymbol{\theta}^{(j)}$ converges to $\hat{\boldsymbol{\theta}}$, the maximum likelihood estimator, if the likelihood function has only a single maximum. Convergence can be established by criteria such as $|\boldsymbol{\theta}^{(j)} - \boldsymbol{\theta}^{(j-1)}| < \delta$ or $|\mathbf{Q}(\boldsymbol{\theta}^{(j)}; \boldsymbol{\theta}^{(j-1)}) - \mathbf{Q}(\boldsymbol{\theta}^{(j-1)}; \boldsymbol{\theta}^{(j-1)})| < \epsilon$. The convergence can be slow in some cases especially if the missing information of $\boldsymbol{Z}$ is substantial. The adaptation for posterior mode evaluation involves replacement of the likelihood $l(\boldsymbol{\theta}; \boldsymbol{Y})$ by the posterior $p(\boldsymbol{\theta}|\boldsymbol{Y})$ in the **E** step.

Example 5.3 *(Rao, 1973) A classical application in the statistical literature refers to a genetic study stating that the 4-dimensional vector of animal counts $\boldsymbol{X} = (X_1, X_2, X_3, X_4)$ has multinomial distribution with parameters n and $\boldsymbol{\pi}$, where $\boldsymbol{\pi} = (1/2 + \theta/4, (1 - \theta)/4, (1 - \theta)/4, \theta/4)$. So the probability function of $\boldsymbol{X}$ is*

$$p(\boldsymbol{x}|\theta) = \frac{(x_1 + x_2 + x_3 + x_4)!}{x_1! x_2! x_3! x_4!} \left(\frac{1}{2} + \frac{\theta}{4}\right)^{x_1} \left(\frac{1 - \theta}{4}\right)^{x_2 + x_3} \left(\frac{\theta}{4}\right)^{x_4}.$$

Direct maximization of the above expression is awkward due to the presence of the term $(1/2) + (\theta/4)$. *To avoid it, the EM method described above will be applied.*

To do that, let $X_1 = Y_0 + Y_1$ *and* $Y_i = X_i$, $i \geq 2$, *where the augmented vector* $\boldsymbol{Y} = (Y_0, Y_1, Y_2, Y_3, Y_4)$ *has multinomial distribution with parameters* n *and* $\pi^* = (1/2, \theta/4, (1-\theta)/4, (1-\theta)/4, \theta/4)$. *To complete the notation, define* $Z = Y_0$, *so that* $\boldsymbol{Y} = (\boldsymbol{X}, Z)$. *Therefore,*

$$p(\boldsymbol{y}|\theta) = \frac{n!}{y_0! y_1! y_2! y_3! y_4!} \left(\frac{1}{2}\right)^{y_0} \left(\frac{\theta}{4}\right)^{y_1} \left(\frac{1-\theta}{4}\right)^{y_2+y_3} \left(\frac{\theta}{4}\right)^{y_4} \quad and$$

$$\log p(\boldsymbol{y}|\theta) = k_1(\boldsymbol{y}) + y_0 \log\left(\frac{1}{2}\right) + (y_1 + y_4) \log\left(\frac{\theta}{4}\right)$$

$$+ (y_2 + y_3) \log\left(\frac{1-\theta}{4}\right)$$

$$= k_2(\boldsymbol{y}) + (y_1 + y_4) \log\theta + (y_2 + y_3) \log(1-\theta)$$

where $k_1(\boldsymbol{y})$ *and* $k_2(\boldsymbol{y})$ *are functions of* $\boldsymbol{y}$ *but not of* θ. *Therefore,*

$$Q(\theta; \theta^{(j)}) = E[k_2(\boldsymbol{Y}) + (Y_1 + Y_4) \log\theta + (Y_2 + Y_3) \log(1-\theta)|\boldsymbol{X}, \theta^{(j)}]$$

$$= k(\boldsymbol{X}, \theta^{(j)}) + E(Y_1 + Y_4|\boldsymbol{X}, \theta^{(j)}) \log\theta$$

$$+ E(Y_2 + Y_3|\boldsymbol{X}, \theta^{(j)}) \log(1-\theta)$$

$$= k(\boldsymbol{X}, \theta^{(j)}) + [E(Y_1|\boldsymbol{X}, \theta^{(j)}) + X_4] \log\theta + (X_2 + X_3) \log(1-\theta),$$

since $Y_i = X_i$, $i = 2, 3, 4$. *We only need to evaluate the expectation of* Y_1 *because* $k(\boldsymbol{X}, \theta^{(j)})$ *does not depend on* θ *and will therefore be irrelevant in the* $\boldsymbol{M}$ *step. From the construction of* $\boldsymbol{Y}$ *it follows that* $(Z|\boldsymbol{X}, \theta) \sim (Z|X_1, \theta) \sim Bin(X_1, p)$ *where* $p = (1/2)/[(1/2) + (\theta/4)] = 2/(2 + \theta)$. *Therefore,* $E(Y_0|\boldsymbol{X}, \theta) = X_1 p$ *and*

$$Q(\theta; \theta^{(j)}) = E[k_2(\boldsymbol{Y})|\boldsymbol{X}, \theta^{(j)}] + (X_1 p^{(j)} + X_4) \log\theta + (X_2 + X_3) \log(1-\theta),$$

where $p^{(j)} = 2/(2 + \theta^{(j)})$. *The* $\boldsymbol{M}$ *step involves finding the value of* θ *that maximizes* $Q(\theta; \theta^{(j)})$. *This is easily obtained by differentiation of* Q *and gives*

$$\theta^{(j+1)} = \frac{X_1 p^{(j)} + X_4}{X_1 p^{(j)} + X_2 + X_3 + X_4}$$

$$= \frac{(X_1 + X_4)\theta^{(j)} + 2X_4}{(X_1 + X_2 + X_3 + X_4)\theta^{(j)} + 2(X_2 + X_3 + X_4)}.$$

To illustrate the results, assume that the counts $\boldsymbol{x} = (125, 18, 20, 34)$ *were observed and the EM algorithm was started at* $\theta^{(0)} = 0.5$. *Then,*

$$\theta^{(j+1)} = \frac{159\theta^{(j)} + 68}{197\theta^{(j)} + 144}.$$

The first 5 iterations of the algorithm give the values 0.608, 0.624, 0.626, 0.627, 0.627.

Example 5.4 *(Randomized response) The proportion θ of individuals belonging to a certain stigmatized category must be estimated. To avoid the non-response (and its consequent loss of information), a new sampling scheme is proposed. An alternative question, not related to the main one, with known proportion of YES responses is introduced together with the guarantee that the selected question will be known exclusively by the respondent. The idea is to increase his/her confidence in providing the correct response without revealing its true status. The probability of a YES response will be $\lambda(\theta) = \pi\theta + (1-\pi)\theta_A$, where θ is the probability of the original question of interest, θ_A is the known probability of a YES answer to the alternative question and π is the probability of selection of the question of interest.*

Using the EM algorithm we get that the observed data is X, the number of YES responses, and $X|\theta \sim Bin(n,\lambda)$. Also, the unobserved data is Z, the number of individuals who will select the question of interest. Then, $Z|X,\theta \sim Bin(X,p)$, where $p = \pi\theta/\lambda$. The joint density of the observed and unobserved data is given by

$$
\begin{aligned}
p(x,z|\theta) &= p(z|x,\theta)p(x|\theta) \\
&= \binom{x}{z}\left(\frac{\pi\theta}{\lambda}\right)^z\left(1-\frac{\pi\theta}{\lambda}\right)^{x-z}\binom{n}{x}\lambda^x(1-\lambda)^{n-x} \\
&= \frac{n!}{z!(x-z)!(n-x)!}(\pi\theta)^z[(1-\pi)\theta_A]^{x-z}[1-\lambda]^{n-x} \\
\log p(x,z|\theta) &= k(x,z) + z\log\theta + (n-x)\log(1-\lambda).
\end{aligned}
$$

Then, the jth iteration of the EM algorithm will have

1. *E step: $Q(\theta;\theta^{(j)}) = E[\log p(X,Z|\theta)|X,\theta^{(j)}] = Xp^{(j)}\log\theta + (n-X)\log(1-\lambda) + k(X,\theta^{(j)})$;*

2. *M step: maximization of Q involves finding the solution of $\partial Q(\theta;\theta^{(j)})/\partial\theta = 0$ but*

$$
\frac{\partial Q(\theta;\theta^{(j)})}{\partial\theta} = \frac{Xp^{(j)}}{\theta} - \frac{n-X}{1-\lambda}\pi.
$$

Solving for θ gives

$$
\theta^{(j+1)} = \frac{Xp^{(j)}[1-(1-\pi)\theta_A]}{Xp^{(j)} + (n-X)\pi}.
$$

The EM algorithm is illustrated in a sample of 150 individuals where 60

YES responses were obtained, based on a procedure with $\pi = 0.7$ and $\theta_A = 0.6$. Initializing the procedure with $\theta^{(0)} = 0.4$, provides the following sequence of estimates: 0.338, 0.322, 0.317, 0.315, 0.314, 0.314, ...

Example 5.5 *(Cancho, Louzada-Neto and Barriga, 2011) Let Y be a non-negative random variable denoting the lifetime of an component in some population. A random variable is said to have a Poisson-Exponential distribution with parameters $\lambda > 0$ and $\beta > 0$ if its density is given by*

$$p(x|\boldsymbol{\theta}) = \frac{\lambda e^{-\lambda x - \beta e^{-\lambda x}}}{1 - e^{-\beta}}, \text{ for } x > 0,$$

where $\boldsymbol{\theta} = (\beta, \lambda)$, with λ controlling the scale and β controlling the shape of the distribution. Its genesis is based on the following. Assume that M has a zero truncated Poisson distribution with probability function given by $P(M = m|\beta) = e^{-\beta}\beta^m/[m!(1 - e^{-\beta})]$, for $m = 1, 2, \ldots$, and $\theta > 0$. Let T_j be a positive random variable independent of M, for $j = 1, 2, \ldots$. Given $M = m$, the random variables T_j, for $j = 1, \ldots, m$, are assumed to be independent and identically distributed according to an exponential distribution with a common density $f(t|\lambda) = \lambda \exp(-\lambda t)$, for $\lambda > 0$. Then, $Y = \max(T_1, T_2, \ldots, T_M)$ has a Poisson-Exponential distribution.

The EM algorithm to obtain the maximum likelihood estimator of $\boldsymbol{\theta}$ is as follows. Let $\mathbf{Y} = (Y_1, \ldots, Y_n)$ and $\mathbf{M} = (M_1, \ldots, M_n)$ have complete likelihood function $L(\boldsymbol{\theta}; \mathbf{y}, \mathbf{m}) = \prod_{i=1}^{n} p(y_i, m_i|\boldsymbol{\theta}) = \prod_{i=1}^{n} p(y_i|m_i, \boldsymbol{\theta})p(m_i|\boldsymbol{\theta})$, where

$$p(y_i|m_i\boldsymbol{\theta}) = m_i\lambda[1 - e^{-\lambda y_i}]^{m_i - 1}e^{-\lambda y_i}, \ y_i > 0,$$

and $p(m_i|\boldsymbol{\theta})$ given by the truncated Poisson probability function above, for $m_i = 1, 2, \ldots, \forall i$.

Setting $\widehat{m}_i = E(M_i|\boldsymbol{\theta} = \widehat{\boldsymbol{\theta}}, y_i)$, we can implement the M-step with

$$Q(\boldsymbol{\theta}; \boldsymbol{\theta}^{(k)}) = c + n\log\lambda + \sum_{i=1}^{n}(\widehat{m}_i - 1)\log(1 - e^{-\lambda y_i})$$

$$-\lambda\sum_{i=1}^{n} y_i + \sum_{i=1}^{n}\widehat{m}_i\log\beta - n\log(e^{\beta} - 1),$$

where c is a constant that is independent of $\boldsymbol{\theta}$, $\widehat{m}_i = 1 + \widehat{\beta}(1 - e^{-\widehat{\lambda}y_i})$ and $\boldsymbol{\theta}^{(k)}$ is an updated value of $\boldsymbol{\theta}$.

Then, the E step and the M step of the algorithm are given by

1. *$\boldsymbol{E}$ step: given a current estimate $\boldsymbol{\theta}^{(k)}$, compute $\widehat{m}_i^{(k)}$;*

2. *$\boldsymbol{M}$ step: update $\boldsymbol{\theta}^{(k)}$ by maximizing $Q(\boldsymbol{\theta}; \boldsymbol{\theta}^{(k)})$ over $\boldsymbol{\theta}$, which leads to the following expressions:*

$$\beta^{(k+1)} = \frac{\sum_{i=1}^{n} \widehat{m}_i^{(k)} (1 - e^{-\beta^{(k+1)}})}{n}$$

and

$$\lambda^{(k+1)} = \frac{n}{\left(\sum_{i=1}^{n} y_i - \sum_{i=1}^{n} \left[\frac{(\widehat{m}_i^{(k)} - 1) y_i e^{-\lambda^{(k+1)} y_i}}{1 - e^{-\lambda^{(k+1)} y_i}} \right] \right)}.$$

The EM iteration for both parameters β and λ is not explicit and a Newton-Raphson algorithm is required for the M step.

Consider the lifetimes of 22 ball bearings (in million revolutions per minute) on an endurance test given by 17.9, 28.9, 33.0, 42.5, 42.1, 45.6, 48.4, 51.8, 52.0, 55.6, 67.8, 68.6, 68.6, 68.9, 84.1, 93.1, 98.6, 105.1, 105.8, 127.9, 128.0, 173.4. After initializing the procedure with $\beta^{(0)} = 7$ and $\lambda^{(0)} = 0.35$, the first 4 iterations of the algorithm give the following sequence of estimates of (β, λ): (7.0005, 0.0347), (7.0015 , 0.0347), (7.0024, 0.0348), (7.0032 , 0.0348). This is an example of slow convergence of the EM algorithm. Only after more than 30 iterations the maximum likelihood estimates, (7.0193, 0.0348), are reached.

5.2.3   Determining likelihood intervals

We can apply the Newton-Raphson algorithm or the Fisher scoring algorithm for determining a likelihood interval. Let $L_r(\theta)$ the relative log-likelihood function be defined as $L_r(\theta) = L(\theta) - L(\hat{\theta})$ and $g(\theta)$ defined as $g(\theta) = L_r(\theta) - \log k$, with $k \in [0, 1]$. The end points of a $100k\%$ likelihood interval can be obtained by solving the equation $g(\theta) = 0$ and applying the Newton-Raphson algorithm or the Fisher scoring algorithm twice but with different starting values $\hat{\theta} - \epsilon$ and $\hat{\theta} + \epsilon$, where $\hat{\theta}$ is the maximum likelihood estimator and $\epsilon > 0$. The convergence of the method is assured as long as starting values are not far from the mode. In the case of the presence of nuisance parameter, one can consider the same specification above but for the substitution of the relative log-likelihood function $L_r(\theta)$ by the the relative profile log-likelihood $L_{pr}(\theta)$.

Example 5.6 *(Example 5.1. continued) Letting $X_1, \ldots, X_n$ be iid random quantities from the Weibull distribution, denoted by $Wei(\alpha, \beta)$ $(\alpha, \beta > 0)$, with relative profile log-likelihood function for α given by $L_{pr}(\alpha, \hat{\beta}(\alpha)) = L_p(\alpha, \hat{\beta}(\alpha)) - L_p(\hat{\alpha}, \hat{\beta}(\hat{\alpha}))$, $g(\theta) = L_{pr}(\alpha, \hat{\beta}(\alpha))) - \log k$.*

Considering the same 25 observations artificially generated from a Weibull distribution with $\alpha = 1.5$ and $\theta = 2.0$, and $k = 0.10$, applying the Newton-Raphson algorithm, the end points of a 10% likelihood interval are given by 1.14

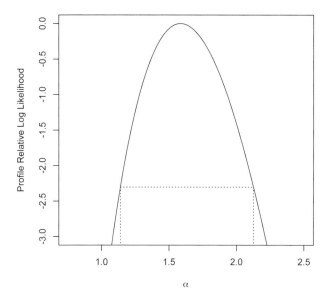

Figure 5.3 *Profile relative log-likelihood function $L_{pr}(\alpha, \hat{\beta}(\alpha))$ for different values of* α.

and 2.13. The profile relative log-likelihood function $L_{pr}(\alpha, \hat{\beta}(\alpha))$ is graphically illustrated in Figure 5.3.

Recall from Chapter 4 that the HPD interval $[\theta_l, \theta_u]$ in the uniparameter case is the collection of all $\theta \in \Theta$ such that $p(\theta|\boldsymbol{x}) > k$, where k is related to γ, the interval credibility, through equation $P(\theta_l \leq \theta \leq \theta_u|\boldsymbol{x}) = \gamma$. Note that we are assuming here that the confidence region is in fact an interval. Defining $g(\theta) = \log p(\theta|\boldsymbol{x}) - k$, apply the Newton-Raphson algorithm or the Fisher scoring algorithm twice but with different starting values $\hat{\theta} - \epsilon$ and $\hat{\theta} + \epsilon$, where $\hat{\theta}$ is the posterior mode and $\epsilon > 0$. The convergence of the method is assured as long as starting values are not far from the mode.

Many difficulties can occur with numerical optimization methods. The solution may not be stable when numerical derivatives are used and the choice of an inappropriate initial value can lead the solution to a local (instead of global) optimum. Nevertheless, many solutions to these problems have been proposed in the literature. There are methods that avoid the numerical evaluation of the second derivative, others that allow for different randomly selected initial values and so forth. The interested reader is referred to Thisted (1976) for further discussion.

5.3 Asymptotic theory

Some analytical methods used in statistical inference will be presented in this section from the standpoint of a potentially infinite sample size. Clearly, as in practice n is never infinite, the results presented here can be applied only when n can be thought of as sufficiently large to ensure that the results stated are good approximations of the exact result.

The results presented here provide methods to obtain approximate solutions to problems for which the exact solution is not feasible or cannot be conveniently obtained. There are various ways to obtain approximate solutions based on asymptotic theory but almost all of them are variations on the methods described in this section.

5.3.1 *Towards asymptotic confidence intervals*

The main idea in this section is to perform calculations under the asymptotic distribution of some suitable statistic. The most common statistic is the maximum likelihood estimator but the score function is another popular choice. Let $\mathbf{X}_n = (X_1, \ldots, X_n)$ be a random sample from $p(x \mid \theta)$ for a scalar θ, $\mathbf{U}(\mathbf{X}; \theta) = \partial \log p(\boldsymbol{x}|\theta)/\partial \theta$ be the score function and $\hat{\theta}_n$ be the maximum likelihood estimator of θ obtained from $\mathbf{X}_n$. Suppose that the Cramer-Rao regularity conditions are true and also that $|\partial^2 U(\mathbf{X}_n; \theta)/\partial \theta^2| < k$. Under these conditions, it follows that

$$\sum_{i=1}^n U(X_i; \hat{\theta}_n) = \sum_{i=1}^n U(X_i; \theta) + (\hat{\theta}_n - \theta) \sum_{i=1}^n \frac{\partial U(X_i; \theta)}{\partial \theta}$$
$$+ \frac{(\hat{\theta}_n - \theta)^2}{2} \sum_{i=1}^n \frac{\partial^2 U(X_i; \xi)}{\partial \theta^2}$$

for any ξ between θ and $\hat{\theta}_n$. By the definition of $\hat{\theta}_n$, the term on the left-hand side is null. Isolating the term $(\hat{\theta}_n - \theta)$ implies that $\sqrt{n}(\hat{\theta}_n - \theta)$ is equal to

$$-\frac{\sqrt{n}/n \left[\sum_{i=1}^n U(X_i; \theta)\right]}{(1/n) \left[\sum_{i=1}^n \partial U(X_i; \theta)/\partial \theta + 0.5(\hat{\theta}_n - \theta) \sum_{i=1}^n \partial U(X_i; \xi)/\partial \theta^2\right]}.$$

Remember that $E[U(X_i; \theta) \mid \theta] = 0$ and $V[U(X_i; \theta) \mid \theta] = I(\theta)$. Then, from the central limit theorem, the numerator $\left[\sum_{i=1}^n U(X_i; \theta)\right]/\sqrt{n}$ converges in distribution to an $N(0, I(\theta))$. The first term of the denominator converges almost surely to $E[\partial U(X_i; \theta)/\partial \theta \mid \theta] = -I(\theta)$. The second term converges almost surely to zero by the (strong) consistence of the maximum likelihood estimator and by bounds imposed on the second derivatives of $U(\mathbf{X}; \boldsymbol{\theta})$, by hypothesis. So, the denominator converge to $-I(\theta)$ with probability one.

Now, remember that if a sequence $Z_n \xrightarrow{\mathcal{D}} Z$ and another sequence W_n converges almost surely to a constant w, then the ratio sequence $Z_n/W_n \xrightarrow{\mathcal{D}}$ Z/w. Thus,

$$\sqrt{n}\,(\hat{\theta}_n - \theta) \xrightarrow{\mathcal{D}} N(0, I^{-1}(\theta)) \text{ when } n \to \infty.$$

It is usually said that for large n the distribution of $\hat{\theta}_n$ is approximately $N(\theta, I^{-1}(\theta)/n)$. It is worth pointing out that the information measure considered is based on a single observation.

The above result is also true for multivariate parameters $\boldsymbol{\theta}$. In this case, the asymptotic result is

$$\sqrt{n}\,(\hat{\boldsymbol{\theta}}_n - \boldsymbol{\theta}) \xrightarrow{\mathcal{D}} N(\mathbf{0}, \mathbf{I}^{-1}(\boldsymbol{\theta})) \text{ when } n \to \infty.$$

The result is used to say that the distribution of $\hat{\boldsymbol{\theta}}_n$ is approximately $N(\boldsymbol{\theta}, \mathbf{I}^{-1}(\boldsymbol{\theta})/n)$, for large n. These results are also valid for maximum likelihood estimators obtained from marginal likelihoods in the empirical Bayes approach. Approximate classical inference can then be made on the basis of those results. Moreover, asymptotic confidence regions can be constructed.

Definition 5.1 *Let $\boldsymbol{\theta}$ be an unknown quantity defined in $\boldsymbol{\Theta}$, $\boldsymbol{U}$ be a function $\boldsymbol{U}_n = \boldsymbol{G}(\boldsymbol{X}_n, \boldsymbol{\theta})$ with values in $\mathcal{U}$ and $\boldsymbol{A}$ be a region in $\mathcal{U}$ such that $\lim_{n \to \infty} Pr(\boldsymbol{U}_n \in \boldsymbol{A}) \geq 1 - \alpha$. A region $\boldsymbol{C} \subset \boldsymbol{\Theta}$ is an asymptotic $100(1-\alpha)\%$ confidence region for $\boldsymbol{\theta}$ if*

$$\boldsymbol{C} = \{\boldsymbol{\theta} : \boldsymbol{G}(\boldsymbol{x}_n, \boldsymbol{\theta}) \in \boldsymbol{A}\}.$$

In this case, $1 - \alpha$ is called the asymptotic confidence level. In the scalar case, the inversion in terms of θ usually leads to an interval, $C = [c_1, c_2]$, say.

It is easy to obtain from the asymptotic results above that $E(\hat{\boldsymbol{\theta}}_n | \boldsymbol{\theta}) \to \boldsymbol{\theta}$; that is, the maximum likelihood estimator is asymptotically unbiased. This means that although the maximum likelihood estimator could be biased for a fixed n, its expected value always tends to the parameter being estimated. Besides, it follows that

$$\lim_{n \to \infty} nV[\hat{\boldsymbol{\theta}}_n \mid \boldsymbol{\theta}] \to \mathbf{I}^{-1}(\boldsymbol{\theta}).$$

So, the variance of the maximum likelihood estimator asymptotically reaches the Cramer-Rao lower bound. As $\hat{\boldsymbol{\theta}}_n$ is asymptotically unbiased, the above limit could be thought of as an indication of the asymptotic efficiency of $\hat{\theta}_n$. Estimators satisfying this property are said to be asymptotically efficient.

Sometimes, $\mathbf{I}(\boldsymbol{\theta})$ depends on $\boldsymbol{\theta}$ making it harder to obtain confidence intervals. It is common in these cases to substitute $\mathbf{I}(\boldsymbol{\theta})$ by $\mathbf{I}(\hat{\boldsymbol{\theta}}_n)$ or, in cases where the expected value is hard to obtain, by $J(\hat{\boldsymbol{\theta}}_n)$, the observed information matrix. The consistency of the maximum likelihood estimator and the strong law of the large numbers justify, respectively, the substitutions made to obtain both of the above results.

The asymptotic result obtained for the score function may also be used to make approximate inference about $\boldsymbol{\theta}$. This is particularly useful when there is no explicit form for the maximum likelihood estimator. Then, for example, the asymptotic $100(1-\alpha)\%$ confidence region for a scalar θ can be constructed using the percentiles of the Normal distribution and is given by

$$\left\{ \theta : \left| \frac{1}{\sqrt{nI(\theta)}} \sum_{i=1}^{n} U(x_i; \theta) \right| < z_{\alpha/2} \right\}.$$

This result can also be extended to the case of parametric vectors.

The same considerations made previously about invariance over 1-to-1 parametric transformations are again true here. The above results are entirely based on the maximum likelihood estimator and on Fisher information measures, and both satisfy the invariance conditions.

Example 5.7 *Let $X \sim Bin(n, \theta)$ with unknown θ. Then*

$$U(X; \theta) = \frac{X}{\theta} - \frac{n - X}{1 - \theta} = \frac{X - n\theta}{\theta(1 - \theta)} \quad and \quad \hat{\theta}_n = \frac{X}{n}.$$

It is also known that $I(\theta) = 1/\theta(1 - \theta)$. Applying the results above gives $\sqrt{n}(\hat{\theta}_n - \theta)$ approximately distributed as an $N(0, \theta(1 - \theta))$ distribution, for large n. Then an asymptotic confidence region with coverage probability of $100(1 - \alpha)\%$ for θ will be given by

$$\left\{ \theta : \left| \frac{\sqrt{n}(\hat{\theta}_n - \theta)}{\sqrt{\theta(1 - \theta)}} \right| < z_{\alpha/2} \right\}.$$

A simple quadratic inequality must be solved to get an explicit solution for the confidence region. In this case, it is easier to proceed with the suggested modifications of θ by its maximum likelihood estimator in the expression of the Fisher information. Thus, observed and expected information coincide at the maximum likelihood estimator since

$$J(\hat{\theta}_n) = \frac{1}{\hat{\theta}_n(1 - \hat{\theta}_n)} = I(\hat{\theta}_n).$$

The above confidence region may be replaced by the interval

$$\left(\hat{\theta}_n - z_{\alpha/2} \sqrt{\frac{\hat{\theta}_n(1 - \hat{\theta}_n)}{n}} \,,\, \hat{\theta}_n + z_{\alpha/2} \sqrt{\frac{\hat{\theta}_n(1 - \hat{\theta}_n)}{n}} \right).$$

An alternative can be to use the asymptotic distribution of $U(X; \theta)/\sqrt{n}$ given by an $N(0, 1/[\theta(1 - \theta)])$ distribution. It is not hard to see that the $100(1-\alpha)\%$ asymptotic confidence region for θ is again given by

$$\left\{ \theta : \left| \frac{\sqrt{n}(\hat{\theta}_n - \theta)}{\sqrt{\theta(1 - \theta)}} \right| < z_{\alpha/2} \right\}.$$

In this case, the confidence region coincides with the region obtained from the asymptotic distribution of $\hat{\theta}_n$.

5.3.2   The delta method

In many practical problems, the interest is focused on a function of θ, say, $g(\theta)$. Assume g is a smooth differentiable function with continuous derivatives. So,

$$\sqrt{n}\left(g(\hat{\theta}_n) - g(\theta)\right) \xrightarrow{D} N\left(0, \left(\frac{\partial g(\theta)}{\partial \theta}\right)^2 I^{-1}(\theta)\right) \text{ when } n \to \infty.$$

The above result is commonly known as the delta method. It forms the basis for obtaining asymptotic confidence intervals for $g(\theta)$ by following exactly the route used for θ. The result follows directly from the Taylor expansion of $g(\hat{\theta}_n)$ around θ,

$$g(\hat{\theta}_n) = g(\theta) + \frac{\partial g(\theta)}{\partial \theta}(\hat{\theta}_n - \theta) + o_p(1),$$

where $o_p(1)$ indicates a term that converges in probability to zero in a stochastic order notation.

The delta method is a general approach with different meanings. It can be seen as an approximation for the variance of a function of a random variable, a bias correction for the expectation of a function of a random variable and also the limiting distribution of a function of a random variable. Note that the asymptotic variance of $\hat{\theta}_n$ can be consistently estimated by evaluating $I^{-1}(\theta)$ at $\theta = \hat{\theta}_n$.

Example 5.8 *(Example 5.7 continued) Let $X \sim Bin(n, \theta)$ with unknown θ, such that the maximum likelihood estimator is given by $\hat{\theta}_n = X/n$ and $\psi = g(\theta) = \text{logit } \theta$ is a parametric transformation of interest. Since $\partial \log(\theta)/\partial \theta = 1/\theta$, $\sqrt{n}\left(\log(\hat{\theta}_n) - \log(\theta)\right) \xrightarrow{D} N(0, (1 - \theta)/\theta)$. Let now*

$\hat{\psi} = g(\hat{\theta}_n) = log(\hat{\theta}_n/(1 - \hat{\theta}_n))$ with $\partial\psi/\partial\theta = [\theta(1 - \theta)]^{-1}$. Then, $\sqrt{n}(\hat{\psi} - \psi) \xrightarrow{\mathcal{D}} N(0, [\theta(1 - \theta)]^{-1})$, or $\hat{\psi} \sim N(\psi, [n\theta(1 - \theta)]^{-1})$ approximately, and this asymptotic distribution may be used to obtain approximate confidence intervals for ψ.

The delta method is easily extended to the multivariate case. Let $\hat{\boldsymbol{\theta}}_n$ be the maximum likelihood estimator of the k-dimensional parameter $\boldsymbol{\theta}$. Then,

$$\sqrt{n}(\hat{\boldsymbol{\theta}}_n - \boldsymbol{\theta}) \xrightarrow{\mathcal{D}} N(\mathbf{0}, \mathbf{I}^{-1}(\boldsymbol{\theta})),$$

where $\mathbf{I}(\boldsymbol{\theta})$ is a $k \times k$ symmetric positive definite matrix denoting the expected Fisher information matrix. Let $\boldsymbol{g}(\boldsymbol{\theta})$ be a smooth differentiable function of $\boldsymbol{\theta}$ with continuous derivative at $\boldsymbol{\theta}$. Then,

$$\sqrt{n}[\boldsymbol{g}(\hat{\boldsymbol{\theta}}_n) - \boldsymbol{g}(\boldsymbol{\theta})] \xrightarrow{\mathcal{D}} N\left(0, \boldsymbol{g}(\boldsymbol{\theta})^T \mathbf{I}^{-1}(\boldsymbol{\theta})\boldsymbol{g}(\boldsymbol{\theta})\right).$$

The above result is the basis for obtaining asymptotic confidence regions for $\boldsymbol{\theta}$. Once again, the same asymptotic normality is retained after replacement of the expected information matrix by the observed information matrix or by the expected information matrix evaluated at the maximum likelihood estimator.

5.3.3   First order asymptotics

In this section we explicitly use a local linearization approach based on Taylor series expansion as an approximation method but use it to approximate the posterior distribution instead of the likelihood function. First, it is important to note that, from the Bayesian point of view, when the sample size increases, its influence on the inference also increases thus minimizing the importance of the chosen prior distribution. However, this will only be true if the prior distribution is non-degenerate, that is, $p(\boldsymbol{\theta}) > 0, \forall \boldsymbol{\theta} \in \Theta$, as we have seen in Section 3.2. As n increases, the posterior distribution will be more and more concentrated around its mode. Note that

$$\begin{aligned} p(\boldsymbol{\theta} \mid \mathbf{x}) \quad &\propto \quad l(\boldsymbol{\theta}; \mathbf{x})p(\boldsymbol{\theta}) \\ &\propto \quad \exp\{L(\boldsymbol{\theta}) + \log p(\boldsymbol{\theta})\} \text{ where } L(\boldsymbol{\theta}) = \log l(\boldsymbol{\theta}; \mathbf{x}). \end{aligned}$$

Since $L(\boldsymbol{\theta}; \mathbf{x}) = \sum_{i=1}^{n} \log l(\boldsymbol{\theta}; x_i)$, the number of terms in $L(\boldsymbol{\theta}; \mathbf{x})$ will increase with n, but $p(\boldsymbol{\theta})$ is fixed, and consequently, the influence of the prior $p(\boldsymbol{\theta})$ becomes less and less relevant. Then, as n gets larger and larger,

$$p(\boldsymbol{\theta} \mid \mathbf{x}) \dot{\propto} \exp\{L(\boldsymbol{\theta}; \mathbf{x})\}.$$

Therefore, the (normalized) likelihood and the posterior will be approximately equal.

Then, using the Taylor expansion of L around $\hat{\boldsymbol{\theta}}$, the maximum likelihood estimator of $\boldsymbol{\theta}$, it follows that

$$L(\boldsymbol{\theta};\mathbf{x}) = L(\hat{\boldsymbol{\theta}};\mathbf{x}) + (\boldsymbol{\theta} - \hat{\boldsymbol{\theta}})' \frac{\partial L(\hat{\boldsymbol{\theta}};\mathbf{x})}{\partial \boldsymbol{\theta}} + \frac{1}{2!}(\boldsymbol{\theta} - \hat{\boldsymbol{\theta}})' \frac{\partial^2 L(\hat{\boldsymbol{\theta}};\mathbf{x})}{\partial \boldsymbol{\theta} \partial \boldsymbol{\theta}'}(\boldsymbol{\theta} - \hat{\boldsymbol{\theta}}) + R(\boldsymbol{\theta},\hat{\boldsymbol{\theta}}),$$

where $R(\boldsymbol{\theta},\hat{\boldsymbol{\theta}})$ contains the terms of higher order, which can be eliminated when $\boldsymbol{\theta}$ is supposed to be close enough to $\hat{\boldsymbol{\theta}}$. As $\partial L(\hat{\boldsymbol{\theta}};\mathbf{x})/\partial \boldsymbol{\theta} = 0$, it follows that

$$L(\boldsymbol{\theta};\mathbf{x}) \doteq k - (\boldsymbol{\theta} - \hat{\boldsymbol{\theta}})' \frac{\mathbf{J}(\hat{\boldsymbol{\theta}})}{2}(\boldsymbol{\theta} - \hat{\boldsymbol{\theta}}) \text{ where } \mathbf{J}(\boldsymbol{\theta}) = -\frac{\partial^2 L(\boldsymbol{\theta};\mathbf{x})}{\partial \boldsymbol{\theta} \partial \boldsymbol{\theta}'}.$$

Therefore, $p(\boldsymbol{\theta} \mid \mathbf{x}) \doteq k \exp\left\{-(\boldsymbol{\theta} - \hat{\boldsymbol{\theta}})' \mathbf{J}(\hat{\boldsymbol{\theta}})(\boldsymbol{\theta} - \hat{\boldsymbol{\theta}})/2\right\}$, for some normalizing constant k and so

$$\mathbf{J}^{1/2}(\hat{\boldsymbol{\theta}})(\boldsymbol{\theta} - \hat{\boldsymbol{\theta}}) \mid \mathbf{x} \sim N(\mathbf{0},\mathbf{I}_p).$$

Then, if $\mathbf{X}_n = (X_1, ..., X_n)$ is a random sample from $p(x \mid \boldsymbol{\theta})$ and $\hat{\boldsymbol{\theta}}$ is the maximum likelihood estimator of $\boldsymbol{\theta}$, it follows under mild regularity conditions that

$$\mathbf{J}^{1/2}(\hat{\boldsymbol{\theta}})(\boldsymbol{\theta} - \hat{\boldsymbol{\theta}}) \mid \mathbf{x}_n \xrightarrow{\mathcal{D}} N(\mathbf{0},\mathbf{I}_p) \text{ when } n \to \infty.$$

In the scalar case,

$$J^{1/2}(\hat{\theta})(\theta - \hat{\theta}) \mid \mathbf{x}_n \xrightarrow{\mathcal{D}} N(0,1) \text{ when } n \to \infty$$

and $J(\theta)$ is given by $-\partial^2 L(\theta)/\partial \theta^2$.

The regularity conditions required are basically the same involved in the statement of the Cramer-Rao inequality (see Section 4.5.2). Observe that $\mathbf{J}(\boldsymbol{\theta}) = \sum_{i=1}^n \partial^2 \log f(x_i \mid \boldsymbol{\theta})/\partial \boldsymbol{\theta}^2$ and so $\mathbf{J}(\boldsymbol{\theta})$ will typically increase with n. These results may still be valid in some cases where the sample does not consist of independent observations.

The above result indicates that $\boldsymbol{\theta} \mid$ x has an approximately $N(\hat{\boldsymbol{\theta}},\mathbf{J}^{-1}(\hat{\boldsymbol{\theta}}))$ distribution, when n is large enough. This means that as long as the regularity conditions are satisfied, it is possible to draw approximate inference about the parameters. In particular, it is possible to construct approximate credibility regions based on the above results.

Definition 5.2 *Let $\boldsymbol{\theta}$ be an unknown quantity defined in Θ. A region $C \subset \Theta$ is an asymptotic $100(1 - \alpha)\%$ credibility region for $\boldsymbol{\theta}$ if $\lim_{n\to\infty} Pr(\boldsymbol{\theta} \in C|\mathbf{x}_n) \geq 1 - \alpha$. In this case, $1 - \alpha$ is called the asymptotic credibility level. In the scalar case, the region C is usually given by an interval, $[c_1, c_2]$, say.*

Note that for validity of the results, the prior $p(\boldsymbol{\theta})$ must be strictly positive and continuous in Θ. If $p(\boldsymbol{\theta})$ is proportional to a constant, it follows that the posterior coincides with the likelihood.

It has just been shown that the distribution of $\boldsymbol{\theta}$ converges to the Normal distribution when n increases. What can be stated about transformations of $\boldsymbol{\theta}$? The answer is not easy because since $\boldsymbol{\theta}$ is Normal we know that the only transformations that preserves normality are the linear transformations of $\boldsymbol{\theta}$. Since all above developments do not depend on any special property associated with $\boldsymbol{\theta}$, any transformation of $\boldsymbol{\theta}$ preserving the regularity conditions will produce the same results. So, these results are valid for any transformation of $\boldsymbol{\theta}$. The only difference will be on the speed of the convergence to the Normal distribution.

The delta method presented in Section 5.3.2 can also be directly applied here. The Taylor series expansion applied directly to some transformation of the parameter is another approximation that leads to the same result. Suppose that $E(\hat{\boldsymbol{\theta}}) = \boldsymbol{\theta}$, $V(\hat{\boldsymbol{\theta}}) = \boldsymbol{\Sigma}$ and $\mathbf{g}$ is a 1-to-1 transformation of $\boldsymbol{\theta}$ with well defined derivatives. Then,

$$\boldsymbol{\Psi} = \mathbf{g}(\boldsymbol{\theta}) = \mathbf{g}(\hat{\boldsymbol{\theta}}) + (\boldsymbol{\theta} - \hat{\boldsymbol{\theta}})' \frac{\partial \mathbf{g}(\hat{\boldsymbol{\theta}})}{\partial \boldsymbol{\theta}} + o(\boldsymbol{\theta} - \hat{\boldsymbol{\theta}}),$$

where $|o(\boldsymbol{u})|/|\boldsymbol{u}| \to 0$ when $\boldsymbol{u} \to \mathbf{0}$. If $\boldsymbol{\theta}$ is close to $\hat{\boldsymbol{\theta}}$, then the last term on the right-hand side can be omitted and ψ will have an approximately linear relationship with $\boldsymbol{\theta}$ where

$$E(\boldsymbol{\Psi}) \doteq \mathbf{g}(\hat{\boldsymbol{\theta}}) \text{ and } V(\boldsymbol{\Psi}) \doteq \left(\frac{\partial \mathbf{g}(\hat{\boldsymbol{\theta}})}{\partial \boldsymbol{\theta}} \right)' \boldsymbol{\Sigma} \frac{\partial \mathbf{g}(\hat{\boldsymbol{\theta}})}{\partial \boldsymbol{\theta}}.$$

Also, if $\boldsymbol{\theta}$ is normally distributed, so shall $\boldsymbol{\Psi}$ be.

An interesting question concerns the choice of the optimal reparametrization in the sense of inducing fast convergence. This theme is still under investigation and some preliminary empirical results seem to indicate that the strongest candidates are the parameter transformations that appear in the definition of the exponential family and the transformations leading to constant non-informative priors. It is worth pointing out that the mean of normally distributed data, with known variance, which has a posterior Normal distribution for any value of n, belongs to both groups of transformations. Note also that the first class of transformations encompasses the class of parameters with constant Fisher information and therefore is, in some sense, stable.

Example 5.9 *(Example 5.7 continued) Let $X \sim Bin(n, \theta)$ with unknown θ.*

It then follows that

$$L(\theta) = \log p(x \mid \theta) = \log \binom{n}{x} + x \log \theta + (n - x) \log(1 - \theta),$$

and its derivatives are given by

$$\frac{\partial L(\theta)}{\partial \theta} = \frac{x}{\theta} - \frac{n - x}{1 - \theta} \Rightarrow \hat{\theta} = \frac{x}{n}$$

$$\frac{\partial^2 L(\theta)}{\partial \theta^2} = -\frac{x}{\theta^2} - \frac{n - x}{(1 - \theta)^2} < 0 \ \ and$$

$$J(\hat{\theta}) = \frac{n\hat{\theta}}{\hat{\theta}^2} + \frac{n - n\hat{\theta}}{(1 - \hat{\theta})^2}$$

$$= \frac{n}{\hat{\theta}} + \frac{n}{1 - \hat{\theta}} = \frac{n}{\hat{\theta}(1 - \hat{\theta})}.$$

So, for large enough n, θ has approximate posterior distribution $N(\hat{\theta}, \hat{\theta}(1 - \hat{\theta})/n)$.

The non-informative prior for this model is given by $p(\theta) \propto \theta^{-1/2}(1 - \theta)^{-1/2}$, as seen in Section 3.4, and the transformation producing a constant non-informative prior is

$$\phi \propto \int_0^\theta u^{-1/2}(1 - u)^{-1/2} \, du \propto sen^{-1}(\sqrt{\theta}).$$

Defining $\phi = sen^{-1}(\sqrt{\theta})$, it follows that

$$\frac{\partial \phi}{\partial \theta} = \frac{\theta^{-1/2}(1 - \theta)^{-1/2}}{2} \quad and \quad J^{-1}(\hat{\phi}) = \frac{\hat{\theta}(1 - \hat{\theta})}{n} \frac{1}{4\hat{\theta}(1 - \hat{\theta})} = \frac{1}{4n}.$$

So, the posterior distribution of ϕ for large enough n is approximately $N(\hat{\phi}, 1/4n)$.

The parameter obtained in the definition of the exponential family is $\psi = logit\ \theta = \log[\theta/(1 - \theta)]$. Then,

$$\hat{\psi} = \log\left(\frac{\hat{\theta}}{1 - \hat{\theta}}\right) \quad and \quad \left|\frac{\partial \psi}{\partial \theta}\right| = \frac{1}{\theta(1 - \theta)},$$

and so the posterior distribution of ψ is approximately $N(\hat{\psi}, 1/[n\hat{\theta}(1 - \hat{\theta})])$.

5.3.4   Higher order asymptotics

More accurate approximations than the one discussed in the Section 5.3.2 also involve the local linearization approach based on Taylor series expansion but with one further term. Then,

$$p(\boldsymbol{\theta}|\mathbf{x}) \simeq p(\hat{\boldsymbol{\theta}}|\mathbf{x}) \exp\left[-\frac{1}{2}(\boldsymbol{\theta}-\hat{\boldsymbol{\theta}})'\mathbf{J}(\hat{\boldsymbol{\theta}})(\boldsymbol{\theta}-\hat{\boldsymbol{\theta}})\right]\left[1+\frac{t(\boldsymbol{\theta})}{3!}\right],$$

where

$$t(\boldsymbol{\theta}) = \sum_{i,j,k}\frac{\partial^3 L(\hat{\boldsymbol{\theta}})}{\partial\theta_i\partial\theta_j\partial\theta_k}(\theta_i-\hat{\theta}_i)(\theta_j-\hat{\theta}_j)(\theta_k-\hat{\theta}_k).$$

These approximations were studied by Lindley (1980). The errors associated with these approximations are, in general, $O(n^{-2})$ in comparison with the $O(n^{-1})$ errors obtained when first order approximations are considered.

Example 5.10 *The expressions stated above applied to the posterior mean of a scalar θ simplify to*

$$E(\theta|\boldsymbol{x}_n) \simeq \hat{\theta} + \int (\theta-\hat{\theta})^4 \frac{1}{6}\frac{\partial^3 L(\hat{\theta})}{\partial\theta^3}\; f_N(\theta|\hat{\theta}, J^{-1}(\hat{\theta}))d\theta,$$

where $f_N(\cdot|a,b)$ denotes the density of the $N(a,b)$ distribution. Then

$$E(\theta|\boldsymbol{x}_n) \simeq \hat{\theta} + \frac{1}{6}\frac{\partial^3 L(\hat{\theta})}{\partial\theta^3}E_N(\theta-\hat{\theta})^4 = \hat{\theta} + \frac{1}{2}\frac{\partial^3 L(\hat{\theta})}{\partial\theta^3}J^{-2}(\hat{\theta}).$$

Example 5.11 *Let $X_1,\ldots,X_n$ be a random sample from the $Pois(\theta)$ and suppose that the prior for θ is a $G(a_0, b_0)$ distribution. So the posterior of θ will be $G(a_1, b_1)$, with $a_1 = a_0 + t$ and $b_1 = b_0 + n$, and $t = \sum_{i=1}^n x_i$.*

It is easy to verify that the posterior mode and mean are $(a_1 - 1)/b_1$ and $a_1/b_1 \simeq \hat{\theta} = t/n$, and for large n,

$$J(\hat{\theta}) = \frac{\partial^2 \log p(\hat{\theta}|\boldsymbol{x}_n)}{\partial\theta^2} = \frac{a_1 - 1}{\hat{\theta}^2}$$

and

$$t(\theta) = \frac{\partial^3 \log p(\hat{\theta}|\boldsymbol{x}_n)}{\partial\theta^3}(\theta-\hat{\theta})^3 = \frac{2(a_1 - 1)}{\hat{\theta}^3}(\theta-\hat{\theta})^3.$$

Then the approximations of second and third order will be

$$\theta|\boldsymbol{x}_n \;\dot\sim\; N\left(\hat{\theta}, \frac{\hat{\theta}^2}{a_1 - 1}\right) \text{ or } \theta|\boldsymbol{x}_n \;\dot\sim\; N\left(\hat{\theta}, \frac{\hat{\theta}^2}{a_1 - 1}\right)\left\{1 + \frac{1}{3}\frac{a_1 - 1}{\hat{\theta}^3}(\theta-\hat{\theta})^3\right\},$$

where the last equation means that the distribution is proportional to the product of a Normal distribution and the correction term in brackets. Using the result from the previous example, it follows that $E(\theta|\boldsymbol{x}_n) \simeq \hat{\theta}$ and $E(\theta|\boldsymbol{x}_n) \simeq \hat{\theta} + 1/b_1$.

On a general ground, let $\mathbf{X} = (X_1, X_2, \ldots, X_n)$ be a sample of independent, identically distributed random variables with log-likelihood given by $L(\boldsymbol{\theta}; \mathbf{X}) = \log l(\boldsymbol{\theta}; \mathbf{X})$ and maximum likelihood estimator $\hat{\boldsymbol{\theta}}$. The higher order approximation to the probability density function of $p(\hat{\boldsymbol{\theta}}|\boldsymbol{\theta})$ of the maximum likelihood estimator is given by

$$\tilde{p}(\hat{\boldsymbol{\theta}}|\boldsymbol{\theta}) \doteq c(\boldsymbol{\theta}, \mathbf{X}) |\mathbf{J}(\boldsymbol{\theta})|^{1/2} \exp(L(\boldsymbol{\theta}; \mathbf{X}) - L(\hat{\boldsymbol{\theta}}; \mathbf{X})),$$

where $c(\boldsymbol{\theta}, \mathbf{X})$ is a normalizing constant, which can be typically written as $(2\pi)^{-d/2}\{1 + O(n^{-1})\}$. According to Brazzale, Davison and Reid (2007), the exact probability density function is approximated by $\tilde{p}(\hat{\boldsymbol{\theta}}|\boldsymbol{\theta})$ with relative error of $O(n^{-3/2})$. The approximation $\tilde{p}(\hat{\boldsymbol{\theta}}|\boldsymbol{\theta})$ is known as the Barndorff-Nielsen approximation.

Example 5.12 *Let $X_1, \ldots, X_n$ be a random sample from an exponential distribution with rate $\psi = 1/\theta$. The log-likelihood $L(\theta)$ is equal to $-n(\log \theta + \hat{\theta}/\theta)$, where $\hat{\theta} = \sum_{i=1}^{n} X_i/n$. The observed information matrix is given by $J(\theta) = n/\theta^2$. Then, the probability density function of $\hat{\theta}$ is approximated by*

$$\tilde{p}(\hat{\theta}|\theta) = c(n)\frac{\hat{\theta}^{n-1}}{\theta^n}e^{-n\hat{\theta}/\theta}, \text{ for } \hat{\theta} > 0,$$

where $c(n) = (2\pi)^{1/2}n^{n-(1/2)}\exp(-n)$. It can be shown that the exact probability density function of $\hat{\theta}$ is given by $\tilde{p}(\hat{\theta}|\theta)$ except for the normalizing constant, whose exact value is $n^n/\Gamma(n)$. This value is known through Stirling's formula to be well approximated by $c(n)$.

We point out that the approximation $\tilde{p}(\hat{\boldsymbol{\theta}}|\boldsymbol{\theta})$ is invariant under one-to-one data transformation. If $\boldsymbol{\psi}$ and $\boldsymbol{\theta}$ are alternative parametrizations, then

$$\tilde{p}(\hat{\boldsymbol{\psi}}|\boldsymbol{\psi}) = \tilde{p}(\hat{\boldsymbol{\theta}}|\boldsymbol{\theta}) \left|\frac{\partial\hat{\boldsymbol{\theta}}}{\partial\hat{\boldsymbol{\psi}}}\right|.$$

5.4 Other analytical approximations

Some of the other analytical methods used in statistical inference for approximations will be presented in this section. Results based on the Kullback-Liebler divergence and a technique for numerical integration named the Laplace method will be discussed.

5.4.1 Approximation by Kullback-Liebler divergence

A measure of the distance or divergence between the densities $p(\theta)$ and $q(\theta)$ is defined by

$$\delta[p(\theta); q(\theta)] = \int p(\theta) \log \left(\frac{p(\theta)}{q(\theta)} \right) \, d\theta = E \left[\log \left(\frac{p(\theta)}{q(\theta)} \right) \right].$$

In the discrete case, all one needs to do is to substitute the integration by a summation. Smaller values for δ indicate closer distributions in Kullback-Leibler sense. In the limit, when $p = q$, the divergence is null. This distance could be used to evaluate the quality of an approximation $q(\theta)$ for $p(\theta)$. Two central questions in the applications are concerned with the determination of the better Normal approximation and the most convenient reparametrization to accommodate such approximation.

These two concerns will be answered only for particular cases in the exponential family. The first concern posed has an easy and general response. The best Normal approximation to the distribution $p(\theta)$ is with mean and variance given respectively by $\mu = E(\theta)$ and $\sigma^2 = V(\theta)$, the mean and variance of the original distribution.

The second concern involves finding the best transformation $\zeta = \zeta(\theta)$ to induce normality in the sense of minimizing the divergence measure between $p(\zeta)$ and its Normal approximation. This is equivalent to finding ζ that minimizes the expected value

$$\int p(\zeta) \log \left(\frac{p(\zeta)}{q(\zeta)} \right) d\zeta,$$

where $q(\zeta)$ is the density function of the $N[E(\zeta), V(\zeta)]$ distribution and $p(\zeta) = p(\theta)|d\zeta/d\theta|$. This is a difficult problem for which the solution will be presented for only two particular distributions, namely Beta and Gamma, as follows:

1. If $\theta \sim Beta(\alpha, \beta)$, then $\zeta(\theta) = \log[\theta/(1-\theta)]$ is the best transformation to induce normality, and the approximating Normal distribution will have mean and variance given by

$$E(\zeta) \simeq \log \left(\frac{\mu}{1 - \mu} \right) = \log \left(\frac{\alpha}{\beta} \right), \text{ where } \mu = \frac{\alpha}{\alpha + \beta} \text{ and}$$

$$V(\zeta) \simeq \frac{1}{\mu(1 - \mu)} \frac{1}{\alpha + \beta + 1} = \frac{(\alpha + \beta)^2}{\alpha\beta(\alpha + \beta + 1)}.$$

2. If $\theta \sim G(\alpha, \beta)$, then $\zeta(\theta) = \log \theta$ is the best transformation and the mean and variance of the approximating Normal distribution will be

$$E(\zeta) \simeq \log \mu \text{ and } V(\zeta) \simeq \frac{1}{\mu\beta} = \frac{1}{\alpha} \text{ where } \mu = \frac{\alpha}{\beta}.$$

The above approximations are based on the delta method. It is worth pointing out that these transformations are exactly those appearing in the exponential family for the Bernoulli and Poisson models, respectively. These are the observational models to which the Beta and Gamma distributions are conjugate, respectively.

Example 5.13 *Suppose we want to elicit the hyperparameters of the conjugate prior for θ in a Poisson model. Assume that $\theta \sim G(\alpha, \beta)$, and now suppose that its mode was assessed as 0.5 and also that $P(\theta < 0.25) = 0.05$. Then $(\alpha - 1)/\beta = 0.5$ or $\beta = 2(\alpha - 1)$ and $P(\theta < 0.25) = P(\zeta < \log 0.25) \simeq \Phi[(\log 0.25 - E(\zeta))/\sqrt{V(\zeta)}]$ where $E(\zeta) \simeq \log[\alpha/(2\alpha - 2)]$ and $V(\zeta) \simeq 1/\alpha$. So $\beta = 2\alpha - 2$ and $1.96 = \{\log 0.25 - \log[\alpha/(2\alpha - 2)]\}/\sqrt{1/\alpha}$. Solving this implicit function for α gives $\alpha = 3.35$ and $\beta = 4.7$.*

5.4.2 Laplace approximation

This class of approximation methods is very useful to evaluate integrals of the type $I = \int f(\boldsymbol{\theta})d\boldsymbol{\theta}$ by rewriting them as

$$\int g(\boldsymbol{\theta}) \exp[-nh(\boldsymbol{\theta})]d\boldsymbol{\theta},$$

where $g : R^p \to R$ and $h : R^p \to R$ are smooth functions, that is, at least three times differentiable. Let $\hat{\boldsymbol{\theta}}$ be the value of $\boldsymbol{\theta}$ which minimizes h. The Laplace method approximates I by

$$\hat{I} = g(\hat{\boldsymbol{\theta}})(2\pi/n)^{p/2}|\hat{\boldsymbol{\Sigma}}|^{1/2} \exp[-nh(\hat{\boldsymbol{\theta}})] \text{ where } \hat{\boldsymbol{\Sigma}} = \left[\frac{\partial^2 h(\hat{\boldsymbol{\theta}})}{\partial\boldsymbol{\theta}\partial\boldsymbol{\theta}'}\right]^{-1}.$$

The Laplace approximation is based on the Taylor approximation for h and g around $\hat{\boldsymbol{\theta}}$. Only the univariate case will be presented here for ease of exposition. As in the last section, it will be assumed that θ and $\hat{\theta}$ are close.

Using Taylor expansion up to the third order, it follows that

$$nh(\theta) = nh(\hat{\theta}) + \frac{n}{2\hat{\Sigma}}(\theta - \hat{\theta})^2 + \frac{nt(\theta)}{3!} + o(n^{-1}), \text{ where } t(\theta) = \frac{\partial^3 h(\hat{\theta})}{\partial\theta^3}(\theta - \hat{\theta})^3.$$

Exponentiating the last expression and applying a linear expansion to $\exp(-nt(\theta))$, it follows that

$$\exp[-nh(\theta)] = \exp[-nh(\hat{\theta})] \exp[-\frac{n}{2\hat{\Sigma}}(\theta - \hat{\theta})^2]\left[1 - \frac{nt(\theta)}{6} + o(n^{-1})\right].$$

The same expansion in Taylor series around $\hat{\theta}$ can be applied to

$$g(\theta) = g(\hat{\theta}) + \frac{\partial g(\theta)}{\partial\theta}(\theta - \hat{\theta}) + o(n^{-1}).$$

Recognizing that

a. $\int \exp[-(n/2\hat{\Sigma})(\theta - \hat{\theta})^2]d\theta = (2\pi)^{1/2}(\hat{\Sigma}/n)^{1/2}$,

b. $\int (\theta - \hat{\theta})^{2k+1} \exp[-(n/2\hat{\Sigma})(\theta - \hat{\theta})^2]d\theta = 0, \forall k$ integer,

c. $\int nt(\theta)(\theta - \hat{\theta}) \exp[-(n/2\hat{\Sigma})(\theta - \hat{\theta})^2]d\theta = o(n^{-1})$

and applying in the expression for I leads to a scalar version of the following proposition.

Proposition 5.1 When $n \to \infty$, $\hat{I} = I[1 + o(n^{-1})]$.

In Bayesian applications, generally $-nh(\theta) = L(\theta) + \log p(\theta)$ which is the expression of the posterior density except for the proportionality constant. If $g(\theta)$ is non-negative, the integral can be redefined by

$$E[g(\theta)] = I = \int \exp[-nh^*(\theta)]d\theta \text{ where } nh^*(\theta) = nh(\theta) - \log g(\theta).$$

Denoting by $\hat{\theta}^*$ the value that minimizes $h^*(\theta)$ and by $\hat{\Sigma}^* = \partial^2 h^*(\hat{\theta})/\partial \theta^2$, it follows an alternative approximation for I given by

$$\tilde{I} = (2\pi)^{1/2}|\hat{\Sigma}^*|^{1/2} \exp[-nh^*(\hat{\theta}^*)].$$

In the case of a multivariate $\boldsymbol{\theta}$, the expression becomes

$$\tilde{I} = (2\pi)^{p/2}|\hat{\boldsymbol{\Sigma}}^*|^{1/2} \exp[-nh^*(\hat{\boldsymbol{\theta}}^*)].$$

Following the same steps as earlier, it is easy to see that $\tilde{I} = I[1 + o(n^{-1})]$. Tierney and Kadane (1986) proposed to evaluate $E[g(\theta)] = [\int g(\boldsymbol{\theta}) \exp[-nh(\boldsymbol{\theta})]d\boldsymbol{\theta}]/ [\int \exp[-nh(\boldsymbol{\theta})]d\boldsymbol{\theta}]$ by approximating separately the numerator and the denominator. They have shown that by doing so, the $o(n^{-1})$ terms cancel out and an improved approximation of order $o(n^{-2})$ is obtained.

The final expression for their approximation can be obtained by combining the above results to give

$$\hat{E}[g(\boldsymbol{\theta})] = \frac{g(\hat{\boldsymbol{\theta}}^*)|\hat{\boldsymbol{\Sigma}}^*|^{1/2} \exp[-nh^*(\hat{\boldsymbol{\theta}}^*)]}{|\hat{\boldsymbol{\Sigma}}|^{1/2} \exp[-nh(\hat{\boldsymbol{\theta}})]}.$$

Example 5.14 (Example 5.11 continued) Let $X_1, \ldots, X_n$ be a random sample from a $Pois(\theta)$ and suppose that a conjugate prior $\theta \sim G(a_0, b_0)$ is used. Taking $g(\theta) = \theta$, it follows that

$$\hat{E}(\theta|\boldsymbol{x}) = \frac{a_1}{b_1} \left(\frac{a_1}{a_1 - 1} \right)^{a_1 - 1/2} e^{-1}, a_1 > 1,$$

where $a_1 = a_0 + \sum_{i=1}^{n} x_i$ and $b_1 = b_0 + n$.

The exact posterior mean in this example is a_1/b_1, and so we can easily evaluate relative errors $(\hat{E}-E)/E$ involved in the approximation. For example, for $a_1 = 6$ the relative error is 0.0028 and for $a_1 = 10$ it will be only 0.00097.

Example 5.15 *(Example 5.4 continued; Migon and Tachibana, 1997) A variation of the randomized response model consists of asking, as the alternative question, the negation of the original one. In this case, $\theta_A = 1 - \theta$ and the probability of a YES response is*

$$\lambda = \pi\theta + (1 - \pi)(1 - \theta) = (2\pi - 1)\theta + (1 - \pi).$$

The Laplace approximation for the evaluation of the posterior mean of θ can be applied. The derivatives and points of maxima can all be obtained analytically or numerically. With the same data of the example and with a unit uniform prior for θ, the exact posterior mode is 0.25. The posterior distribution of θ is a mixture of $n + 1$ Beta distributions, and the calculations become tedious as the sample size increases. The performance of the Laplace approximations can be assessed as a function of the sample size by keeping the YES proportion fixed at $60/150 = 0.40$ and letting the sample size n vary. For this example, the exact and approximated posterior means are shown in Table 5.1.

Table 5.1: *Exact and Approximated Posterior Means.*

n	Exact	Laplace
50	0.280	0.270
150	0.251	0.255
450	0.251	0.251

Even though the exact posterior mean itself depends on the sample size, the Laplace approximation gets better as the value of n increases.

Example 5.16 *(Example 5.1 continued) Using the same data of the example presented in Section 5.2.1, we obtain the following posterior estimates via Laplace methods with a non-informative prior: $E(\alpha|\boldsymbol{x}) = 1.59$, $V(\alpha|\boldsymbol{x}) = 0.064$ and $E(\theta|\boldsymbol{x}) = 2.05$, $V(\theta|\boldsymbol{x}) = 0.079$. These values compare well with the maximum likelihood results, as $\sqrt{V(\theta|\boldsymbol{x})} = 0.281$.*

5.5   Numerical integration methods

Numerical integration, also known as quadrature technique, is a collection of methods convenient to solve some useful problems in inference, mainly when the dimension of the parameter space is moderate. They become useful when analytical solution is not available analytically.

First, we will take care of the general problem of obtaining the value of the integral $I = \int_a^b f(x)dx$, where $f : R \to R$ is a smooth function. Let $w : R \to R^+$ be a well-defined weight function. Quadrature methods are essentially approximations of I obtained by evaluating f at points x_i, $i = 1, \ldots, n$. The simplest solution is given by the weighted sum

$$\hat{I} = \sum_1^n w_i f(x_i), \text{ where } w_i = w(x_i),\, i = 1, \ldots, n.$$

The quadrature methods are fully characterized by choosing the points of evaluation or nodes $x_1, \ldots, x_n$ in the interval (a, b) and the corresponding weights involved. An integration rule must have easily obtainable weights and nodes. The nodes should lie in the region of integration and the weights should all be positive.

5.5.1 Newton-Cotes type methods

The interval of integration (a, b) with finite a, b is divided into n equal parts, the function $f(x)$ is evaluated in the middle point of each interval and the weights are then applied. Then

$$\hat{I}_{NC} = h \sum_{i=1}^n f(a + (2i - 1)h/2)$$

approximates I, with $h = (b - a)/n$. These methods are generically named Newton-Cotes rules.

This is an approximation by the area of the rectangles with equal base $(b - a)/n$. Often a good approximation is obtained with n of the order of 10^2, which seems reasonable for the unidimensional case.

A slight variation is the trapezoidal rule involving unit weights except in the extremes of the interval, when they are set to $1/2$. The rule gives the approximation

$$\hat{I}_T = h \left[\frac{f(a)}{2} + \sum_{i=1}^n f(a + (2i - 1)h/2) + \frac{f(b)}{2} \right].$$

The Simpson rule is another variation described by weights alternating between $4/3$ and $2/3$, except in the extreme where they assume the value $1/3$. In this case the approximation is given by

$$\hat{I}_S = \frac{h}{3} \left[f(a) + 4 \sum_{i=1}^{n/2} f\left(a + (4i+1)\frac{h}{2}\right) + 2 \sum_{i=1}^{n/2} f\left(a + (4i+3)\frac{h}{2}\right) + f(b) \right].$$

The p-dimensional case is slightly more demanding. A general solution follows from an iterative application of the Cartesian product rule. Let $\mathbf{x} = (x_1, x_2)$ be a bidimensional vector. A quadrature in two dimensions is based on the Cartesian product rule

$$\int f(\mathbf{x})dx = \int \left[\int f(x_1, x_2)dx_2\right] dx_1 = \int f_1(x_1)dx_1.$$

The last term on the right-hand side is obtained by integration with respect to x_2 using the unidimensional quadrature rule

$$\int f_2(x_2)dx_2 \simeq \sum_{j=1}^{m} w_j f_2(x_{2,j}),$$

where $f_2(x_2) = f(x_1, x_2)$. As the last integral is also unidimensional it can be approximated again by quadrature $\int f_1(x_1)dx_1 \simeq \sum_{i=1}^{n} w_i f_1(x_{1,i})$. Joining the two weights it follows that

$$\int f(\mathbf{x})dx \simeq \sum_{i=1}^{n} \sum_{j=1}^{m} w_i w_j f(x_{1,i}, x_{2,j}).$$

This is a bidimensional rule based on $n \times m$ evaluation points $(x_{1,i}, x_{2,j})$ and with weights (w_i, w_j), $i = 1, \ldots, n$, $j = 1, \ldots, m$. Note that these rules will quickly become unfeasible when the dimensionality increases.

From now on, a general integration method more adequate to statistical problems will be introduced. We will be concerned with an integral over the whole real line.

5.5.2   Gauss-Hermite rules

The method introduced in this section is specifically useful for integration over the whole real line. If the domain of the integral is over positive values, which sometimes happens when we are dealing with a precision or a scale parameter, then the logarithmic reparametrization is often convenient.

Assuming that the integrand $f(x)$ can be expressed in the form of $g(x)\exp(-x^2)$, a general unidimensional rule is

$$\int_{-\infty}^{\infty} f(x)dx = \int_{-\infty}^{\infty} g(x)e^{-x^2}dx \simeq \sum_{1}^{n} h_i f(x_i),$$

where the x_i's are the zeros of a Hermite polynomial of degree n, $H_n(x)$, and the weights h_i depend on n and on the Hermite polynomial $H_{n-1}(x)$, evaluated at x_i, $i = 1, \ldots, n$. When $g(x)$ is a polynomial of maximum degree

$2n - 1$ the formula is exact. There are other Gaussian integration rules that can be applied when the domain of integration is finite. The values of the zeroes and heights are tabulated in many mathematical tables and can be found, for example, in Abramowitz and Stegun (1965).

The accuracy of the approximation depends on $f(x)$ being well approximated by a polynomial of degree $2n-1$ or less times a Normal weight function. For the multivariate case some sort of parameter orthogonality must be guaranteed for application of the Cartesian rule.

The integration rule introduced earlier can be extended to a more general approximating function $f(x) = g(x)(\sqrt{2\pi\sigma^2})^{-1}\exp\{-0.5[(x-\mu)/\sigma]^2\}$ leading to the solution

$$\int f(x)dx = \int g(x)\frac{1}{\sqrt{2\pi\sigma^2}}\exp\left[-\frac{1}{2}\left(\frac{x-\mu}{\sigma}\right)^2\right]dx.$$

Making the variable transformation $z = (x - \mu)/\sqrt{2\sigma^2}$, it follows that

$$\begin{aligned}
\int f(x)dx &= \int g(\sqrt{2\sigma^2}z + \mu)\frac{1}{\sqrt{2\pi\sigma^2}}e^{-z^2}\sqrt{2\sigma^2}dz \\
&= \pi^{-1/2}\int g(\sqrt{2\sigma^2}z + \mu)e^{-z^2}dz
\end{aligned}$$

and the Gauss-Hermite approximation will be

$$\begin{aligned}
\int f(x)dx &\simeq \pi^{-1/2}\sum_{i=1}^{n}h_i f(\sqrt{2\sigma^2}z_i + \mu) \\
&= \pi^{-1/2}\sum_{i=1}^{n}h_i\sqrt{2\pi\sigma^2}e^{z_i^2}g(\sqrt{2\sigma}z_i + \mu) \\
&= \sum_{i=1}^{n}m_i g(x_i),
\end{aligned}$$

where $m_i = \sqrt{2\sigma^2}e^{z_i^2}h_i$ and $x_i = \sqrt{2\sigma}z_i + \mu$, $i = 1,\ldots,n$.

In general, the mean μ and variance σ^2 of the approximating Normal are unknown. The following strategy, proposed by Naylor and Smith (1982), can be used:

a. Let μ_0 and σ_0^2 be initial values for μ and σ^2.

b. Apply the above expression with $g(\theta) = \theta$ and $g(\theta) = \theta^2$, respectively. This will provide approximating values for the mean and variance.

c. Repeat the last step until the mean and variance obtained are stabilized.

In the multiparameter case, the Cartesian product rule is applied after some orthogonalization is operated over the parameters. This can be done through a Cholesky decomposition of the (approximated) variance-covariance matrix Σ. Let $\Sigma = \mathbf{H}\,\mathbf{D}\,\mathbf{H}'$, where $\mathbf{D}$ is a diagonal matrix and $\mathbf{H}$ is a lower triangular matrix, and define the transformation $\boldsymbol{z} = \mathbf{D}^{-\frac{1}{2}}\mathbf{H}^{-1}(\mathbf{x} - \boldsymbol{\mu})/\sqrt{2}$. Therefore,

$$
\int f(\mathbf{x})d\mathbf{x} = \pi^{-p/2}\int g(\boldsymbol{\mu} + \sqrt{2}\mathbf{H}\,\mathbf{D}^{1/2}\boldsymbol{z})\exp(-\boldsymbol{z}'\boldsymbol{z})d\boldsymbol{z}
$$

$$
\simeq \pi^{-p/2}\sum_{i_p=1}^{n_p}\ldots\sum_{i_1=1}^{n_1} h_{i_1}\ldots h_{i_n}f(\boldsymbol{\mu} + \sqrt{2}\mathbf{H}\,\mathbf{D}^{1/2}\boldsymbol{z}),
$$

where n_i is the number of nodes involved in the approximation at the ith coordinate and $\boldsymbol{z} = (z_{i_1}, \ldots, z_{i_p})$, $i_j = 1, \ldots, n_j$, $j = 1, \ldots p$.

If the mean vector $\boldsymbol{\mu}$ and variance-covariance matrix Σ of the approximating Normal density are unknown, an extension of the iterative method of Naylor and Smith presented previously can be applied.

5.6   Simulation methods

In this section we will discuss a collection of techniques useful for solving many of the relevant statistical problems. From a classical point of view we are interested in simulating a sampling distribution of a statistic arising from a possibly complex model, and in the Bayesian case we are interested in obtaining some characteristics of the posterior distribution. All the techniques described in this section share a common characteristic: they involve the random generation of samples from a distribution of interest. This distribution is the sampling distribution of an estimator in the classical approach and the posterior distribution in the Bayesian case. Interested readers can complement their study by reading the books by Efron (1982), Gamerman and Lopes (2006) and Ripley (1987).

5.6.1   *Monte Carlo method*

The basic idea of the Monte Carlo method consists of writing the desired integral as an expected value with respect to some probability distribution. To motivate our discussion we will begin with a very simple problem. Assume we wish to calculate the integral of a smooth function in a known interval (a, b), that is,

$$
I = \int_a^b g(\theta)d\theta.
$$

The above integral can be rewritten as

$$I = \int_a^b [(b-a)g(\theta)]\frac{1}{b-a}d\theta.$$

This problem can be thought of as the evaluation of the expectation of $[(b-a)g(\theta)]$ with respect to the uniform distribution over (a,b) and

$$I = E_{U(a,b)}[(b-a)g(\theta)],$$

where $U(a,b)$ represents the uniform distribution in (a,b).

A method of moments estimator of this quantity is

$$\hat{I} = \frac{1}{n}\sum_{i=1}^{n}(b-a)g(\theta_i),$$

where $\theta_1,\ldots,\theta_n$ is a random sample selected from the uniform distribution on (a,b).

An algorithm for construction of $\hat{I}$ can be described by the following steps:

1. Generate $\theta_1, \theta_2, \ldots, \theta_n$ from a $U(a,b)$ distribution.
2. Calculate $g(\theta_1), g(\theta_2), \ldots, g(\theta_n)$.
3. Obtain the sample mean: $\bar{g} = (1/n)\sum_{i=1}^{n} g(\theta_i)$.
4. Determine: $\hat{I} = (b-a)\bar{g}$.

A generalization can be obtained straightforwardly. Let $I = E_p[g(\theta)]$ be the expected value of $g(\theta)$ with respect to a distribution with density $p(\theta)$. The algorithm is similar to that described above with modifications of the sampling in step 1 from the $U(a,b)$ to a density $p(\cdot)$. In this general case, the integral approximation is given by

$$\hat{I} = \frac{1}{n}\sum_{i=1}^{n}\frac{g(\theta_i)}{p(\theta_i)}.$$

The multivariate extension is based on evaluation of

$$I = \int_{a_1}^{b_1}\cdots\int_{a_k}^{b_k} g(\boldsymbol{\theta})d\boldsymbol{\theta} = \int_{a_1}^{b_1}\cdots\int_{a_k}^{b_k}\frac{g(\boldsymbol{\theta})}{p(\boldsymbol{\theta})}p(\boldsymbol{\theta})d\boldsymbol{\theta},$$

and the Monte Carlo estimator is given by

$$\hat{I} = \frac{1}{n}\sum_{i=1}^{n}\frac{g(\boldsymbol{\theta}_i)}{p(\boldsymbol{\theta}_i)},$$

where $\boldsymbol{\theta}_1,\ldots,\boldsymbol{\theta}_n$ is a random sample selected from the $p(\boldsymbol{\theta})$ distribution on $(a_1,b_1) \times \ldots \times (a_p,b_p)$.

It is worth noting that

$$E_p[\hat{I}] = \int \cdots \int \frac{g(\boldsymbol{\theta})}{p(\boldsymbol{\theta})} p(\boldsymbol{\theta}) d\boldsymbol{\theta} = I.$$

So, $\hat{I}$ is an unbiased estimator of I.

Some questions need to be answered to implement these techniques. How large n must be? How does the Monte Carlo method compare with quadrature rules? Or, to pose it in another way: when is Monte Carlo preferred to numerical integration? An elementary example will be useful to motivate further developments.

Example 5.17 *Suppose the evaluation of $I = \int_0^1 e^x dx$ is desired. Even though, we know $I = 1.7182$, we will approximate it by Monte Carlo as an exercise. As we have just seen, the simple Monte Carlo estimator of I is $\bar{I} = (1/n) \sum_{i=1}^n \exp(X_i)$, $X_i \sim U(0,1)$, $i = 1, \ldots, n$. Since $\hat{I}$ is a sample mean, its precision to estimate I can be measured by its variance, which is given by*

$$Var(\hat{I}) = \frac{1}{n} V(e^X) = \frac{1}{n} \left(\int_0^1 e^{2x} dx - I^2 \right).$$

Then, the variance of the Monte Carlo estimator is approximated by

$$V(\hat{I}) \cong \frac{1}{n} \left[\frac{1}{n} \sum_{i=1}^n e^{2X_i} - \left(\frac{1}{n} \sum_{i=1}^n e^{X_i} \right)^2 \right].$$

Table 5.2 presents the Monte Carlo approximation of the integral I and the variance of the estimator. As expected, $Var(\hat{I})$ decreases as n increases.

Table 5.2: *Monte Carlo Approximation with Samples from a $U(0,1)$ Distribution.*

n	100	500	1000	10000
$\hat{I}$	1.7701	1.7162	1.6848	1.7198
$Var(\hat{I})$	0.0031	0.0007	0.0003	0.0001

Following Example 5.17, the variance of the estimator can be calculated in the general case as

$$V(\hat{I}) = \frac{1}{n} \left(E_p \left[\frac{g(\boldsymbol{\theta})^2}{p(\boldsymbol{\theta})^2} \right] - I^2 \right).$$

It can be noted that the variance of the estimator converges to zero when n converges to infinity. Moreover, as $\hat{I}$ is an unbiased estimator of I, $\hat{I}$ is also a consistent estimator of I.

5.6.2 Monte Carlo with importance sampling

Monte Carlo with importance sampling is a technique developed to reduce the variance of the estimator based on simple Monte Carlo. Moreover, the technique can also be applied to calculate the expected value of a function $g(X)$ with respect to a density p when the variance of the Monte Carlo estimator is large or when it is is difficult to simulate from the density p, and the application of a direct Monte Carlo integration may be compromised. So, a second density h is chosen to be simulated from, and density is known as the importance sampling density.

Consider that the integral I of interest is $I = E[g(X)]$, the expectation of a given function $g(X)$ with respect to a density p. It can then be written as

$$I = \int g(x)p(x)dx = \int g(x)\frac{p(x)}{h(x)}h(x)dx,$$

where $h(x)$ is a density. An alternative method of moments estimator for I can be obtained as

$$\hat{I} = \frac{1}{n}\sum_{1}^{n} g(X_i)w(X_i), \tag{5.1}$$

where $w(X_i) = p(X_i)/h(X_i)$, $X_i \sim h(.)$, $i = 1,\ldots,n$ and $h(.)$ is the importance sampling density.

The unique difference with respect to simple Monte Carlo is the first step where sampling from the uniform distribution is replaced by sampling from h and the third step where values of $g * w$ instead of values of g are averaged. Therefore, $Var(\hat{I}) = (1/n)\int (g(x)w(x) - I)^2h(x)dx$. Choosing $g(x)w(x)$ approximately constant can make $Var(\hat{I})$ as small as we want. Therefore, whenever possible the importance sampling density should be roughly similar to $g(x)p(x)$. Note that simple Monte Carlo is the particular case of the importance sampling integration when we take $h = p$.

Example 5.18 *(Example 5.17 continued) If we take as importance sampling density $h(x) = \frac{2}{3}(1+x)$, $x \in (0,1)$, then $w(x) = p(x)/h(x) = 3/[2(1+x)]$ and $g(x)w(x) = (3/2)e^x/(1+x) \doteq k$, for $x \in (0,1)$. Then,*

$$I = \int_0^1 e^x \left[\frac{3}{2(1+x)}\right]\left[\frac{2}{3}(1+x)\right] dx = \int_0^1 \frac{3}{2}\frac{e^x}{1+x}h(x)dx.$$

Therefore, $\bar{I} = (1/n)\sum_{i=1}^{n} 3e^{X_i}/[2(1+X_i)]$, where $X_i \sim h(x)$, $i = 1,\ldots,n$ and $Var(\bar{I}) = (1/n)[(3/2)^2 \int_0^1 e^{2x}/(1+x)^2dx - I^2] = 0.027/n$. The variance reduction is quite substantial, approximately one tenth of the value obtained with a simple Monte Carlo using the same number of replications.

The implementation of the algorithm in this case depends on sampling from the density h. The importance sampling distribution function is given by

$$H(x) = \begin{cases} 0 & \text{if } x < 0; \\ \frac{2}{3}(x + \frac{x^2}{2}) & \text{if } 0 \le x < 1; \\ 1 & \text{if } x \ge 1. \end{cases}$$

Using the probability integral transform, we simply have to generate a unit uniform random variable U and solve for $U = (2/3)(X + X^2/2)$. The unique solution satisfying the equation is $X = [(4 + 12U)^{1/2} - 2]/2$.

Example 5.19 *Let $\theta = P(X > 2)$ where X has a standard Cauchy distribution with density*

$$p(x) = \frac{1}{\pi(1 + x^2)}, \; x \in R.$$

In this example, $g(X) = I_X[(2, \infty)]$ and $\theta = \int_{-\infty}^{\infty} g(x)p(x)dx$.

Let $X_1, \ldots, X_n$ be a random sample from a Cauchy distribution. It is easy to obtain that

$$\hat{\theta} = \frac{1}{n} \sum_{i=1}^{n} I_{X_i}(2, \infty) = \frac{\#(X_i > 2)}{n}$$

and $n\hat{\theta} \sim Bin(n, \theta)$.

Note that the distribution function of the Cauchy distribution is $F(x) = \int_{-\infty}^{x} p(t)dt = 0.5 + \pi^{-1}\arctan x$, so that $\theta = 1 - P(2) = 0.1476$. Then, $Var(\hat{\theta}) = \theta(1 - \theta)/n = 0.126/n$.

Let h be a density defined by $h(x) = 2/x^2 I_x[(2, \infty)]$. It is easy to obtain that the distribution function is $H(x) = 1 - 2/x$, if $x \ge 2$. For any $U \sim U(0, 1)$ it then follows that $X = 2/(1 - U)$ has density h and a sample $X_1, \ldots, X_n$ from h can easily be obtained. The importance sampling estimator of θ is given by

$$\tilde{\theta} = \frac{1}{n} \sum_{i=1}^{n} w(X_i), \; \text{where } w(x) = \frac{1}{2\pi} \frac{x^2}{1 + x^2}.$$

It can be shown that the variance of this estimator is smaller than that of $\hat{\theta}$ (see Exercise 5.18).

The multivariate extension is again trivial. Assume the interest is to obtain the expectation value of g with respect to p given by

$$I = \int g(\mathbf{x})p(\mathbf{x})d\mathbf{x}.$$

Using the multivariate density h as the importance sampling density, the importance sampling Monte Carlo estimator is given by

$$\hat{I} = \frac{1}{n} \sum_{i=1}^{n} \sum_{i=1}^{n} g(\mathbf{X}_i) w(\mathbf{X}_i),$$

where $w(\mathbf{X}_i) = p(\mathbf{X}_i)/h(\mathbf{X}_i)$ and $\mathbf{X}_i \sim h(.)$, $i = 1, \ldots, n$.

Example 5.20 *Suppose we want to calculate the value of $P(X > Y)$, where $X \sim Exp(1)$ and $Y|X = x \sim Normal(x, 1)$. So, the joint density of X and Y is given by*

$$p(x, y) = \exp(x) \frac{1}{\sqrt{2\pi}} \exp\left\{ \frac{1}{2}(y - x)^2 \right\} I_{(0,\infty)}(x)$$

and the integral of interest is

$$I = \int_0^\infty \int_{-\infty}^\infty g(x, y) p(x, y) \, dx dy,$$

where $g(x, y) = I(x > y)$.

So, if we consider $h(x, y) = \exp(x) I(x > 0)(1/\sqrt{2\pi}) \exp(-y^2/2)$, i.e., a product of independent exponential and standard Normal densities, which is very easy to be simulated from, we have the following estimator

$$\hat{I} = \frac{1}{n} \sum_{i=1}^{n} g(X_i, Y_i) \frac{p(X_i, Y_i)}{h(X_i, Y_i)} = \frac{1}{n} \sum_{i=1}^{n} \exp\left\{ \frac{2X_i Y_i - X_i^2}{2} \right\} I(X_i > Y_i),$$

where $X_1, \ldots, X_n$ is a random sample of the $Exp(1)$ distribution, and $Y_1, \ldots, Y_n$ is a random sample of the $Normal(0, 1)$ distribution.

Therefore, the Monte Carlo algorithm can be used to solve any of the basic inference problems cited in the introduction of this chapter that can be written as expectations. In the Bayesian case, when one wants to evaluate $E[g(\boldsymbol{\theta})|\mathbf{x}]$, the algorithm can be summarized as follows:

1. Generate $\boldsymbol{\theta}_1, \ldots, \boldsymbol{\theta}_n$ from the posterior density $p(\theta|\mathbf{x})$ (or the importance density $h(\boldsymbol{\theta})$).

2. Calculate $g_i = g(\boldsymbol{\theta}_i)$ (or $g_i = g(\boldsymbol{\theta}_i)p(\boldsymbol{\theta}_i|\mathbf{x})/h(\boldsymbol{\theta}_i)$), $i = 1, \ldots, n$.

3. Obtain the estimator $\hat{E}[g(\boldsymbol{\theta})] = (1/n) \sum_{i=1}^{n} g_i$.

Example 5.21 *Suppose $X_1, \ldots, X_n$ is a random sample of a $Bin(m, \theta)$, where m is known, and $\theta \sim U(0, 1)$. So the posterior distribution is $\theta|\mathbf{x} \sim Beta(a, b)$ and the posterior mean is $a/(a + b)$, where $a = 1 + \sum_{i=1}^{n} X_i$ and $b = 1 + n - \sum_{i=1}^{n} X_i$.*

Suppose interest is focused on calculating the posterior variance numerically using a $U(0,1)$ importance sampling density. So the function of interest is

$$g(\theta) = \left(\theta - \frac{a}{a+b} \right)^2$$

and the g_i's values are given by

$$g_i = \left(\theta_i - \frac{a}{a+b} \right)^2 \frac{\theta_i^{a-1}(1-\theta_i)^{b-1}}{B(a,b)}, \qquad i = 1, \ldots, k,$$

where $\theta_1, \ldots, \theta_k$ is a random sample of the $U(0,1)$ and $B(a,b)$ is the Beta function.

5.6.3 Resampling methods

In this section we will be concerned with sampling and resampling techniques from a classical and a Bayesian point of view. First, the classical bootstrap, which essentially consists of resampling from the empirical distribution of a statistic, will be presented. A weighted version of the bootstrap will be useful to implement the Bayesian argument. Intuitively the argument proceeds as follows: a sample is generated from the prior distribution and a resample is taken using some likelihood-based weights. It will be shown that the points in the resample constitute an approximate sample from the posterior distribution. This approximation becomes better as sample sizes increase. Another classical resampling technique, named jackknife, will be presented and exemplified. Its Bayesian version will be developed for the exponential family. A general account of these resampling methods from a classical perspective is provided by Davison and Hinkley (1997) and Efron (1982).

The main objective of jackknife and bootstrap is to obtain a measure of the accuracy of some complex statistics. From a classical point of view the question arises from the fact that the sampling distribution is hard to be determined in many cases. As some examples we can mention robust statistics, such as trimmed means, the correlation coefficient, concordance measures in probabilistic classification and so on.

On the Bayesian side the interest in techniques such as jackknife and bootstrap is slightly different. One important application of leave-one-out methods is to obtain information about the influence of particular observations or, more generally, to define diagnostic measures. One may also use the bootstrap as a resampling technique useful to implement the Bayesian paradigm, as will be shown later in this section.

This class of procedures is characterized by its computational demand, although they are often very easy to implement even in complex situations or high dimensional problems.

5.6.3.1 Jackknife

The jackknife is a useful technique to build confidence intervals, and it works also as a bias reduction technique, as will be illustrated in an example. This is a generic tool and so in specific problems it could provide less accurate results. The basic idea of splitting the sample was introduced by Quenouille (1949, 1956) to eliminate estimation bias.

Suppose that $X_1, \ldots, X_n$ is a random sample from $p(x|\theta)$ and that $\hat{\theta}(\mathbf{X})$ is an estimator of θ. Denote by $\hat{\theta}_i$ the estimator based on the original sample without the ith observation. Let $\tilde{\theta}_i = n\hat{\theta} - (n-1)\hat{\theta}_i$ be a sequence of pseudo-values and define the jackknife estimator of θ as

$$\hat{\theta}_J = \frac{1}{n} \sum_{i=1}^{n} \tilde{\theta}_i.$$

The name *pseudo-value* derives from the fact that for the special case where $\hat{\theta}(\mathbf{X}) = \bar{X}$, the pseudo-value coincides with the ith observation, that is, $\tilde{\theta}_i = \sum_1^n X_j - \sum_{j \neq i}^n X_j = X_i$. It is not difficult to show that $\hat{\theta}_J$ is unbiased if $\hat{\theta}$ and $\hat{\theta}_i$ are also unbiased. Besides that, the jackknife estimator has the property of eliminating terms of order $1/n$ on the bias of the estimator.

Example 5.22 Let $X_1, \ldots, X_n$ be iid observations from the uniform distribution on $(0, \theta)$. It is well known that $T = \max_i X_i$ is a sufficient statistic for
θ with $E(T) = (1 - 1/n)\theta$, $\forall \theta$. A jackknife estimator is given by

$$\hat{\theta}_J = \frac{1}{n} \sum_{i=1}^{n} \tilde{\theta}_i, \;\; where \; \tilde{\theta}_i = nT - (n-1)\hat{\theta}_i,$$

and $\hat{\theta}_i = \max\{X_1, \ldots, X_{i-1}, X_{i+1}, \ldots, X_n\}$. Then, $E(\hat{\theta}_J) = n(1 - 1/n)\theta - (n-1)[1 - 1/(n-1)]\theta = \theta$.

Let $\tilde{\theta}_i$, $i = 1, \ldots, n$ represent random variables approximately independent and identically distributed with mean θ. A jackknife estimate of the sample variance will be given by

$$\hat{\sigma}_J^2 = \frac{1}{n-1} \sum_{i=1}^{n} (\tilde{\theta}_i - \hat{\theta}_J)^2$$

and therefore the statistic

$$\frac{\hat{\theta}_J - \theta}{(\hat{\sigma}_J^2/n)^{1/2}}$$

has approximately a standard Student-t distribution with $n - 1$ degrees of freedom. An approximate $100(1 - \alpha)\%$ confidence interval for θ is given by

$$\left(\tilde{\theta}_J - t_{n-1,\alpha/2} \, \frac{\hat{\sigma}_J}{n^{1/2}} \, , \, \tilde{\theta}_J + t_{n-1,\alpha/2} \, \frac{\hat{\sigma}_J}{n^{1/2}} \right).$$

Example 5.23 *(Correlation coefficient) A very popular data set in statistics was given by Fisher (1936) and contains measurements on three species of iris. We concentrate on the sepal length and sepal width of the specie iris setosa. The correlation coefficient of length and width is calculated in a sample of 50 observations. The sample correlation coefficient is estimated as $\hat{\rho} = 0.742$. Using the jackknife methodology we obtain the point estimation 0.743, and the 95% confidence interval based on the Student-t distribution was $(0.63, 0.84)$.*

The jackknife is connected to cross-validation which is a general technique for assessing accuracy, measuring the predictive performance of a statistical procedure or model. The cross-validation basic idea is to take two random subsets of the data, fit a model or apply a statistical procedure to the first subset and then test it on the second subset, repeating the procedure several times. For a data of size n, the general case is to take a random subset of $n - k$ elements, which is sometimes referred to as leave-k-out, while a extreme case is to leave-one-out, which has also been called jackknife.

From a Bayesian perspective the notion of jackknife corresponds to obtaining the posterior or predictive distribution leaving one of the observations out and is very useful for model checking. For example, the influence of an observation can be assessed by this procedure. Using a divergence measure such as Kullback-Liebler, the posterior or predictive distributions based on the whole sample could be compared with the distribution based on leaving one observation out to evaluate its influence in the analysis.

For the exponential family with one parameter, defined in Chapter 2, the conjugate analysis leads to the posterior density

$$p(\theta|\mathbf{x}) \propto \exp\{\alpha_1 \phi(\theta) + \beta_1 b(\theta)\},$$

where $\alpha_1 = \alpha + T(\mathbf{x})$, $\beta_1 = \beta + n$ and $T(\mathbf{x}) = \sum_{i=1}^{n} u(x_i)$ and $p(y|\mathbf{x}) = a(y)k(\alpha_1, \beta_1)/k(\alpha_1 + u(y), \beta_1 + n + 1)$. So, leaving one observation out would make the posterior density

$$p(\theta|\mathbf{x}_i) \propto \exp\{\alpha_1' \phi(\theta) + \beta_1' b(\theta)\},$$

where $\alpha_1' = \alpha + T(\mathbf{x}_i)$, $\beta_1' = \beta + n - 1 = \beta_1 - 1$, $T(\mathbf{x}_i) = \sum_{j \neq i}^{n} u(x_j)$.

5.6.3.2   Bootstrap

The concept of bootstrap was introduced by Efron (1979). The method of bootstrap consists of generating a large number of samples based on the empirical distribution obtained from the original sampled data. Confidence intervals with some pre-specified coverage probability can be built up easily under mild assumptions.

Let $X_1, \ldots, X_n$ be the observed data from a random sample of a distribution $p(x|\theta)$; $\theta \in \Theta$ is the unknown parameter. Let $\hat{\theta}(\mathbf{x})$ be an estimator of θ. The empirical distribution function is defined by $\hat{F}_n(x) = (1/n)\#(X_i \leq x)$, $\forall x \in R$, as seen in Chapter 4.

A resample procedure consists of the selection of samples with replacement from a finite population using equal probability. This corresponds to selecting a sample from the empirical distribution $\hat{F}_n(\mathbf{x})$. These sampled values will be denoted by $\{X_1^*, \ldots, X_n^*\}$ and the bootstrap estimator of θ by $\hat{\theta}^*(\mathbf{x}^*)$. The inference will be based on B replications of the above procedure and in the evaluation of the statistic of interest, in this case the estimator $\hat{\theta}^* = \hat{\theta}^*(\mathbf{x}^*)$ for each of the B replications. Denote the resulting values by $\hat{\theta}_1^*, \ldots, \hat{\theta}_B^*$. The bootstrap distribution of $\hat{\theta}^*$ is given by the empirical distribution formed by the resampled values. Summarizing, the bootstrap distribution of the $\hat{\theta}^*$ is used in order to make the inferences about θ as a replacement for the sampling distribution of $\hat{\theta}$.

A central assumption in the method is that $\hat{F}_n$ is a good approximation of F, that is, the bootstrap distribution of $\hat{\theta}^*$ is similar to that of $\hat{\theta}$ or that the distribution of $\hat{\theta}^* - \tilde{\theta}$ is similar to that of $\hat{\theta} - \theta$, where $\tilde{\theta}$ is the value of the corrssponding parameter for $\hat{F}_n$.

The mean and variance of these B replications will be denoted by $\bar{\theta}^* = (1/n)\sum_{i=1}^{B} \hat{\theta}_i^*$ and $\hat{\sigma}^2(\hat{\theta}^*) = [1/(B-1)]\sum_{i=1}^{B}(\hat{\theta}_i^* - \bar{\theta}^*)^2$. From the above suppositions it follows that $Var(\hat{\theta}) \simeq Var(\hat{\theta}^*) \simeq \hat{\sigma}^2(\hat{\theta}^*)$ and $E[\hat{\theta} - \theta] = \bar{\theta}^* - \tilde{\theta}$. A bias adjusted estimate of θ will be $\hat{\hat{\theta}} = \hat{\theta} - [\bar{\theta}^* - \tilde{\theta}]$.

Confidence intervals for θ can be built from the percentiles of the bootstrap distribution. Let θ_γ^* be the $100\gamma\%$ percentile of the bootstrap distribution of $\hat{\theta}^*$, that is, $P(\hat{\theta}^* \leq \theta_\gamma^*) = \alpha$. The interval $(\theta_{\alpha/2}^*, \theta_{1-\alpha/2}^*)$ obtained as described earlier is named the $100(1-\alpha)\%$ bootstrap confidence interval.

Example 5.24 *Let $\mathbf{X} = (X_1, \ldots, X_n)$ be a random sample from the $Pois(\theta)$ distribution, with $n = 250$ and $\theta = 3$. Figure 5.4 shows the exact and the bootstrap distributions for $\hat{\theta}$ considering $B = 1,000$.*

Example 5.25 *(Example 5.23 continued) Returning to the evaluation of the correlation between length and width of samples of iris setosa and applying the*

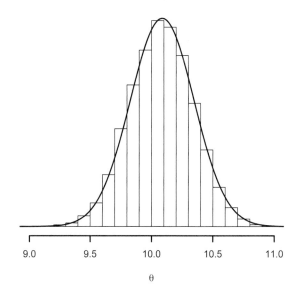

Figure 5.4: *Exact and bootstrap empirical distributions for* $\hat{\theta}$.

bootstrap with values of B equal to $100, 400$ *and* $1,600$ *give* $(L_{5\%}, mean, U_{95\%})$
equal to $(0.64, 0.736, 0.81)$, $(0.71, 0.741, 0.82)$ *and* $(0.71, 0.742, 0.82)$. *It seems*
that $n = 400$ *is a reasonable number of replications to accurately describe*
the bootstrap distribution. It is interesting to note that this is the number of
replications usually recommended in the literature. The point estimates for the
bootstrap and jackknife are almost the same although the confidence intervals
are shorter for the bootstrap.

Overall, while the jackknife uses n new data sets, each containing all the
original data points but leaving-one-out, the bootstrap considers B data sets
each containing n points obtained by random sampling of the original set
of n points. Although the jackknife is a useful tool for estimating standard
errors of estimators, according to Efron (1982), the bootstrap estimate of
standard errors is superior, in the sense that the jackknife may provide an
inconsistent estimate. This is the case of the sample median. Moreover, the
jackknife estimate can be seen as an approximation to the bootstrap, which
is therefore preferred.

5.6.3.3 *Weighted bootstrap*

Sometimes we are not able to sample directly from the distribution of interest, $p(x)$. A useful strategy is to sample from an approximation of this distribution and use the rejection sampling scheme:

1. Generate x from an auxiliary density $h(x)$.

2. Generate u independently from a uniform distribution on $(0, 1)$.

3. If $u \leq \frac{p(x)}{Ah(x)}$, accept x; otherwise return to step 1, where $A = \max p(x)/h(x)$.

The probability of accepting a value x generated from $h(x)$ is

$$P(\text{accept } x) = \frac{1}{A} \int \int I_u \left(0, \frac{p(x)}{Ah(x)} \right) h(x) \, dx du.$$

The expected number of accepted values in n independent runs of the algorithm will be n/A. So, the algorithm is improved by decreasing the value of A as much as possible.

If the determination of A is difficult, the following modification of the algorithm can be applied:

1. Take a sample from $x_1, \ldots, x_n$ from $h(x)$.

2. Evaluate the weights $w(x_i) = p(x_i)/h(x_i)$, $i = 1, \ldots, n$.

3. Select a new sample $x_1^*, x_2^*, \ldots, x_m^*$ from the set $\{x_1, \ldots, x_n\}$ with respective probabilities given by $w_i / \sum_{i=1}^n w_i$, $i = 1, \ldots, n$ with replacement.

Note that

$$P(x^* \leq a) = \sum_{i=1}^n \frac{w_i}{\sum_{j=1}^n w_j} I_{x_i}(\infty, a).$$

Taking the limit as $n \to \infty$,

$$\sum_{i=1}^n \frac{w_i}{\sum_{j=1}^n w_j} I_{x_i} \to \int_\infty^a p(x) dx.$$

It is interesting to note that the algorithm allows approximate sampling from $p(x)$ even when p is known only up to an arbitrary constant. This is particularly useful for Bayesian inference where in many cases the proportionality constant of the posterior distribution is not known.

The above sampling algorithm is known in the literature as the weighted bootstrap. Again, many questions deserve consideration in the applications. For example, how big must n, the initial sample size, be? And m, the resampling size? Is this approximation efficient? Note that if values of x were not generated in some regions, then these values would never be resampled even if the weights were large.

We shall concentrate here on a modification of the algorithm to solve Bayes theorem numerically. Remember that

$$p(\theta \mid \mathbf{x}) = kp(\theta)l(\theta; \mathbf{x}), \ \theta \in \Theta,$$

where $p(\theta)$ is the prior distribution, $l(\theta; \mathbf{x})$ is the likelihood function and $\mathbf{x}$ denotes the available data.

Taking in the algorithm $h(x) = p(\theta)$ gives $w(x) = p(\theta \mid \mathbf{x})p(\theta) = kl(\theta; \mathbf{x})$. Therefore, the algorithm simplifies to the following:

1. Take a sample $\theta_1, \ldots, \theta_n$ from the prior distribution $p(\theta)$.

2. Evaluate the weights $w_i = p(\theta \mid \mathbf{x})/p(\theta) = kl(\theta; \mathbf{x})$, $i = 1, \ldots, n$.

3. Sample $\theta_1^*, \theta_2^*, \ldots, \theta_m^*$ with replacement from the finite population $\{\theta_1, \ldots, \theta_n\}$ with respective weights $l_i / \sum_{i=1}^n l_i$, where $l_i = l(\theta_i; \mathbf{x})$, $i = 1, \ldots, n$.

It is worth noting that often it is necessary to make some adjustments to take care of some numerical difficulties. One must make sure to sample over the relevant region of the parameter space. It may well be that the prior is concentrated over a region of low posterior probability. In this case, the prior is not a suitable candidate for initial sampling and other distributions must be used.

Example 5.26 *Let $Y_i \sim N(\mu_i, \sigma^2)$, where $\mu_i = \beta x_i$ and $\sigma^2 = 1$, $i = 1, \ldots, 5$. Using the fairly vague $N(0, 4)$ prior distribution for β, an adaptation of Theorem 2.1 gives the posterior*

$$\beta \mid \boldsymbol{x}, \boldsymbol{y} \sim N \left(\frac{\sum x_i y_i}{\sum x_i^2 + 1/4}, \frac{1}{\sum x_i^2 + 1/4} \right).$$

If the observed data is $\boldsymbol{y} = (-2, 0, 0, 0, 2)$ and $\boldsymbol{x} = (-2, -1, 0, 1, 2)$, the posterior mean and variance are given by 0.78 and 0.097. Simulating $n = 1000$ samples and $m = 500$ resamples, we can easily obtain numerical summaries. The estimated mean is 0.801 and the estimated variance is 0.087, which is in complete agreement with the above results and with the Bayesian bootstrap resample depicted in Figure 5.5. Note that the range of relevant values is considerably shortened after moving from the prior to the posterior distribution.

5.6.4 Markov chain Monte Carlo methods

The central idea behind the Markov chain Monte Carlo (MCMC, in short) method is to build up a Markov chain from which is easy to simulate and

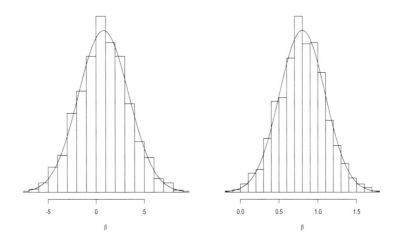

Figure 5.5 *Summary of inference for β in Example 5.26: Left panel: initial sample; right panel: final resample. The true prior and posterior densities are also shown.*

that has equilibrium distribution given by the distribution of interest. These techniques often are more powerful than the quadrature rules and simple Monte Carlo because they can be successfully applied to highly dimensional problems. A general discussion about this topic can be found in Gamerman and Lopes (2006).

Let $X_1, \ldots, X_p$ have the joint density $p(\mathbf{x}) = p(x_1, \ldots, x_p)$ defined in the space $\mathcal{X} \subset R^p$. In fact, derivations below are also valid for the more general case where the X_i's are higher-dimension variables (vectors, matrices, ...). Suppose that a homogeneous, irreducible and aperiodic Markov chain with state-space $\mathcal{X}$ and equilibrium distribution $p(\mathbf{x})$ can be constructed. Denote by $q(\mathbf{x}, \mathbf{y})$ the transition kernel of the chain, which means that $q(\mathbf{x}, \cdot)$ defines a conditional distribution governing the transitions from state $\mathbf{x}$.

In other words, it is possible to build a chain with transition probabilities invariant on time, where each state can be reached from any other state with a finite number of iterations and also without absorbing states. Assume further that it is easy to generate values from these transition probabilities. This means that for any given initial stage, a trajectory from the chain can be generated. For a sufficiently large number of iterations, this trajectory will eventually produce draws from the equilibrium distribution $p(\mathbf{x})$. By constructing a suitable Markov chain, one is able to perform a Monte Carlo simulation of values from p, hence the name MCMC.

There are many possible ways to construct such a chain. One scheme is provided by the Gibbs sampler algorithm, proposed by Geman and Geman (1984) and popularized to the statistical community by Gelfand and Smith (1990). Let $p_i(x_i|\mathbf{x}_{-i})$ denote the conditional density function of X_i given values of all the other X_j's ($j \neq i$) and assume that it is possible to generate from these distribution for each $i = 1, \ldots, p$. The algorithm starts by arbitrarily choosing initial values $\mathbf{x}^0 = (x_1^0, \ldots, x_p^0)$. If in the jth iteration we have the chain at state $\mathbf{x}^{(j)}$, then the position of the chain at iteration $j + 1$ will be denoted by $\mathbf{x}^{(j+1)}$ and will be given after

- generating a random quantity $x_1^{(j+1)}$ from $p_1(x_1|(x_2^{(j)}, \ldots, x_p^{(j)}))$;
- generating a random quantity $x_2^{(j+1)}$ from $p_2(x_2|(x_1^{(j+1)}, x_3^{(j)}, \ldots, x_p^{(j)}))$;
- successively repeating the procedure for $i = 3, \ldots, p$ where at the last step a random quantity $x_p^{(j+1)}$ is generated from $p_p(x_p|(x_1^{(j+1)}, \ldots, x_{p-1}^{(j+1)}))$.

This way, a vector $\mathbf{x}^{(j+1)} = (x_1^{(j+1)}, \ldots, x_p^{(j+1)})$ is formed. Under suitable regularity conditions the limiting distribution of $\mathbf{x}^{(j)}$, as $j \to \infty$ is just $p(\mathbf{x})$.

Another scheme is provided by the Metropolis-Hastings algorithm initially proposed by Metropolis et al. (1953) and later extended by Hastings (1970). It is based on the same idea of using an auxiliary distribution, previously used for importance sampling, accept-reject schemes and weighted bootstrap. Let $q(\mathbf{x}, \mathbf{y})$ denote an arbitrary transition kernel and assume that at iteration j the chain is at state $\mathbf{x}^{(j)}$. Then, the position of the chain at iteration $j + 1$ will be denoted by $\mathbf{x}^{(j+1)}$ and will be given after

- proposing a move to $\mathbf{x}^*$ according to $q(\mathbf{x}^{(j)}, \cdot)$;
- accepting the proposed move with probability

$$\alpha(\mathbf{x}^{(j)}, \mathbf{x}^*) = \min\left\{1, \frac{p(\mathbf{x}^*)/q(\mathbf{x}^{(j)}, \mathbf{x}^*)}{p(\mathbf{x}^{(j)})/q(\mathbf{x}^*, \mathbf{x}^{(j)})}\right\}$$

and thus setting $\mathbf{x}^{(j+1)} = \mathbf{x}^*$, or rejecting the move with probability $1 - \alpha(\mathbf{x}^{(j)}, \mathbf{x}^*)$, thus setting $\mathbf{x}^{(j+1)} = \mathbf{x}^{(j)}$ otherwise.

It is not difficult to show that the Metropolis-Hastings chain has equilibrium distribution given by $p(\mathbf{x})$. Note that the move is made in block for all model parameters. In practice, with highly dimensional models, it is very difficult to find suitable kernels q^* for such spaces and at the same time ensure

large enough acceptance probabilities. A commonly used variation of the algorithm incorporates the blocking strategy used in the Gibbs sampler and per-
forms moves componentwise by defining transition kernels q_i^*, for $i = 1, \ldots, p$.
A transition is then completed after cycling through all p components of $\mathbf{x}$,
and the generations of the components are made according to the Metropolis-
Hastings scheme.

Whatever the scheme used to generate the chain, a stream of values
$\mathbf{x}^{(1)}, \mathbf{x}^{(2)}, \ldots$ is formed. Although consecutive values $\mathbf{x}^{(j)}$ and $\mathbf{x}^{(j+1)}$ are correlated, a random sample of size n from $p(\mathbf{x})$ can be formed by retaining n
successive values after convergence has been ascertained. If approximately independent observations are required, one might hold n values lagged by l units,
for example, $\mathbf{x}^{(m)}, \mathbf{x}^{(m+l)}, \ldots, \mathbf{x}^{(m+(n-1)l)}$, where m is large enough to ensure
convergence has been achieved and l is large enough to carry only residual
correlation over the chain. This will be a random sample of n approximately
iid elements of the joint distribution $p(\mathbf{x})$. This sample is valid for any positive
value of m and l; in particular, $l = 1$ is a common choice since independence
is not really required.

After generating a large random sample, the inference about each x_i can
be done as in any Monte Carlo method. For example, the mean of the ith
component of $\mathbf{x}$ is estimated by $(1/n) \sum_{k=0}^{n-1} x_i^{(m+kl)}$. This idea can be applied
in the Bayesian context to obtain a sample from the posterior distribution of a
parametric vector $\boldsymbol{\theta}$ or in the frequentist context to obtain a sample from the
sampling distribution of an estimator or a test statistic $\mathbf{T}(\mathbf{X})$. Nevertheless,
most of the work and applications in the area are geared towards the Bayesian
approach.

Estimates using the known conditional distribution can also be obtained.
In the case of the mean of X_i, the *Rao-Blackwellized* estimator of $E_p(X_i)$ is

$$\hat{E}_p(X_i) = \frac{1}{n} \sum_{k=0}^{n-1} E_p(X_i | \mathbf{X}_{-i}^{(m+kl)}).$$

This estimator of $E_p(X_i)$ is usually better than $\bar{X}_i$, which is based only on
the generated values. The improvement is justified by a more efficient use
of the (probabilisitic) information available. A very similar idea was used
in the Rao-Blackwell theorem (see Section 4.3) to prove that conditioning
of sufficient statistics improves the estimator, hence the name. It is worth
pointing out that inferences about any quantity related to the X's are easily
done. For example, the mean value of $g(\mathbf{X})$, given by $E_p[g(\mathbf{X})]$, is estimated
as $(1/n) \sum_{k=0}^{n-1} g(\mathbf{x}^{(m+kl)})$, where the $\mathbf{X}^{(j)}$, $\forall j \geq 1$, are the values generated
from the chain.

There are many practical problems that are easily handled by the combination of Bayesian methods and Gibbs sampling or some other MCMC methods, but they are difficult to handle by other means.

Example 5.27 *Let $X_1, \ldots, X_n$ be a random sample from the $N(\theta, \sigma^2)$ distribution and assume independent prior distributions $\theta \sim N(\mu_0, \tau_0^2)$ and $\phi \sim G(n_0/2, n_0 \sigma_0^2/2)$. Note that this distribution is different from the usual conjugate prior used so far in this book but may be a suitable representation of the prior knowledge in some situations. Then the joint posterior is*

$$p(\theta, \phi | \boldsymbol{x}) \propto l(\theta, \sigma^2; \boldsymbol{x}) p(\theta) p(\phi)$$
$$\propto \phi^{n/2} \exp\left\{ -\frac{\phi}{2} \left[ns^2 + n(\overline{x} - \theta)^2 \right] \right\}$$
$$\times \exp\left\{ -\frac{1}{2\tau_0^2}(\theta - \mu_0)^2 \right\} \phi^{n_0/2-1} \exp\left(-\frac{n_0 \sigma_0^2 \phi}{2} \right)$$
$$\propto \phi^{[(n+n_0)/2]-1}$$
$$\times \exp\left\{ -\frac{1}{2} [\phi(n_0 \sigma_0^2 + ns^2 + n(\theta - \bar{x})^2) + \tau_0^{-2}(\theta - \mu_0)^2] \right\}.$$

This distribution has no known form and it is not possible to perform the analytic integration to obtain the proportionality constant. The kernels of the marginal distributions can be obtained but are of no known form which prevents the exact evaluation of their means, variance and so forth. Nevertheless, the conditional posterior distributions of $\theta | \phi$ and $\phi | \theta$ are easy to obtain. They are proportional to the posterior density once the terms that do not depend on the quantity of interest are incorporated to the proportionality constant. So,

$$p(\theta | \phi, \boldsymbol{x}) \propto \exp\left\{ -\frac{1}{2}[n\phi(\theta - \bar{x})^2) + \tau_0^{-2}(\theta - \mu_0)^2] \right\}$$
$$\propto \exp\left\{ -\frac{1}{2} [\theta^2(\tau_0^{-2} + n\phi) - 2\theta(\tau_0^{-2}\mu_0 + n\phi\bar{x})] \right\}, \quad and$$
$$p(\phi | \theta, \boldsymbol{x}) \propto \phi^{[(n+n_0)/2]-1} \exp\left\{ -\frac{\phi}{2}[n_0 \sigma_0^2 + ns^2 + n(\theta - \bar{x})^2] \right\},$$

from where it is clear that $(\theta | \phi, \boldsymbol{x}) \sim N[\mu_1(\phi), \tau_1^2(\phi)]$ and $(\phi | \theta, \boldsymbol{x}) \sim G(n_1/2, n_1 \sigma_1^2(\theta)/2)$, where

$$\mu_1(\phi) = \frac{\tau_0^{-2}\mu_0 + n\phi\bar{x}}{\tau_0^{-2} + n\phi}, \quad \tau_1^2(\phi) = \frac{1}{\tau_0^{-2} + n\phi}, \quad n_1 = n_0 + n$$

and $n_1 \sigma_1^2(\theta) = n_0 \sigma_0^2 + ns^2 + n(\theta - \bar{x})$. Therefore, it is easy to generate from the conditionals, and the Gibbs sampler becomes easy to be implemented.

One difficult problem in the application of MCMC schemes is to ensure the convergence of the chain. Some theoretical results and many diagnostic statistics are available. The practical recommendation is to monitor the trajectory of the chain using output diagnostics. A common approach is to plot the averages of selected quantities, such as the components of the vector $\mathbf{X}$, and assess by visual inspection whether the convergence has occurred using multiple chains generated from different starting values. More formal diagnostic tools have already been derived and should also be used in addition to visual inspection. Convergence of the chain can be slow for many reasons. For example, if the components of $\mathbf{x}$ are highly correlated, or if the joint density has multiple modes with regions of low probability between some of them, then the chain may take a large number of iterations to converge. The interested reader can refer to the books by Gamerman and Lopes (2006) and Gilks et al. (1995) for some theoretical and practical aspects of MCMC methodology.

Example 5.28 *(Example 5.27 continued) Let $X_1, \ldots, X_{20}$ be a random sample of size $n = 20$ from a $N(\theta, \sigma^2)$ distribution, with $\theta = 5$ and $\sigma^2 = 2$. Moreover, assume independent prior distributions $\theta \sim N(0, \tau_0^2)$ and $\phi \sim G(a_0/2, b_0/2)$, where τ_0^2, a_0 and b_0 are known values with $a_0 = n_0$ and $b_0 = n_0 \sigma_0^2$. We can obtain a sample of joint posterior by repeatedly simulating values from the full conditional distributions of θ and σ^2, given in Example 5.27.*

The mean posterior estimates of θ and σ^2 with their respective standard deviations in parentheses are given by 5.132 (0.3233) and 2.114 (0.7480), based on 2,000 iterations of the Gibbs sampling method but discarding the first 100 simulation results, which are taken as the burn-in sample. Figure 5.6 shows the estimates of the marginal posterior distributions of θ and σ^2.

Example 5.29 *(Example 3.14 continued) Consider the model*

$$
\begin{aligned}
X_i \mid \theta_i &\sim N(\theta_i, 1) \\
\theta_i \mid \mu, \tau^2 &\sim N(\mu, \tau^2), \quad for \quad i = 1, \ldots, k.
\end{aligned}
$$

The above model favors similarities between the means by pooling them into the same prior distribution. To complete the model we must specify prior distributions for parameters of the last hierarchical level, generally vague priors. Thus, the prior distributions used are $\mu \sim N(0, c)$, with large values for c, and $\tau^2 \sim U[0, A]$, with, for example, $c = 1.0 \times 10^{12}$ and $A = 100$. These choices allow us to state that our estimates will be based mainly on data information.

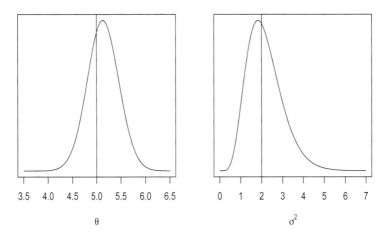

Figure 5.6 *Marginal posterior distributions of the parameters θ and σ^2, with vertical lines indicating the true parameter values.*

The chain trajectories and posterior histograms, obtained via Gibbs sampler in the transformed scale for the first level parameter θ_1 and hyperparameters μ and τ^2 under the hierarchical model, are presented in Figure 5.7. It can be seen that the chains mix very well and indicate convergence of the process. These results were obtained with the software WinBUGS (for details, see Gilks, Richardson and Spiegelhalter (1995)) and are based on two chains of 70,000 iterations, excluding the first 10,000 as burn-in and at intervals of 10 iterations to mitigate sampling autocorrelations. This procedure was adopted for all the results obtained via the Gibbs sampler in this example.

Table 5.3 presents the maximum likelihood estimates $\hat{p}_i = y_i$, the true batting ratio of the season, the James-Stein estimator $\hat{p}_i^{(EB)}$ and the Bayes estimator $\hat{p}_i^B$ obtained with the full Bayesian model above. We use the inverse transformation $y_i = [\sin(x_i n^{-1/2}) + 1]/2$ in all estimates. The effect of the hierarchical model is to shrink the estimates towards a common mean. Moreover, all Bayes and empirical Bayes estimates are closer to the true batting ratios in the season than the maximum likelihood estimates, except for Players 1 and 15 and, to a lesser extent, for the Player 10.

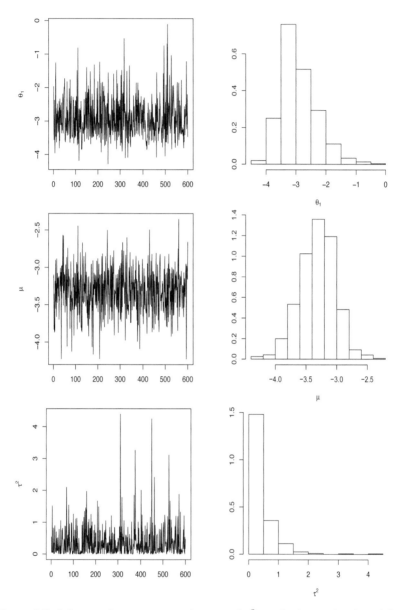

Figure 5.7 *Inference for parameters* θ_1, μ *and* τ^2 *in the hierarchical model of Example 5.29. Left column: trajectories generated via Gibbs sampler; right column: histograms of the marginal posterior distributions.*

Table 5.3: *Estimation Summary for Example 5.29 (1970 Batting Averages).*

Player	$\hat{p}_i$	p_i	$\hat{p}^{EB}$	$\hat{p}_i^B$
1	0.400	0.346	0.290	0.291
2	0.378	0.298	0.296	0.286
3	0.356	0.276	0.282	0.282
4	0.333	0.222	0.277	0.278
5	0.311	0.273	0.273	0.272
6	0.311	0.270	0.273	0.273
7	0.289	0.263	0.268	0.268
8	0.267	0.210	0.264	0.264
9	0.244	0.269	0.259	0.259
10	0.244	0.230	0.259	0.259
11	0.222	0.264	0.254	0.255
12	0.222	0.256	0.254	0.254
13	0.222	0.303	0.254	0.254
14	0.222	0.264	0.254	0.254
15	0.222	0.226	0.254	0.255
16	0.200	0.285	0.249	0.249
17	0.178	0.316	0.244	0.244
18	0.156	0.200	0.239	0.238

Exercises

§5.2

1. Consider the genetic application of Section 5.2 where a 4-dimensional vector of counts $\mathbf{x} = (X_1, X_2, X_3, X_4)$ has multinomial distribution with parameters n and $\boldsymbol{\pi}$, where $\boldsymbol{\pi} = (1/2 + \theta/4, (1-\theta)/4, (1-\theta)/4, \theta/4)$. Assume that the observed data was (125, 18, 20, 34).

   (a) Obtain the equations required for calculation of the maximum likelihood estimator via the Newton-Raphson algorithm and apply them to obtain the maximum likelihood estimate for the given dataset.

   (b) Obtain the equations required for calculation of the maximum likelihood estimator via the Fisher scoring algorithm and apply them to obtain the maximum likelihood estimate for the given dataset.

   (c) Use the expressions given for the successive iterates in the EM algorithm to show that the likelihood is monotonically increasing through the steps of the algorithm.

(d) Compare the three different algorithms for finding the maximum likelihood estimator in terms of computational complexity and time.

(e) Assume now a prior $\theta \sim Beta(a, b)$. Repeat the exercise to obtain the
(generalized) maximum likelihood estimator. Specify numerical values
for a and b and obtain the corresponding posterior modes.

2. Consider a random sample $X_1, \ldots, X_n$ from the $G(\alpha, \beta)$ distribution with
both parameters unknown. Obtain the maximum likelihood equations and
describe an iterative scheme to obtain the maximum likelihood estimator
of α and β.

3. Consider the randomized response model given in Example 5.4. Show that
the maximum likelihood estimator of θ can be calculated directly using
its invariance properties and evaluate its value with data provided in the
example.

4. Consider the EM algorithm with a sequence of iterated values $\boldsymbol{\theta}^{(j)}$, $j \geq$
1. Show that the sequence satisfies $L(\boldsymbol{\theta}^{(j)}; \mathbf{x}) \leq L(\boldsymbol{\theta}^{(j+1)}; \mathbf{x})$, $\forall j$, and is
therefore monotonically increasing in the likelihood $l(\boldsymbol{\theta}; \mathbf{x})$.

5. Consider a random sample $Y_1, \ldots, Y_n$ from the Poisson-exponential distribution (see Example 5.5) with scale parameter $\lambda > 0$, shape parameter
$\theta > 0$ and density given by

$$f(y) = \frac{\theta \lambda e^{-\lambda y - \theta e^{-\lambda y}}}{1 - e^{-\theta}}, \quad y > 0.$$

(a) Show that the maximum likelihood estimator of θ is given by $\hat{\theta} = (\sum_{i=1}^{n} y_i - n\hat{\lambda}^{-1})/\sum_{i=1}^{n} y_i e^{-\hat{\lambda} y_i}$, where $\hat{\lambda}$ is the maximum likelihood estimator for λ.

(b) Derive the equations for the EM algorithm presented in Example 5.5.

§5.3

6. Let $\mathbf{X}_n = (X_1, \ldots, X_n)$ be a random sample from the $N(0, \theta^2)$ distribution.

(a) Obtain the asymptotic posterior distribution of θ as $n \to \infty$.

(b) Obtain the asymptotic posterior mean and variance of θ^2.
Hint: $X \sim N(0, 1) \Rightarrow X^2 \sim \chi_1^2$.

(c) Obtain the asymptotic posterior distribution of θ^2 based on the delta
method and compare its mean and variance with the results obtained in
(b).

7. Let $X \sim Bin(20, \theta)$ and assume that $X = 7$ was observed. Obtain a 90%
credibility interval for θ using a uniform prior and

(a) the fact that if $z \sim Beta(a, b)$, then

$$\frac{b}{a} \frac{z}{1 - z} \sim F(2a, 2b).$$

(b) an asymptotic approximation for $\psi = \theta/1 - \theta$.

(c) an asymptotic approximation for $\phi = \sin^{-1}(\sqrt{\theta})$.

(d) Compare the results.

8. Let $\mathbf{X}_n = (X_1, \ldots, X_n)$ be independent random variables where $X_i \sim Pois(\theta t_i)$, $i = 1, \ldots, n$ and $t_1, \ldots, t_n$ are known times.

(a) Prove that the maximum likelihood estimator of θ is $\hat{\theta} = \overline{X}/\overline{t}$, where $\overline{X} = \sum_{i=1}^{n} X_i/n$ and $\overline{t} = \sum_{i=1}^{n} t_i/n$.

(b) Obtain the asymptotic posterior distribution of $\theta|\mathbf{x}_n$ and construct an asymptotic $100(1-\alpha)\%$ credibility interval for θ assuming that n is large.

(c) Obtain the asymptotic posterior distribution of $\theta^{1/2}|\mathbf{x}_n$ and, based on it, construct an asymptotic $100(1 - \alpha)\%$ confidence interval for θ assuming that n is large.

(d) Compare the confidence intervals obtained in (b) and (c), especially with respect to their lengths.

9. Let $X_1, \ldots, X_n$ be a random sample from the distribution with density

$$f(x \mid \theta) = \theta x^{\theta - 1} I_x([0, 1]).$$

(a) Verify which function(s) of θ (up to linear transformations) can be estimated with highest efficiency and determine their corresponding estimators.

(b) Obtain the asymptotic $100(1 - \alpha)\%$ confidence interval for θ basing calculations on the asymptotic distribution of the score function $U(\mathbf{X}; \theta)$.

(c) Repeat item (b) basing calculations now on the central limit theorem applied to the sample mean $\overline{X}$.

(d) Obtain the asymptotic $100(1 - \alpha)\%$ credibility interval for θ based on approximations for the posterior distribution of θ.

10. Let $X_1, \ldots, X_n$ be a random sample from the $Pois(\theta)$ distribution and define $\lambda = \theta^{1/a}$, $a \neq 0$.

(a) Obtain the likelihood function $l(\lambda; \mathbf{X})$.

(b) Obtain Jeffreys non-informative prior for λ.

(c) Obtain the Taylor expansion of $L(\lambda; \mathbf{X}) = \log l(\lambda; \mathbf{X})$ around the maximum likelihood estimator of λ and determine the value(s) of a for which the 3rd order term vanishes.

(d) Discuss the importance of the result obtained in the previous item in terms of asymptotic theory.

11. Let $X_1, \ldots, X_n$ be a random sample from a Gamma distribution with parameters μ and β with probability density function

$$p(x|\mu, \beta) = \frac{1}{x\Gamma(\beta)\mu^\beta}(\beta x)^\beta e^{-\beta x/\mu}, \text{ for } x > 0.$$

(a) Obtain the likelihood function $l(\mu, \beta; \mathbf{X})$.

(b) Prove that $\hat{\mu} = \bar{\mathbf{X}}$ and $\log \hat{\beta} - \psi(\hat{\beta}) = \log(\hat{\mathbf{X}}/\tilde{\mathbf{X}})$, where $\hat{\mathbf{X}}$ and $\tilde{\mathbf{X}}$ are the arithmetic and geometric means, respectively. Determine the function ψ.

(c) Obtain the Taylor expansion of $L(\mu, \beta; \mathbf{X}) = \log l(\mu, \beta; \mathbf{X})$ around the maximum likelihood estimators of μ and β.

(d) Prove that the joint density of the maximum likelihood estimators denoted by $p(\hat{\mu}, \hat{\beta}|\mu, \beta)$ can be decomposed as $p(\hat{\mu}, \hat{\beta}|\mu, \beta) = p(\hat{\mu}|\mu, \beta)p(\hat{\beta}|\beta)$, where

$$p(\hat{\mu}|\mu, \beta) = \frac{1}{\mu\hat{\mu}}(\beta\hat{\mu})^{n\beta} e^{-n\beta\hat{\mu}/\mu}, \text{ for } \hat{\mu} > 0, \text{ and}$$

$$p(\hat{\beta}|\beta) = \frac{\hat{\beta}(\psi'(\hat{\beta}) - 1)^{1/2}}{(\Gamma(\beta)\Gamma(\hat{\beta}))^{-n}} \exp\left\{ n\left(\hat{\beta} - \beta\log(\hat{\beta}) - (\beta - \hat{\beta})\psi(\hat{\beta})\right) \right\},$$

for $\hat{\beta} > 0$.

12. Let $X_1, \ldots, X_n$ be a random sample from the uniform distribution over the interval $[0, \theta]$ and let $\hat{\theta}_n$ be the maximum likelihood estimator of θ.

(a) Obtain a non-degenerate asymptotic distribution for $\hat{\theta}_n$, or in other words, find functions $h(n)$, $a(\theta)$ and $b(\theta)$ and a non-degenerate asymptotic distribution P such that

$$h(n)\frac{\hat{\theta}_n - a(\theta)}{b(\theta)} \xrightarrow{\mathcal{D}} Z \text{ where } Z \sim P \text{ when } n \to \infty.$$

Hint: Use the density of $\hat{\theta}_n$ to obtain the form of h, a and b and use the result $(1 + s/n)^n \to e^s$ when $n \to \infty$ for $s \in R$.

(b) Comment about the convergence rate and compare it against the standard rate $n^{1/2}$ of the maximum likelihood estimator, under regularity conditions.

(c) Obtain the asymptotic $100(1 - \alpha)\%$ confidence interval for θ of smallest length based on the results of item (a).

(d) Show that the posterior distribution of

$$h(n) \frac{\hat{\theta}_n - a(\theta)}{b(\theta)}$$

converges to P where h, a, b and P are the same ones obtained in item (a).

(e) Obtain the asymptotic $100(1-\alpha)\%$ HPD credibility interval for θ. Compare it against the asymptotic $100(1-\alpha)\%$ confidence interval for θ obtained in item (c).

(f) Compare the intervals obtained in items (c) and (e) with the exact interval based on the exact distribution of $\hat{\theta}$, obtained in Exercise 4.30.

13. Consider again the variation of the randomized response model which consists of asking as the alternative question the negation of the original one.

(a) Show that the posterior distribution of θ is a mixture of $n + 1$ Beta distributions.

(b) Obtain the relevant derivatives and points of maxima required for the evaluation of the posterior mean of θ analytically or numerically.

(c) Apply the results of the previous items to compare against the results in Table 5.1 with exact and approximated posterior means for θ.

§5.4

14. Show that for any distribution $p(\theta)$ in the exponential family, the best Normal approximation in the Kullback-Leibler sense has mean and variance given, respectively, by $\mu = E(\theta)$ and $\sigma^2 = V(\theta)$, the mean and variance of the original distribution.

§5.5

15. Apply the Gauss-Hermite integration rules to obtain approximations for the posterior expectation of α and θ, given the observed values already provided in the Weibull example of Section 5.3, and compare the results with the approximations from the Laplace method.

§5.6

16. Use the simple Monte Carlo method to evaluate $\int_{-\infty}^{\infty} e^{-x^2/2} dx$ and compare it with the known answer $\sqrt{2\pi}$. Also, evaluate the variance of the estimator. Hint: Make a transformation to take the line into the interval $[0, 1]$ and then proceed as earlier.

17. Show that if an integral $I = \int g(x)p(x)dx$ is estimated by importance sampling, then its estimator

$$\bar{I} = \frac{1}{n} \sum_{1}^{n} g(x_i)w(x_i), \text{ where } w(x_i) = \frac{p(x_i)}{h(x_i)} \text{ and } x_i \sim h(x), \, i = 1, \ldots, n.$$

is unbiased and has variance given by

$$Var(\bar{I}) = \frac{1}{n} \int [g(x)w(x) - I]^2 h(x)dx.$$

18. Let $\theta = P(X > 2)$, where X has a standard Cauchy distribution with density

$$p(x) = \frac{1}{\pi(1+x^2)}, \, x \in R.$$

Let h be an importance sampling density defined by $h(x) = 2I_x[(2, \infty)]/x^2$. Show that use of this sampling density reduces the variance of the estimator of θ over the simple Monte Carlo estimator.

19. Show that the Rao-Blackwellized estimator of $E_p(X_i)$ given by

$$\hat{E}_p(X_i) = \frac{1}{n} \sum_{k=0}^{n-1} E_p(X_i|\mathbf{X}_{-i}^{(m+kl)})$$

provides an unbiased and consistent estimator of $E_p(X_i)$. Generalize the result to obtain the Rao-Blackwellized estimator of the marginal density of X_i and show that it is also an unbiased and consistent estimator of the marginal density of X_i, for $i = 1, ..., n$.

20. Let $X_1, ..., X_n$ be a random sample from a Poisson distribution with mean that is either θ or ϕ. The mean is θ up to an unknown break point m from where it becomes ϕ.

   (a) Obtain the likelihood of the unknown parameters θ, ϕ and m.

   (b) Suggest a reasonable family of conjugate prior distributions for θ, ϕ and m.

   Hint: To simplify matters, assume independent priors for θ, ϕ and m.

   (c) Obtain the full conditional distributions required for implementation of the Gibbs sampler.

(d) Generate data $(X_1, ..., X_n)$ for given values of θ, ϕ and m and apply the Gibbs sampler to draw inference about them.

21. (Casella and George, 1992) Let π denote the following discrete distribution over $S = \{0, 1\}^2$

$$
\begin{array}{c c c}
 & \multicolumn{2}{c}{X_2} \\
 & 0 & 1 \\
\hline
X_1 \quad 0 & \pi_{00} & \pi_{01} \\
1 & \pi_{10} & \pi_{11} \\
\end{array}
$$

where $\pi_{00} + \pi_{01} + \pi_{10} + \pi_{11} = 1$ and $\pi_{ij} > 0$, for $i, j = 1, 2$. Assume that instead of drawing samples directly from π, one decides to draw values from π through the Gibbs sampler.

(a) Show that the transition probabilities for X_1 are given by the conditional distribution π_1 of $X_1 | X_2 = j$,

$$\pi_1(0|j) = \frac{\pi_{0j}}{\pi_{+j}} \text{ and } \pi_1(1|j) = \frac{\pi_{1j}}{\pi_{+j}}$$

where $\pi_{+j} = \pi_{0j} + \pi_{1j}$, $j = 0, 1$.

(b) Show that the transition probabilities for X_2 are given by the conditional distribution π_2 of $X_2 | X_1 = i$,

$$\pi_2(0|i) = \frac{\pi_{i0}}{\pi_{i+}} \text{ and } \pi_2(1|i) = \frac{\pi_{i1}}{\pi_{i+}}$$

where $\pi_{i+} = \pi_{i0} + \pi_{i1}$, $i = 0, 1$.

(c) Show that the 4×4 transition matrix P of the chain formed by the Gibbs sampler has elements

$$
\begin{aligned}
P((i, j), (k, l)) &= Pr((X_1, X_2)^{(n)} = (k, l) | (X_1, X_2)^{(n-1)} = (i, j)) \\
&= \frac{\pi_{kl}}{\pi_{k+}} \frac{\pi_{kj}}{\pi_{+j}}, \text{ for } (i, j), (k, l) \in S.
\end{aligned}
$$

(d) Show that π is the only stationary distribution of this chain.

(e) Extend the results for cases when X_1 can take n_1 values and X_2 can take n_2 values.

22. Show that the Metropolis-Hastings chain has equilibrium distribution given by $p(\mathbf{x})$.

Chapter 6

Hypothesis testing

In this chapter we still consider statistical problems involving an unknown quantity θ known to belong to a parametric space Θ. In many instances, the inferential process may be summarized in the verification of some assertions or conjectures about θ. For example, one may be interested in verifying whether a coin is fair, if a collection of quantities is independent or if distinct populations are probabilistically equal. Each one of the assertions above constitutes a hypothesis and can be associated with a model. This means that it can be associated with parameter values in some form. Considering the simple case of two alternative hypotheses, two disjoint subsets Θ_0 and Θ_1 belonging to Θ are formed with the corresponding values of θ in each of the 2 hypotheses.

The objective here is to test them, namely to decide which one (if any) of the hypotheses should be accepted. This chapter presents the relevant theory required to adequately address this decision problem. As previously, this task will be performed from both classical and Bayesian points of view.

6.1 Introduction

Denote by H_0 the hypothesis that $\theta \in \Theta_0$ and by H_1 the hypothesis that $\theta \in \Theta_1$. Our statistical problem now is to decide whether H_0 or H_1 is accepted, or in other words, whether θ is in Θ_0 or in Θ_1. If the subset of the parameter space defining a hypothesis contains a single element, the hypothesis is said to be simple. Otherwise, it is said to be composite. Under a simple hypothesis, the observational distribution is completely specified, whereas under a composite hypothesis, it is only specified that the observational distribution belongs to a family. From now on, the hypotheses H_0 and H_1 will be uniquely associated with disjoint subsets Θ_0 and Θ_1 of the parameter space. Whenever they are simple, the notation $\Theta_0 = \{\theta_0\}$ and $\Theta_1 = \{\theta_1\}$ will be used. Note that in some cases, only one of the hypotheses is simple.

Typically, a test of hypotheses is a decision problem with a number of possible actions. If the researcher makes the wrong decision, he incurs a penalty or suffers a loss. Once again, his/her objective is to minimize the loss in some form. For example, under the Bayesian approach he/she would try to minimize the expected loss. A rule to decide the hypothesis to be accepted is called a test procedure or simply a test and will be denoted by ψ. One may define for example that $\psi = i$ if the hypothesis accepted is H_i, $\forall i$.

From the Bayesian perspective, one may have many alternative hypotheses $H_1, \ldots, H_k$ that can be compared through $P(H_i \mid \mathbf{x})$, $i = 1, \ldots, k$. Under the classical perspective, it is important to have only two hypotheses H_0 and H_1 at each time. The theory of this chapter is developed for this case in order to ease comparisons between the two approaches.

Usually there is a hypothesis that is more important. This will be denoted by H_0 and called the null hypothesis. The other hypothesis against which H_0 is tested is denoted by H_1 and called the alternative hypothesis. If H_0 and H_1 exhaust all possibilities, then necessarily one of them must be true. In this case, rejecting one of them necessarily implies accepting the other one.

In this chapter, we start with the presentation of the classical procedures and later present the Bayesian procedures. We then move on to establish a connection between hypothesis tests and confidence intervals. Finally, tests based on asymptotic theory results are described from both classical and Bayesian perspectives. A systematic introductory account of this topic is presented in Bickel and Doksum (1977) and DeGroot (1970). The classical theory is presented at a more formal level in Lehmann (1986).

Assume that the statistician is offered the choice of observing a sample $X_1, \ldots, X_n$ from a distribution that depends on the unknown parameter $\boldsymbol{\theta}$ previously deciding which hypothesis to accept. In a problem of this kind, the statistician can specify a test procedure by splitting the sample space into two subsets. Under the frequentist viewpoint, this is the only available option because the only source of information comes from the data. One subset of the sample space will contain the values of $\mathbf{X}$ that will lead to acceptance of H_0, and the other one will lead to rejection of H_0. This latter set is called the critical region and a test procedure gets completely specified by the critical region. Of course, the complement of the critical region contains sample results that lead to the acceptance of H_0.

6.2   Classical hypothesis testing

The general theory of classical hypothesis testing comes from the pioneering work of Neyman and Pearson (1928). The probabilistic characteristics of a classical test can be described by specification of $\pi(\boldsymbol{\theta})$, the probability that the test leads to the rejection of H_0, for each value of $\boldsymbol{\theta} \in \boldsymbol{\Theta}$. The function π is called the power of the test. If $\mathbf{C}$ is the critical region, then π is defined by

$$\pi(\boldsymbol{\theta}) = P(\mathbf{X} \in C \mid \boldsymbol{\theta}), \quad \forall \boldsymbol{\theta} \in \boldsymbol{\Theta}.$$

Some textbooks define the power function only for $\boldsymbol{\theta} \notin \boldsymbol{\Theta}_0$. The size or significance level α of a test procedure is defined as

$$\alpha \geq \sup_{\boldsymbol{\theta} \in \boldsymbol{\Theta}_0} \pi(\boldsymbol{\theta}).$$

Just as in the case of confidence levels seen in Section 4.4, the inequality above is a technical requirement. It is more useful in discrete sample spaces where not all values in $[0, 1]$ are possible probability values. As will shortly be seen, one wishes to use as small a value for α as possible. In practice, this means that an equality is used.

For any given test procedure ψ, two types of errors can be committed. A type I error is committed when the test indicates rejection of H_0 when it is true. Note that the largest possible value for the probability of this error is α. Similarly, the type II error is committed when the test indicates acceptance of H_0 when it is false. The probability of type II error is usually denoted by β. Note that $\beta(\boldsymbol{\theta}) = 1 - \pi(\boldsymbol{\theta})$, for $\boldsymbol{\theta} \in \boldsymbol{\Theta}_1$. In the case of simple hypotheses, the probability of type I error is $\alpha = \pi(\boldsymbol{\theta}_0)$ and the probability of type II error is $\beta = 1 - \pi(\boldsymbol{\theta}_1)$.

6.2.1 Simple hypotheses

It is useful to start the study of the theory with the case of two simple hypotheses $H_0 : \boldsymbol{\theta} = \boldsymbol{\theta}_0$ and $H_1 : \boldsymbol{\theta} = \boldsymbol{\theta}_1$. Ideally, one wishes to find a test procedure for which the two error probabilities are as small as possible. In practice, it is impossible to find a test for which these probabilities are simultaneously minimized. As an alternative, one may seek to construct a test that minimizes linear combinations of α and β.

Theorem 6.1 *(Optimal test) Assume that* $\mathbf{X} = (X_1, \ldots, X_n)$ *is a random sample from* $p(x|\boldsymbol{\theta})$, $H_0 : \boldsymbol{\theta} = \boldsymbol{\theta}_0$ *and* $H_1 : \boldsymbol{\theta} = \boldsymbol{\theta}_1$. *Let* ψ^* *be a test of* H_0 *versus* H_1 *such that* H_0 *is accepted if* $p_0/p_1 > k$ *and* H_0 *is rejected if* $p_0/p_1 < k$, *where* $p_i = p(\boldsymbol{x} \mid \boldsymbol{\theta}_i)$, $i = 0, 1$ *and* $k > 0$. *(If* $p_0/p_1 = k$, *nothing can be decided.) Then, any other test* ψ *will be such that*

$$a\alpha(\psi^*) + b\beta(\psi^*) \leq a\alpha(\psi) + b\beta(\psi),$$

where $\alpha(\psi)$ and $\beta(\psi)$ respectively denote the probabilities of errors of type I and II of test ψ, for any $a, b \in R^+$.

Proof (Continuous case) Let $\mathbf{C}$ be the critical region of any arbitrary test ψ and define $p_i = p(\mathbf{x} \mid \boldsymbol{\theta}_i)$, $i = 0, 1$. Then, for $a, b \in R^+$,

$$
\begin{aligned}
a\alpha(\psi) + b\beta(\psi) &= a \int_{\mathbf{C}} p(\mathbf{x} \mid \boldsymbol{\theta}_0) \, d\mathbf{x} + b \int_{\overline{\mathbf{C}}} p(\mathbf{x} \mid \boldsymbol{\theta}_1) \, d\mathbf{x} \\
&= a \int_{\mathbf{C}} p(\mathbf{x} \mid \boldsymbol{\theta}_0) \, d\mathbf{x} + b \left[1 - \int_{\mathbf{C}} p(\mathbf{x} \mid \boldsymbol{\theta}_1) \, d\mathbf{x} \right] \\
&= b + \int_{\mathbf{C}} (a p_0 - b p_1) \, d\mathbf{x}.
\end{aligned}
$$

So, minimization of $a\alpha(\psi) + b\beta(\psi)$ is equivalent to choosing the critical region $\mathbf{C}$ in such a way that the value of the integral is minimal. This will occur if the integration is performed over a set that includes every point $\mathbf{x}$ such that $a p_0 - b p_1 < 0$ and that does not include points $\mathbf{x}$ such that $a p_0 - b p_1 > 0$.

Therefore, minimization of $a\alpha(\psi) + b\beta(\psi)$ is achieved by having the critical region $\mathbf{C}$ include only points $\mathbf{x}$ such that $a p_0 - b p_1 < 0$. (If the sampling distribution is continuous and $a p_0 - b p_1 = 0$, this point has 0 contribution and is irrelevant). This completes the demonstration because $a p_0 - b p_1 < 0$ iff $p_0/p_1 < k = b/a$, which corresponds to the description of the test ψ^*.   □

The ratio p_0/p_1 is called the likelihood ratio (LR, in short). The theorem establishes that a test that minimizes $a\alpha(\psi) + b\beta(\psi)$, rejects H_0 when the LR is small and accepts H_0 when the LR is large. Usually, the null hypothesis H_0 and error of type I are privileged. Therefore, one considers only tests ψ such that $\alpha(\psi)$ cannot be larger than a prespecified level α_0 and searches for the test that minimizes $\beta(\psi)$ among them. This is a variation of the problem solved with the theorem and the solution is provided by the following lemma.

Lemma 6.1 *(Neyman-Pearson) Assume that $\mathbf{X} = (X_1, \ldots, X_n)$ is a random sample from $p(x|\boldsymbol{\theta})$, $H_0 : \boldsymbol{\theta} = \boldsymbol{\theta}_0$ and $H_1 : \boldsymbol{\theta} = \boldsymbol{\theta}_1$. Let ψ^* be a test of H_0 versus H_1 such that H_0 is accepted if $p_0/p_1 > k$ and H_0 is rejected if $p_0/p_1 < k$, where $p_i = p(\boldsymbol{x} \mid \boldsymbol{\theta}_i)$, $i = 0, 1$. (If $p_0/p_1 = k$, nothing can be decided.) Then, for any other test ψ such that $\alpha(\psi) \leq \alpha(\psi^*)$, $\beta(\psi) \geq \beta(\psi^*)$. Also, $\alpha(\psi) < \alpha(\psi^*)$ implies $\beta(\psi) > \beta(\psi^*)$.*

Proof Following the definition of the optimal test ψ^* in the theorem, it follows that for any other test ψ

$$
\alpha(\psi^*) + k\beta(\psi^*) \leq \alpha(\psi) + k\beta(\psi),
$$

for $k > 0$. If $\alpha(\psi) \leq \alpha(\psi^*)$, then necessarily $\beta(\psi) \geq \beta(\psi^*)$. Also, if $\alpha(\psi) < \alpha(\psi^*)$, it follows that $\beta(\psi) > \beta(\psi^*)$, completing the demonstration.   □

Special attention must be given to the wording of the lemma. It considers only acceptance or rejection of H_0 with no reference to H_1. This is consistent with the preferential status given to H_0 and also to the fact that H_0 and H_1 do not exhaust the parameter space. This point will be readdressed below.

In the lemma, α_0 plays the role of significance level. Recalling that $\pi(\theta_1) = 1 - \beta(\psi)$, minimization of β implies in maximization of π. Hence, the Neyman-Pearson lemma shows that of all tests with a given significance level, the one that is based on the likelihood ratio has largest power or is more powerful.

Example 6.1 *In the $N(\theta, \sigma^2)$ with known σ^2, consider the test of $H_0: \theta = \theta_0$ versus $H_1: \theta = \theta_1$ with $\theta_0 < \theta_1$. Then,*

$$\frac{p_0}{p_1} = \frac{p(\boldsymbol{x}|\theta_0)}{p(\boldsymbol{x}|\theta_1)} = \frac{(2\pi\sigma^2)^{-n/2} \exp\left\{-\frac{1}{2\sigma^2} \sum_{i=1}^n (x_i - \theta_0)^2\right\}}{(2\pi\sigma^2)^{-n/2} \exp\left\{-\frac{1}{2\sigma^2} \sum_{i=1}^n (x_i - \theta_1)^2\right\}}$$

$$= \exp\left\{\frac{1}{2\sigma^2}\left[-2\theta_1 \sum_{i=1}^n x_i + n\theta_1^2 + 2\theta_0 \sum_{i=1}^n x_i - n\theta_0^2\right]\right\}$$

$$= \exp\left\{\frac{1}{2\sigma^2}\left[2(\theta_0 - \theta_1)\sum_{i=1}^n x_i\right]\right\} \exp\left\{\frac{1}{2\sigma^2}n(\theta_1^2 - \theta_0^2)\right\}$$

$$\propto \exp\left\{\frac{n(\theta_0 - \theta_1)\bar{x}}{\sigma^2}\right\},$$

where the proportionality constant involves the constants θ_0, θ_1 and σ^2.

The likelihood ratio test accepts H_0 when $p_0/p_1 > k$. Then,

$$\frac{p_0}{p_1} > k \iff \exp\left\{\frac{n(\theta_0 - \theta_1)\bar{x}}{\sigma^2}\right\} > c_3 \iff \frac{n(\theta_0 - \theta_1)\bar{x}}{\sigma^2} > c_2 \iff \bar{x} < c_1$$

since $\theta_0 < \theta_1$ for constants c_1, c_2 and c_3. As the best estimator of θ is $\overline{X}$, the sample mean, when testing H_0 against H_1, one expects the test to accept H_0 for small values of $\overline{X}$. That is exactly the result of the optimal likelihood ratio test. The next step is to determine the value of c_1. To do this, note that the test has level α and therefore $\alpha = P(\text{rejection of } H_0 \mid \theta = \theta_0) = P(\overline{X} > c_1 \mid \theta = \theta_0)$. But

$$\overline{X} \mid \theta_0 \sim N\left(\theta_0, \frac{\sigma^2}{n}\right) \quad \text{or} \quad Z = \frac{\overline{X} - \theta_0}{\sigma/\sqrt{n}} \sim N(0, 1).$$

Since $\alpha = P\left(Z > (c_1 - \theta_0)\sqrt{n}/\sigma\right)$, then

$$\frac{(c_1 - \theta_0)}{\sigma}\sqrt{n} = z_\alpha \Rightarrow c_1 = \theta_0 + z_\alpha \frac{\sigma}{\sqrt{n}}.$$

The test with significance level α accepts H_0 if $\overline{X} < \theta_0 + \sigma z_\alpha / \sqrt{n}$.

Note that the test is completely specified and does not depend on the value of θ_1 but on the fact that $\theta_1 > \theta_0$. This means that the test is the same for any value of θ_1 such that $\theta_1 > \theta_0$. Therefore, the LR test is also more powerful to test H_0 versus $H_1 : \theta_1 > \theta_0$.

The fact that the test does not depend on the value of θ_1 may also cause problems. Consider the case when H_0 is rejected but $\bar{x}$ is much closer to θ_0 than to θ_1; that is, $\bar{x} - \theta_0 \ll \theta_1 - \bar{x}$. In this case, common sense suggests that given the choices of H_0 and H_1 one should choose H_0. This is another reason to avoid commitments towards acceptance or rejection of H_1 since the test does not provide information about it. Similar comments apply if the distances of $\bar{x}$ to θ_0 and θ_1 are much larger than the distance between θ_0 and θ_1. The intuitive reasoning based on the frequentist argument is that if sampling of $\boldsymbol{X}$ is repeated many times, in only $100\alpha\%$ of them, will H_0 be erroneously rejected.

The power of the test $\pi(\theta) = P(\text{rejection of } H_0 \mid \theta)$ is given by

$$\pi(\theta) = P\left[\overline{X} > \theta_0 + \frac{\sigma}{\sqrt{n}} z_\alpha \mid \theta > \theta_0\right].$$

But $\overline{X} \mid \theta > \theta_0 \sim N(\theta, \sigma^2/n)$ and therefore $\sqrt{n}(\overline{X} - \theta)/\sigma \sim N(0,1)$. Then,

$$
\begin{aligned}
\pi(\theta) &= P\left[\sqrt{n}\frac{(\overline{X} - \theta)}{\sigma} > \frac{\theta_0 + (\sigma/\sqrt{n})z_\alpha - \theta}{\sigma/\sqrt{n}} \mid \theta > \theta_0\right] \\
&= P\left[\sqrt{n}\frac{(\overline{X} - \theta)}{\sigma} > \frac{(\theta_0 - \theta)}{\sigma}\sqrt{n} + z_\alpha \mid \theta > \theta_0\right] \\
&= 1 - \Phi\left(z_\alpha - \frac{(\theta - \theta_0)}{\sigma}\sqrt{n}\right),
\end{aligned}
$$

that is, an increasing function of θ. So, the more distant the parameter value in the alternative, the smaller the chances of a type II error.

This test is not only the most powerful test for H_0 versus H_1: $\theta > \theta_0$, but also for $H_0 : \theta \le \theta_0$ versus H_1: $\theta > \theta_1$ because the level of the test with the new hypothesis is

$$\max_{\theta \le \theta_0} \pi(\theta) = \max_{\theta \le \theta_0} P\left[\overline{X} > \theta_0 + \frac{\sigma}{\sqrt{n}} z_\alpha\right].$$

As just seen, π is an increasing function of θ, the maximum in the region $\{\theta : \theta \le \theta_0\}$ is given at the value θ_0 and the value of π at this point is $1 - \Phi(z_\alpha) = \alpha$.

Also, in the example we could evaluate the size γ at which we would reject H_0 after observing $\boldsymbol{X} = \mathbf{x}$; that is, we could evaluate γ such that the limit $\theta_0 + \sigma z_\gamma/\sqrt{n}$ of the critical region is given by $\bar{x}$. This gives a more precise account of the strength of the data evidence in favor of or against the hypotheses. Smaller values of γ indicate that the sample mean is consistent with a lower probability of type I error and therefore there is more data evidence against H_0. Likewise, larger values of γ indicate a higher probability of type I error and therefore there is more data evidence in favor of H_0.

More generally, suppose a test where H_0 is rejected when a statistic T belongs to a region of the form $[T > c]$, and let t be the observed value of T. The statistic T used to define the critical region is called a test statistic. Then, evaluation of $Pr(T > t|H_0)$ gives an idea of how extreme the observed value is under H_0. This probability is usually known as the p-value. In the previous example, the p-value is given by $1 - \Phi(\sqrt{n}(\bar{x} - \theta_0)/\sigma)$. The notion of a p-value is useful for allowing the knowledge of the size at which one would reject H_0 based on the information actually obtained for

$$H_0 \text{ is rejected} \iff p\text{-value} < \alpha,$$

where α is a prespecified level of the test. It should be stressed that under the frequentist treatment, no probabilities can be associated to the hypothesis as θ is not random. Therefore, no association between the p-value and the probability of H_0 can be made because such probability simply cannot be defined. The notion of p-value can be placed in a general setting whenever it makes sense to specify the border of the critical region in terms of observed values of the test statistic.

Returning now to the case of discrete populations, it is not always possible to obtain tests of any prespecified level exactly. By exactly, we mean to have $\alpha = \eta$ where $\eta = \sup_{\theta \in \Theta_0} P(\text{rejection of } H_0 | \theta \in \Theta_0)$. The notion of p-value becomes even more important here.

There is an alternative approach that allows one to obtain tests of an exact level even for discrete distribution. This alternative is known as randomized tests where any prespecified level is obtained after realization of an additional independent Bernoulli experiment with success probability conveniently chosen to complete the difference between α and η.

6.2.2   Composite hypotheses

Consider again the test ψ of $H_0\colon \theta \in \Theta_0$ versus $H_1\colon \theta \in \Theta_1$. Let α be the fixed significance level and $\pi_\psi(\theta)$ the power function of ψ. Then, $\pi_\psi(\theta) \leq \alpha$ for every $\theta \in \Theta_0$.

Definition 6.1 *A test ψ^* is uniformly more powerful (UMP, in short) for $H_0\colon \theta \in \Theta_0$ versus $H_1\colon \theta \in \Theta_1$ at the significance level α if*

1. $\alpha(\psi^*) \leq \alpha$;

2. $\forall \psi$ with $\alpha(\psi) \leq \alpha$, $\pi_\psi(\theta) \leq \pi_{\psi^*}(\theta)$, $\forall \theta \in \boldsymbol{\Theta}_1$.

The test of the example above is UMP to test $\theta = \theta_0$ versus $\theta > \theta_0$ and also to test $\theta \leq \theta_0$ versus $\theta > \theta_0$.

Theorem 6.2 *Let $\boldsymbol{X} = (X_1, \ldots, X_n)$ be a random sample from $p(x|\theta)$ and $p(x|\theta)$ belong to the one-parameter exponential family with density*

$$p(\boldsymbol{x} \mid \theta) = a(\boldsymbol{x}) \exp\{\phi(\theta)T(\boldsymbol{x}) + b(\theta)\}$$

and ϕ is a strictly increasing function of θ, then the UMP test of level α to test $H_0: \theta \leq \theta_0$ versus $H_1: \theta > \theta_0$ is given by the critical region $T(\boldsymbol{X}) > c$ where c is such that $\alpha \geq P(T(\boldsymbol{X}) > c \mid \theta_0)$ (with equality in the continuous case). The power of this test is an increasing function of θ. If the hypotheses are interchanged or ϕ is a strictly decreasing function of θ, then the UMP test of level α rejects H_0 if $T(\boldsymbol{X}) < c$ where c is such that $\alpha \geq P(T(\boldsymbol{X}) < c \mid \theta_0)$ and the power of this test is again an increasing function of θ. If the two conditions above are simultaneously true, the UMP test remains unaltered.

Proof Consider the standard case where ϕ is strictly increasing and the hypotheses are $H_0 : \theta = \theta_0$ and $H_1 : \theta = \theta_1 > \theta_0$. In this case, Neyman-Pearson's lemma ensures that the most powerful test rejects H_0 when

$$\frac{p(\mathbf{x}|\theta_0)}{p(\mathbf{x}|\theta_1)} < c_4 \quad \Longleftrightarrow \quad \frac{\exp\{\phi(\theta_0)T(\mathbf{x}) + b(\theta_0)\}}{\exp\{\phi(\theta_1)T(\mathbf{x}) + b(\theta_1)\}} < c_3$$

$$\Longleftrightarrow \quad \exp\{[\phi(\theta_0) - \phi(\theta_1)]T(\mathbf{x})\} < c_2$$

$$\Longleftrightarrow \quad [\phi(\theta_0) - \phi(\theta_1)]T(\mathbf{x}) < c_1$$

$$\Longleftrightarrow \quad T(\mathbf{x}) > c,$$

where $c_4, \ldots, c_1, c$ are constants such that $\alpha = P(T(\boldsymbol{X}) > c | \theta_0)$. For the most powerful test, it can be shown that $\pi(\theta_0) \leq \pi(\theta_1)$. So, the power function is an increasing function of θ and

$$\pi(\theta_0) = \sup_{\{\theta : \theta < \theta_0\}} \pi(\theta).$$

So, the test is UMP for $H_0 : \theta \leq \theta_0$. As in the calculation above it was only used that $\theta_1 > \theta_0$, the results must be equally true for any such value of θ_1. Therefore, the test is UMP for $H_1 : \theta > \theta_0$.

In the case of a strictly decreasing ϕ,

$$[\phi(\theta_0) - \phi(\theta_1)]T(\mathbf{x}) < c_1 \quad \Longleftrightarrow \quad T(\mathbf{x}) < c$$

and the critical region becomes $\{\mathbf{X} : T(\mathbf{X}) < c\}$.

In the case of interchanged hypotheses, all inequalities must be reversed because one must work with the ratio $p(\mathbf{x}|\theta_1)/p(\mathbf{x}|\theta_0)$ instead of $p(\mathbf{x}|\theta_0)/p(\mathbf{x}|\theta_1)$. This leads to the critical region in the form $\{\mathbf{X} : T(\mathbf{X}) < c\}$.

Finally, in the case of a strictly decreasing ϕ and interchanged hypotheses, double reversal of the inequality preserves it and the critical region remains in the form $\{\mathbf{X} : T(\mathbf{X}) > c\}$. $\square$

Example 6.2 *Let* $X_1, \ldots, X_n$ *be a random sample from the* $Ber(\theta)$ *distribution. We know from Section 2.5 that* $\phi(\theta) = \log[\theta/(1-\theta)]$ *which is an increasing function of* θ *and that* $T(\mathbf{X}) = \sum_{i=1}^{n} X_i$. *Therefore, the UMP test for* $H_0 : \theta \leq \theta_0$ *versus* $H_1 : \theta > \theta_0$ *has critical region of the form* $\sum_{i=1}^{n} X_i > c$.

The property that guarantees the existence of UMP tests in the exponential family is in fact more general. It finds an appropriate setting under families with monotone likelihood ratio.

Definition 6.2 *The family of distributions* $\{p(x \mid \theta), \theta \in \Theta\}$ *is said to have monotone likelihood ratio if there is a statistic* $T(\mathbf{X})$ *such that* $\forall \theta_1, \theta_2 \in \Theta$ *with* $\theta_1 < \theta_2$ *the likelihood ratio*

$$\frac{p(\mathbf{X} \mid \theta_2)}{p(\mathbf{X} \mid \theta_1)}$$

is a monotone function of $T(\mathbf{X})$.

The uniform distribution over the interval $[0, \theta]$ does not belong to the exponential family. Nevertheless, it has monotone likelihood ratio because the ratio of sampling densities is a monotonically increasing function of $T(\mathbf{X}) = \max_i X_i$.

The results just proved for exponential families can be extended for families with monotone likelihood ratio. So, if the likelihood ratio is an increasing function of $T(\mathbf{X})$, then the UMP test of level α for $H_0: \theta \leq \theta_0$ versus $H_1: \theta > \theta_0$ is given by the critical region of the form $T(\mathbf{X}) > c$ where c is such that $\alpha = P(T(\mathbf{X}) > c \mid \theta_0)$ and the power of this test is an increasing function of θ. Likewise, if the hypotheses are interchanged or the likelihood ratio is a decreasing function of $T(\mathbf{X})$, the UMP test of level α rejects H_0 if $T(\mathbf{X}) < c$ where c is such that $\alpha \geq P(T(\mathbf{X}) < c \mid \theta_0)$ and the power of this test is again an increasing function of θ. If the two conditions above are simultaneously true, the UMP test remains unaltered.

These results make intuitive sense. The larger the likelihood ratio, the more plausible the value θ_0 relative to θ_1. If the likelihood ratio is an increasing function of $T(\mathbf{X})$, the same reasoning is true for $T(\mathbf{X})$. Therefore, a reasonable rejection region for H_0 would be given by small values of $T(\mathbf{X})$.

Example 6.3 *(Example 6.2 continued) Let $X_1, \ldots, X_n$ be a random sample from the $Ber(\theta)$ distribution. Then, $p(\boldsymbol{X} \mid \theta) = \theta^T (1-\theta)^{n-T}$ with $T = \sum X_i$. If $\theta_1 < \theta_2$, the LR is*

$$\frac{\theta_2^T (1-\theta_2)^{n-T}}{\theta_1^T (1-\theta_1)^{n-T}} = \left[\frac{\theta_2 (1-\theta_1)}{\theta_1 (1-\theta_2)} \right]^T \left(\frac{1-\theta_2}{1-\theta_1} \right)^n = \xi^T \, \eta^n,$$

with

$$\xi = \frac{\theta_2 (1-\theta_1)}{\theta_1 (1-\theta_2)} \ \text{and} \ \eta = \frac{1-\theta_2}{1-\theta_1}.$$

Since $1 - \theta_1 > 1 - \theta_2$, $\xi > 0$ and the likelihood ratio is increasing in T, confirming results obtained earlier in the example.

So far, only one-sided tests have been considered. These are tests where the parametric regions defining the hypotheses are given by a single strict inequality. An example of interest of a test of hypotheses that is not one-sided is obtained when H_0: $\theta = \theta_0$ versus H_1: $\theta \neq \theta_0$. Such test may be useful when comparing two competing treatments.

Assuming now the observation of a sample from the $N(\theta, \sigma^2)$ distribution with known σ^2, consider the three tests of $H_0 : \theta = \theta_0$ below based on $\bar{X}$:

Test 1. Reject H_0 if $\mid \bar{X} - \theta_0 \mid > 1.645\sigma/\sqrt{n}$.

Test 2. Reject H_0 if $\bar{X} - \theta_0 > 1.282\sigma/\sqrt{n}$.

Test 3. Reject H_0 if $\mid \bar{X} - \theta_0 \mid < 0.126\sigma/\sqrt{n}$.

Calculation of the rejection probability of H_0 shows that the three tests have level 0.1. The next step is to proceed with evaluation of the power of each of the tests. It can be easily seen from Figure 6.1 that none of the tests is UMP over the other two. Nevertheless, the first test is the only one with $\min_{\Theta_1} \pi(\theta) > \pi(\theta_0)$. This means that the rejection probability is larger under the alternative than under the null hypothesis. This way, one guarantees that the chances of rejecting H_0 are larger when H_0 is false. This seems like a reasonable property to require from tests.

Definition 6.3 *A test ψ for $H_0 : \boldsymbol{\theta} \in \boldsymbol{\Theta}_0$ against $H_1 : \boldsymbol{\theta} \in \boldsymbol{\Theta}_1$ is said to be unbiased if for every pair $(\boldsymbol{\theta}, \boldsymbol{\theta}')$ where $\boldsymbol{\theta} \in \boldsymbol{\Theta}_0$ and $\boldsymbol{\theta}' \in \boldsymbol{\Theta}_1$, $\pi_\psi(\boldsymbol{\theta}) \leq \pi_\psi(\boldsymbol{\theta}')$. The power function is at least as large in $\boldsymbol{\Theta}_1$ as it is in $\boldsymbol{\Theta}_0$. If the test does not satisfy the condition above, it is said to be biased.*

One can then try to construct UMP tests for H_0: $\boldsymbol{\theta} = \boldsymbol{\theta}_0$ versus H_1: $\boldsymbol{\theta} \neq \boldsymbol{\theta}_0$ within the class of unbiased tests. In the one-parameter exponential family, it can be shown that if ϕ is a strictly increasing function of θ, the UMP unbiased test of level α for H_0: $\theta = \theta_0$ versus H_1: $\theta \neq \theta_0$ accepts H_0 when

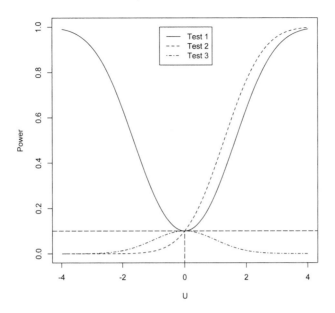

Figure 6.1: *Power functions for the above tests, where $U = \sqrt{n}(\theta - \theta_0)/\sigma$.*

$c_1 < T(\mathbf{X}) < c_2$ with $P(c_1 < T(\mathbf{X}) < c_2 \mid \theta_0) = 1 - \alpha$ and (c_1, c_2) is an interval of highest sampling density of $T(\mathbf{X})$. The sampling distributions of $T(\mathbf{X})$ are not necessarily symmetric (around θ_0) as in the Normal case above. For such tests, we are led to two distinct p-values. In general, the smaller one is used.

It may not be possible in general to find unbiased tests. A general procedure for testing H_0: $\boldsymbol{\theta} \in \boldsymbol{\Theta}_0$ versus H_1: $\boldsymbol{\theta} \in \boldsymbol{\Theta}_1$ is based on the maximum likelihood ratio (MLR, in short) statistic given by

$$\lambda(\mathbf{X}) = \frac{\sup_{\boldsymbol{\theta} \in \boldsymbol{\Theta}_0} p(\mathbf{X} \mid \boldsymbol{\theta})}{\sup_{\boldsymbol{\theta} \in \boldsymbol{\Theta}_1} p(\mathbf{X} \mid \boldsymbol{\theta})}.$$

The most common case for use of this procedure is when $\boldsymbol{\Theta}_0$ and $\boldsymbol{\Theta}_1$ are exclusive and exhaustive and $\boldsymbol{\Theta}_0$ is of smaller dimension than $\boldsymbol{\Theta}_1$. In these cases, the denominator is replaced by the supremum over the whole parametric space $\boldsymbol{\Theta}$, which is easier to evaluate. Formally, the statistic $\lambda(\mathbf{X})$ is being replaced by $\max\{\lambda(\mathbf{X}), 1\}$. In any case, $\lambda(\mathbf{X})$ is a random variable depending on the sample. The maximum (or generalized) likelihood ratio test for H_0 of level α accepts H_0 if $\lambda(\mathbf{X}) > c$ where c satisfies

$$\alpha \geq \sup_{\boldsymbol{\theta} \in \boldsymbol{\Theta}_0} P(\lambda(\mathbf{X}) < c \mid \boldsymbol{\theta}).$$

Once again, α is taken as equal to the supremum of the above probabilities whenever possible. The power of the test is given by $\pi(\boldsymbol{\theta}) = P(\lambda(\mathbf{X}) < c \mid \boldsymbol{\theta})$. This test rejects the null hypothesis H_0 if the maximized value of the likelihood under H_0 is distant from the global maximized value. This is an indication that there is great improvement in likelihood by consideration of points outside H_0, and H_0 is not providing a good description of the data. In this case, it makes sense to reject H_0.

It is important to distinguish between the above test and the test based on monotone likelihood ratios. Although the maximum likelihood ratio test enjoys good asymptotic properties, it is not always unbiased or UMP. The main difficulties associated with it are the calculation of the maximized likelihood in closed form and the determination of its sampling distribution. The first point was dealt with in previous chapters and the second one will be addressed below when asymptotic tests are treated in Section 6.5.

Other desirable properties in test procedures are similarity and invariance. Consider the problem of testing $H : \boldsymbol{\theta} = \boldsymbol{\theta}_0$ in the presence of a nuisance parameter ϕ. A test is said to be similar if the level of the test is the same whatever the value of the nuisance parameter. An example of a similar test is the t test, to be seen later in this section. A test is said to be invariant (under a specific family of transformations) if the distribution of any transformation of the observations inside the family remains in the same family. This will allow the hypotheses to remain unaltered with any of the transformations operated over the data. The tests presented below in this section are all invariant under linear transformations of the observations.

6.2.3   Hypothesis testing with the Normal distribution

The most common tests for samples from a Normal distribution are presented here. Once again, let $X_1, \ldots, X_n$ be iid with $X_i \sim N(\theta, \sigma^2)$ and suppose one wishes to test $H_0: \theta = \theta_0$ versus $H_1: \theta \neq \theta_0$. Assume initially that σ^2 is known. In this case, we have shown that the UMP unbiased test of level α is given by the critical region $|\sqrt{n}(\bar{X} - \theta_0)/\sigma| > z_{\alpha/2}$. We will now obtain the MLR test.
    Since $\Theta_0 = \{\theta_0\}$,

$$\sup_{\theta \in \Theta_0} p(\mathbf{x} \mid \theta) = p(\mathbf{x} \mid \theta_0).$$

For $\theta \in \Theta_1 = \Theta - \{\theta_0\}$, the maximum of $p(\mathbf{x} \mid \theta)$ is obtained at $\hat{\theta}$, the maximum likelihood estimator of θ, which in this case is the sample mean $\bar{X}$. Note that for every value of θ, $P(\hat{\theta} = \theta_0) = 0$, and therefore considering maximization of Θ instead of Θ_1 in the expression of the MLR does not produce any change. Then, the MLR statistic is given by

$$
\begin{aligned}
\lambda(\mathbf{X}) &= \frac{p(\mathbf{X} \mid \theta_0)}{p(\mathbf{X} \mid \hat{\theta})} \\
&= \frac{(2\pi\sigma^2)^{-n/2} \exp\left\{-\left[\sum(X_i - \bar{X})^2 + n(\bar{X} - \theta_0)^2\right]/2\sigma^2\right\}}{(2\pi\sigma^2)^{-n/2} \exp\left\{-\sum(X_i - \bar{X})^2/2\sigma^2\right\}} \\
&= \exp\left\{-\frac{n}{2\sigma^2}(\bar{X} - \theta_0)^2\right\}.
\end{aligned}
$$

Observe that under H_0, $\bar{X} \sim N(\theta_0, \sigma^2/n)$, and therefore $Z = \sqrt{n}(\bar{X} - \theta_0)/\sigma \sim N(0, 1)$. The MLR statistic is given by $\lambda(\mathbf{X}) = \exp(-Z^2/2)$. The MLR test rejects H_0 if

$$
\lambda(\mathbf{X}) < c_2 \iff Z^2/2 > c_1 \iff |Z| > c.
$$

Given a significance level α, $\alpha = P(|Z| > c \mid H_0)$ or $c = z_{\alpha/2}$. The power of the test is $\pi(\theta) = P(|Z| > z_{\alpha/2} \mid H_1)$.

Under H_1, $\bar{X} \sim N(\theta, \sigma^2/n)$. So,

$$
\bar{X} - \theta_0 \sim N\left(\theta - \theta_0, \frac{\sigma^2}{n}\right) \iff W = \frac{\sqrt{n}}{\sigma}[\bar{X} - \theta_0 - (\theta - \theta_0)] \sim N(0, 1).
$$

Then, $W = Z - \sqrt{n}\frac{(\theta - \theta_0)}{\sigma}$, and therefore the power of the test is

$$
\begin{aligned}
\pi(\theta) &= 1 - P\left(-z_{\alpha/2} < Z < z_{\alpha/2} \mid \theta\right) \\
&= 1 - P\left(-z_{\alpha/2} - \sqrt{n}\frac{(\theta - \theta_0)}{\sigma} < W < z_{\alpha/2} - \sqrt{n}\frac{(\theta - \theta_0)}{\sigma} \mid \theta\right) \\
&= 1 + \Phi\left(-z_{\alpha/2} - \sqrt{n}\frac{(\theta - \theta_0)}{\sigma}\right) - \Phi\left(z_{\alpha/2} - \sqrt{n}\frac{(\theta - \theta_0)}{\sigma}\right) > \alpha.
\end{aligned}
$$

The MLR test is unbiased as $\pi(\theta) > \alpha$, $\forall \theta \neq \theta_0$. The rate of increase of the power depends on σ. The smaller σ (more precise distribution, more concentrated population), the faster the rate of growth of the power towards 1.

In the case where σ^2 is unknown, $\Theta = \{(\theta, \sigma^2) : \theta \in R, \sigma^2 > 0\}$ and $\Theta_0 = \Theta\{(\theta, \sigma^2) : \theta = \theta_0, \sigma^2 > 0\}$. Since the dimension of Θ_0 is smaller than the dimension of Θ_1 and of Θ, we will work with the latter. As earlier, this change can be proved to affect only a zero probability set. As seen previously,

$$
\sup_{(\theta,\sigma^2) \in \Theta_0} p(\mathbf{X} \mid \theta, \sigma^2) = p(\mathbf{X} \mid \theta_0, \hat{\sigma}_0^2), \text{ where } \hat{\sigma}_0^2 = \Sigma(X_i - \theta_0)^2/n \text{ and}
$$

$$
\sup_{(\theta,\sigma^2) \in \Theta} p(\mathbf{X} \mid \theta, \sigma^2) = p(\mathbf{X} \mid \hat{\theta}, \hat{\sigma}^2), \text{ where } \hat{\theta} = \bar{X} \text{ and } \hat{\sigma}^2 = \Sigma(X_i - \bar{X})^2/n.
$$

Therefore, the MLR statistic becomes

$$
\begin{aligned}
\lambda(\mathbf{X}) &= \frac{p(\mathbf{X} \mid \theta_0, \hat{\sigma}_0^2)}{p(\mathbf{X} \mid \hat{\theta}, \hat{\sigma}^2)} \\
&= \frac{(2\pi\hat{\sigma}_0^2)^{-n/2} \exp\left\{-\frac{1}{2\hat{\sigma}_0^2}\sum(X_i - \theta_0)^2\right\}}{(2\pi\hat{\sigma}^2)^{-n/2} \exp\left\{-\frac{1}{2\hat{\sigma}^2}\sum(X_i - \bar{X})^2\right\}} \\
&= \left(\frac{\hat{\sigma}^2}{\hat{\sigma}_0^2}\right)^{n/2} \frac{\exp\left(-\frac{1}{2\hat{\sigma}_0^2}n\hat{\sigma}_0^2\right)}{\exp\left(-\frac{1}{2\hat{\sigma}^2}n\hat{\sigma}^2\right)} \\
&= \left(\frac{\hat{\sigma}^2}{\hat{\sigma}_0^2}\right)^{n/2}.
\end{aligned}
$$

One can also write

$$
\begin{aligned}
\frac{\hat{\sigma}_0^2}{\hat{\sigma}^2} &= \frac{\sum(X_i - \bar{X})^2 + n(\bar{X} - \theta_0)^2}{\sum(X_i - \bar{X})^2} \\
&= 1 + \frac{n(\bar{X} - \theta_0)^2}{(n-1)S^2} \\
&= 1 + \frac{T^2}{n-1} \quad \text{where } T = \sqrt{n}\frac{(\bar{X} - \theta_0)}{S}
\end{aligned}
$$

and the MLR statistic can be rewritten as $\lambda(\mathbf{X}) = [1 + T^2/(n-1)]^{-n/2}$. Therefore, the MLR test accepts H_0 if $T^2 < c_1$ or $|T| < c$. As $T|H_0 \sim t_{n-1}(0,1)$, the value of c of the level α test is $t_{\alpha/2,n-1}$. This test is usually known as the t test and is possibly the most used test in Statistics. It can be shown that the power of the test is a strictly increasing function of $|\theta - \theta_0|$ (see Exercise 6.5). An immediate consequence is the unbiasedness of the test since the smallest value of the power occurs when $\theta = \theta_0$.

This test is similar because none of the properties of the test is affected by the value of the nuisance parameter σ^2. This was achieved by replacement of σ^2 by its estimator S^2 and the existence of the pivotal quantity T. This test is also invariant under linear transformations.

Another common test is the test of equality of two means. Consider two random samples $\mathbf{X}_1 = (X_{11}, \ldots, X_{1n_1})$ from the $N(\theta_1, \sigma^2)$ distribution and $\mathbf{X}_2 = (X_{21}, \ldots, X_{2n_2})$ from the $N(\theta_2, \sigma^2)$ distribution. Assume also for simplicity that the variances are equal. Then, the hypothesis of interest is defined by the parameter space $\Theta_0 = \{(\theta_1, \theta_2, \sigma^2) : \theta_1 = \theta_2 = \theta \in R, \ \sigma^2 > 0\}$. Observe that under H_0 the problem contains only a single sample of size $n_1 + n_2$ from the $N(\theta, \sigma^2)$ distribution. The parameters can be more usefully described by the parameter of interest $\theta_2 - \theta_1$, any transformation of θ_1 and θ_2 that is not a multiple of $\theta_2 - \theta_1$ and σ^2. The last two are nuisance parameters.

Once again, the MLR statistic is based on maximizations over Θ and Θ_0. These operations give

$$\sup_{(\theta_1,\theta_2,\sigma^2)\in\Theta_0} p(\mathbf{X}_1,\mathbf{X}_2|\theta,\sigma^2)=\left(\frac{1}{2\pi\hat{\sigma}_0^2}\right)^{n_1+n_2}\exp\left\{-\frac{1}{2\hat{\sigma}_0^2}\left[\sum_{i=1}^{2}\sum_{j=1}^{n_i}(X_{ij}-\hat{\theta})^2\right]\right\},$$

where

$$\hat{\theta}=\frac{1}{n_1+n_2}\left[\sum_{i=1}^{n_1}X_{1i}+\sum_{i=1}^{n_2}X_{2i}\right]$$

and

$$\hat{\sigma}_0^2=\frac{1}{n_1+n_2}\left[\sum_{i=1}^{n_1}(X_{1i}-\hat{\theta})^2+\sum_{i=1}^{n_2}(X_{2i}-\hat{\theta})^2\right],$$

and $\sup_{(\theta_1,\theta_2,\sigma^2)\in\Theta} p(\mathbf{X}_1,\mathbf{X}_2|\theta_1,\theta_2,\sigma^2)$ equals to

$$\left(\frac{1}{2\pi\hat{\sigma}^2}\right)^{n_1+n_2}\exp\left\{-\frac{1}{2\hat{\sigma}^2}\left[\sum_{i=1}^{2}\sum_{j=1}^{n_i}(X_{ij}-\hat{\theta}_i)^2\right]\right\},$$

where

$$\hat{\theta}_i=\bar{X}_i,\ i=1,2 \text{ and } \hat{\sigma}^2=\frac{1}{n_1+n_2}\left[\sum_{i=1}^{n_1}(X_{1i}-\hat{\theta}_1)^2+\sum_{i=1}^{n_2}(X_{2i}-\hat{\theta}_2)^2\right].$$

This gives $\lambda(\mathbf{X}_1,\mathbf{X}_2)=(\hat{\sigma}^2/\hat{\sigma}_0^2)^{(n_1+n_2)/2}$. Note that

$$\bar{X}_1-\hat{\theta}=\frac{n_2}{n_1+n_2}(\bar{X}_1-\bar{X}_2) \text{ and } \bar{X}_2-\hat{\theta}=\frac{n_1}{n_1+n_2}(\bar{X}_2-\bar{X}_1),$$

which implies that

$$\sum_{i=1}^{2}\sum_{j=1}^{n_i}(X_{ij}-\hat{\theta})^2 = \sum_{i=1}^{2}\sum_{j=1}^{n_i}(X_{ij}-\bar{X}_i)^2+n_i(\bar{X}_i-\hat{\theta})^2$$

$$= \sum_{i=1}^{2}\sum_{j=1}^{n_i}(X_{ij}-\bar{X}_i)^2+\frac{n_1 n_2}{n_1+n_2}(\bar{X}_1-\bar{X}_2)^2.$$

Therefore,

$$\lambda(\mathbf{X}_1,\mathbf{X}_2) = \left(1+\frac{(n_1^{-1}+n_2^{-1})(\bar{X}_1-\bar{X}_2)^2}{\sum_{i=1}^{2}\sum_{j=1}^{n_i}(X_{ij}-\bar{X}_i)^2}\right)^{-(n_1+n_2)/2}$$

$$= \left(1+\frac{T^2}{n_1+n_2-2}\right)^{-(n_1+n_2)/2}$$

where

$$T = \frac{\bar{X}_1 - \bar{X}_2}{S\sqrt{n_1^{-1} + n_2^{-1}}} \quad \text{with } S^2 = \frac{n_1 + n_2}{n_1 + n_2 - 2}\hat{\sigma}^2.$$

So, the MLR test accepts H_0 when $\lambda(\mathbf{X}_1, \mathbf{X}_2) > c_1$ or when $|T| < c$. As seen previously, $T|H_0 \sim t_{n_1+n_2-2}(0,1)$. For a level α test, $c = t_{\alpha/2,n_1+n_2-2}$. Analogously, it can be shown that the power as this test is a strictly increasing function of $|\theta_1 - \theta_2|$. Therefore, this test is unbiased for the same reasons as the previous t test. It is also similar because none of its properties is affected by the two nuisance parameters and invariant under linear transformations of the observations.

6.3   Bayesian hypothesis testing

In the Bayesian context, the problem of deciding about which hypothesis to accept is conceptually simpler. Typically, one would compare the hypotheses $H_1, \ldots, H_k$ through their respective posterior probabilities, obtained via Bayes theorem as

$$p(H_i \mid \mathbf{x}) \propto p(\mathbf{x} \mid H_i)\, p(H_i).$$

Once again, this setup can be framed as a decision problem. In addition to the posterior probabilities attached to the hypotheses (or states of nature), a loss structure associated with the possible actions can be incorporated.

Returning to the special case of two hypotheses, suppose one wishes to test H_0: $\boldsymbol{\theta} \in \boldsymbol{\Theta}_0$ versus H_1: $\boldsymbol{\theta} \in \boldsymbol{\Theta}_1$. It suffices to examine the posterior probabilities $p(H_0 \mid \mathbf{x})$ and $p(H_1 \mid \mathbf{x})$. Assuming equal error loss, if $p(H_0 \mid \mathbf{x}) > p(H_1 \mid \mathbf{x})$, then H_0 should be accepted as the most plausible hypothesis for $\boldsymbol{\theta}$. In this case, it can be said that H_0 is preferable to H_1. Otherwise, H_1 is preferred to H_0. There is a clearcut rule for choice between the hypotheses, which is not always true under the frequentist framework.

As

$$p(H_0 \mid \mathbf{x}) \quad \propto \quad p(\mathbf{x} \mid H_0)\, p(H_0)$$
$$p(H_1 \mid \mathbf{x}) \quad \propto \quad p(\mathbf{x} \mid H_1)\, p(H_1)$$

and reiterating that the proportionality constant is the same in both expressions,

$$\frac{p(H_0 \mid \mathbf{x})}{p(H_1 \mid \mathbf{x})} = \frac{p(H_0)}{p(H_1)}\frac{p(\mathbf{x} \mid H_0)}{p(\mathbf{x} \mid H_1)}.$$

The ratio $p(H_0)/p(H_1)$ is called the prior odds between H_0 and H_1 and the ratio $p(H_0 \mid \mathbf{x})/p(H_1 \mid \mathbf{x})$ is called the posterior odds between H_0 and H_1. The ratio

$$\frac{p(\mathbf{x} \mid H_0)}{p(\mathbf{x} \mid H_1)}$$

is called the Bayes factor and is denoted by $BF(H_0; H_1)$. This concept was introduced by Jeffreys (1961). Note that it is the ratio of marginal likelihoods of the hypotheses.

So, the posterior odds are given by the product of the prior odds and the Bayes factor. Once again, the likelihood ratio introduces the influence of the observations in the setting of hypothesis testing. In general, the likelihoods here are marginal in the sense that they are obtained after integrating out some of the parameters not associated with the specification of the hypotheses.

Despite their notational simplicity, it is not easy in many cases to specify $p(H_j)$ when H_j is a simple hypothesis, for some j, and $\boldsymbol{\theta}$ is continuous. If a prior density f is specified for $\boldsymbol{\theta} \in \Theta$, one will have that $p(H_i) = p(H_i \mid \mathbf{x}) = 0$. In these cases, one solution is to attribute a lump prior probability p to H_j, for $p \in (0, 1)$. So, if H_1 is the complement of a simple hypothesis H_0, $p(H_1) = 1-p$, and this probability is distributed over the different values of $\boldsymbol{\theta}$ under H_1, according to the prior distribution for $\boldsymbol{\theta} \mid H_1$. This distribution will have density f over Θ_1.

As H_0 is a simple hypothesis, say, $\boldsymbol{\theta} = \boldsymbol{\theta}_0$, it follows that $p(\mathbf{x} \mid H_0) = p(\mathbf{x} \mid \boldsymbol{\theta}_0)$, the marginal density of $\mathbf{X}$ given H_0. This can also be referred to as the marginal likelihood of H_0 based on $\mathbf{X}$. The marginal likelihood of H_1 based on $\mathbf{X}$ is

$$
\begin{aligned}
p(\mathbf{x} \mid H_1) &= \int_{\Theta - \{\boldsymbol{\theta}_0\}} p(\mathbf{x}, \boldsymbol{\theta} \mid H_1) \, d\boldsymbol{\theta} \\
&= \int_{\Theta - \{\boldsymbol{\theta}_0\}} p(\mathbf{x} \mid \boldsymbol{\theta}, H_1) \, p(\boldsymbol{\theta} \mid H_1) \, d\boldsymbol{\theta} \\
&= \int_{\Theta} p(\mathbf{x} \mid \boldsymbol{\theta}) f(\boldsymbol{\theta}) \, d\boldsymbol{\theta},
\end{aligned}
$$

observing that the last integration is performed over the entire parameter space Θ because a single point does not alter its value. The Bayes factor is reduced to

$$BF(H_0; H_1) = \frac{p(\mathbf{x} \mid \boldsymbol{\theta}_0)}{\int_{\Theta} p(\mathbf{x} \mid \boldsymbol{\theta}) f(\boldsymbol{\theta}) \, d\boldsymbol{\theta}}.$$

Note that it provides the relative odds between H_0 and H_1 without taking into account the prior odds. It is a Bayesian measure of the relative goodness of the fit of any given pair of models to the data set. A Bayes factor larger than 1 indicates that H_0 fits the data better than H_1 and thus has larger likelihood. Also, the marginal prior for $\boldsymbol{\theta}$ is mixed, and therefore the marginal distribution of $\mathbf{X}$ is

$$
\begin{aligned}
p(\mathbf{x}) &= \int_{\Theta} p(\mathbf{x}|\boldsymbol{\theta})dF(\boldsymbol{\theta}) \\
&= p\,p(\mathbf{x}|\boldsymbol{\theta}_0) + (1-p)\int_{\Theta} p(\mathbf{x}|\boldsymbol{\theta})f(\boldsymbol{\theta})\,d\boldsymbol{\theta} \\
&= p\,p(\mathbf{x}|\boldsymbol{\theta}_0) + (1-p)\,p(\mathbf{x}|H_1).
\end{aligned}
$$

Let $\mathbf{X} = (X_1, \ldots, X_n)$ be a random sample from the $N(\theta, \sigma^2)$ distribution and assume one wishes to test $H_0: \theta = \theta_0$ versus $H_1: \theta \neq \theta_0$. Consider initially, the case of known variance σ^2. Then, with a prior probability p for H_0 and assuming $\theta \mid H_1 \sim N(\mu, \tau^2)$ gives

$$
\begin{aligned}
p(\mathbf{x} \mid H_1) &= \int p(\mathbf{x} \mid \theta)p(\theta)\,d\theta \\
&= \left(\frac{1}{2\pi\sigma^2}\right)^{n/2} \exp\left\{-\frac{1}{2\sigma^2}\sum_{i=1}^{n}(x_i - \bar{x})^2\right\} \\
&\quad \times \int \exp\left\{-\frac{n}{2\sigma^2}(\theta - \bar{x})^2\right\} \frac{1}{\sqrt{2\pi}\tau}\exp\left\{-\frac{1}{2\tau^2}(\theta - \mu)^2\right\}d\theta.
\end{aligned}
$$

After a few substitutions and algebraic transformations,

$$
\begin{aligned}
p(\mathbf{x} \mid H_1) &= \left(\frac{1}{2\pi\sigma^2}\right)^{n/2} \exp\left\{-\frac{ns^2}{2\sigma^2}\right\} \\
&\quad \times \left(\frac{\sigma^2/n}{\tau^2 + \sigma^2/n}\right)^{1/2} \exp\left\{-\frac{1}{2}\frac{(\bar{x} - \mu)^2}{\tau^2 + \sigma^2/n}\right\},
\end{aligned}
$$

where $s^2 = \sum_{i=1}^{n}(x_i - \bar{x})^2/n$. The Bayes factor is

$$
\begin{aligned}
BF(H_0; H_1) &= \frac{\left(\frac{1}{2\pi\sigma^2}\right)^{-n/2}\exp\left\{-\frac{ns^2}{2\sigma^2}\right\}\exp\left\{-\frac{n}{2\sigma^2}(\bar{x} - \theta_0)^2\right\}}{\left(\frac{1}{2\pi\sigma^2}\right)^{-n/2}\exp\left\{-\frac{ns^2}{2\sigma^2}\right\}\left(\frac{\sigma^2/n}{\tau^2+\sigma^2/n}\right)^{1/2}\exp\left\{-\frac{1}{2}\frac{(\bar{x}-\mu)^2}{\tau^2+\sigma^2/n}\right\}} \\
&= \left(\frac{\tau^2 + \sigma^2/n}{\sigma^2/n}\right)^{1/2}\exp\left\{\frac{n}{2}\left[\frac{(\bar{x} - \mu)^2}{\sigma^2 + n\tau^2} - \frac{(\bar{x} - \theta_0)^2}{\sigma^2}\right]\right\}.
\end{aligned}
$$

As expected, the Bayes factor depends only on the sample through $\bar{x}$.

To obtain the sample value that maximizes the Bayes factor, it suffices to solve the maximization problem for $\bar{x}$. Taking the logarithm of the previous expression and differentiating with respect to $\bar{x}$ gives

$$
\begin{aligned}
\frac{\partial \log BF}{\partial \bar{x}} &= \frac{n}{2}\left[\frac{2(\bar{x} - \mu)}{\sigma^2 + n\tau^2} - \frac{2(\bar{x} - \theta_0)}{\sigma^2}\right] = 0 \\
\frac{\partial^2 \log BF}{\partial \bar{x}^2} &= \frac{n}{2}\left[\frac{2}{\sigma^2 + n\tau^2} - \frac{2}{\sigma^2}\right] = \frac{n}{2}\left[-\frac{2n\tau^2}{\sigma^2(\sigma^2 + n\tau^2)}\right] < 0,
\end{aligned}
$$

and solving the first equation provides the maximum. The solution is

$$\bar{x}_{\max} = \theta_0 + \frac{\sigma^2}{n\tau^2}(\theta_0 - \mu),$$

and the maximized value of the Bayes factor is

$$\exp\left\{\frac{(\theta_0 - \mu)^2}{2\tau^2}\right\}\left(1 + \frac{n\tau^2}{\sigma^2}\right)^{1/2} > 1.$$

The larger the sample value n, the larger the chances of H_0; the larger τ^2, the larger the maximized value of the Bayes factor. The prior uncertainty plays a crucial role in the Bayesian comparison of sharp null hypotheses. In the limit, when $\tau^2 \to \infty$, the Bayes factor $BF(H_0; H_1)$ also increases indefinitely. This limit is the result obtained when a non-informative prior for $\theta \mid H_1$ is used. This was first noted by Lindley (1957) and is known as Lindley's paradox. It has been the object of study of comparisons between the Bayesian and frequentist approaches to hypothesis testing.

To ease comparisons between the approaches, take $\mu = \theta_0$ and $\tau^2 = \sigma^2$. The first assumption centers the alternative prior distribution over the single value of H_0, and the second takes the prior variance in the alternative as equal to the observational variance. Then,

$$
\begin{aligned}
BF(H_0; H_1) &= \left(\frac{\sigma^2 + \sigma^2/n}{\sigma^2/n}\right)^{1/2} \exp\left\{\frac{n}{2}\left[\frac{(\bar{x} - \theta_0)^2}{\sigma^2 + n\sigma^2} - \frac{(\bar{x} - \theta_0)^2}{\sigma^2}\right]\right\} \\
&= (n+1)^{1/2} \exp\left\{-\frac{n}{2}\left[\frac{n(\bar{x} - \theta_0)^2}{\sigma^2(n+1)}\right]\right\} \\
&= (n+1)^{1/2} \exp\left\{-\frac{n}{n+1}\frac{z^2}{2}\right\} \text{ where } z = \sqrt{n}\,\frac{\bar{x} - \theta_0}{\sigma}.
\end{aligned}
$$

If $p(H_0) = p$, then

$$\frac{p(H_0|\mathbf{x})}{p(H_1|\mathbf{x})} = \frac{p}{1-p}BF(H_0; H_1) \text{ if and only if}$$

$$p(H_0|\mathbf{x}) = \left\{1 + \left[\frac{p}{1-p}BF(H_0; H_1)\right]^{-1}\right\}^{-1}.$$

Assuming prior indifference $(p(H_0) = p(H_1) = 1/2)$ gives

$$p(H_0 \mid \mathbf{x}) = \left\{1 + \left[(1 + n)^{1/2}\exp\left\{-\frac{n}{n+1}\frac{z^2}{2}\right\}\right]^{-1}\right\}^{-1}.$$

The posterior probability of H_0 can be calculated for different values of

Table 6.1: *Values of $p(H_0|x)$.*

n	$|z|$ (p-value)	
	1.96 (0.05)	2.576 (0.01)
1	0.35	0.21
10	0.37	0.14
100	0.60	0.27
1000	0.80	0.53

n and z since both classical and Bayesian tests are based on z^2. Table 6.1 shows the $p(H_0 \mid \mathbf{x})$ working with the most common values, associated with the p-values 0.05 and 0.01, for different sample sizes.

The probabilities of H_0 get smaller as the value of $|z|$ gets larger, which is reasonable. What is not reasonable is the values that are obtained for these probabilities. But they reinforce the idea that p-values should not be taken as probabilities that can be associated with H_0. Another interesting result is that for any given value of z, posterior probabilities can vary substantially from a very low value that would lead to rejection of H_0 to a very large value that would lead to acceptance of H_0. One possible way to reconcile these findings with the significance level is that levels should also be changed with sample size. Large sample size should call for a reduced level of significance and vice versa. A more detailed discussion of the subject can be found in Berger and Sellke (1987).

Assume now the same situation but with an unknown observational variance σ^2 and take $\phi = \sigma^{-2}$. A prior for ϕ must also be specified. Since the hypotheses do not involve ϕ, it is reasonable to take the same marginal prior for ϕ under both hypotheses. Adopting conjugate priors under both hypotheses gives

$$n_0 \sigma_0^2 \phi \sim \chi_{n_0}^2 \text{ under } H_0 \text{ and } H_1 \text{ and } \theta \mid \phi, H_1 \sim N(\mu, (c\phi)^{-1}).$$

The relevant quantities for the calculation of the Bayes factor are

$$p(\mathbf{x} \mid H_0) = \int p(\mathbf{x} \mid \theta_0, \phi)\, p(\phi \mid H_0)\, d\phi$$

and

$$p(\mathbf{x} \mid H_1) = \int \int p(\mathbf{x} \mid \theta, \phi)\, p(\theta \mid \phi, H_1)\, p(\phi \mid H_1)\, d\theta d\phi,$$

where all the above densities are known. Substituting their expressions gives

$$BF(H_0; H_1) = \left(\frac{c+n}{c}\right)^{1/2}$$

$$\times \left\{ \frac{n_0\sigma_0^2 + (n-1)s^2 + [cn/(c+n)](\bar{x} - \mu)^2}{n_0\sigma_0^2 + (n-1)s^2 + n(\bar{x} - \theta_0)^2} \right\}^{(n_0+n)/2},$$

where now $s^2 = [1/(n-1)] \sum (x_i - \bar{x})^2$.

It is interesting to study the behavior of the Bayes factor in extreme situations such as with non-informative priors. Taking $n_0 \to 0$ and assuming as previously that $\mu = \theta_0$ gives

$$
\begin{aligned}
BF(H_0; H_1) &= \left(\frac{c+n}{c}\right)^{1/2} \left\{ \frac{(n-1)s^2 + [c/(c+n)]n(\bar{x} - \theta_0)^2}{(n-1)s^2 + n(\bar{x} - \theta_0)^2} \right\}^{n/2} \\
&= k^{-1/2} \left\{ \frac{n-1+kt^2}{n-1+t^2} \right\}^{n/2},
\end{aligned}
$$

where $k = c/(c+n)$ and $t = \sqrt{n}(\bar{x} - \theta_0)/s$. The above expression is graphed in Figure 6.2. The Bayes factor is a symmetric function of the sample values through the statistic t. It varies from the highest value of support of H_0 when $t = 0$ to a minimum value when $|t| \to \infty$, as expected.

In the general case, the hypotheses can be incorporated to the prior distribution and the problem of hypothesis testing can be thought of as comparison of possible alternative distributions for $\boldsymbol{\theta}$. For example, in the above test, the priors corresponding to H_0 and H_1 would be $P(\theta = \theta_0 | \phi) = 1$ and $\theta | \phi \sim N(\mu, (c\phi)^{-1})$. The prior corresponding to H_0 is degenerate. This will be the case whenever one of the hypotheses corresponds to a parameter space with smaller dimension than the complete parameter space Θ.

A common problem with the Bayes Factor is observed when we have a uniform non-informative prior in a non-limited parametric space. So the Bayes factor cannot be directly computed once the priori distribution is not a proper density function. Indeed, the prior probabilities of the hypothesis suppositions are not defined. Such problems may also occur even in the presence of proper uniform priors, as illustrated in the example below.

Example 6.4 *Consider $\boldsymbol{X} = (X_1, \ldots, X_n)$ a random sample of the $N(\mu, \sigma^2)$, with σ^2 known and prior of μ is $50 - 50$ mixed distribution of the degenerate distribution in μ_0 and a uniform distribution in $(-c, c)$, i.e., the prior distribution for μ is given by $\pi(\mu) = 0.5 I_{\mu_0}(\mu) + 0.5(1/2c) I_\mu(-c, c)(1 - I_{\mu_0}(\mu))$. If we consider the hypotheses $H_0 : \mu = \mu_0$ vs $H_1 : \mu \neq \mu_0$, the Bayes factor is*

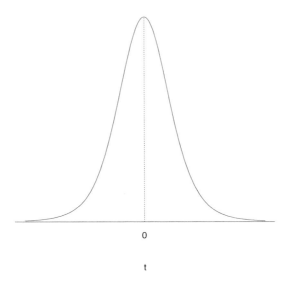

Figure 6.2: *Bayes factor for H_0: $\theta = \theta_0$ versus H_1: $\theta \neq \theta_0$ as a function of t.*

given by

$$BF(H_0; H_1) = \frac{2c\sqrt{n}\phi(z)/\sigma}{\left[\Phi\left(\sqrt{n}\,\frac{\bar{x}+c}{\sigma}\right) - \Phi\left(\sqrt{n}\,\frac{\bar{x}-c}{\sigma}\right)\right]},$$

where $z = \sqrt{n}(\bar{x} - \mu_0)/\sigma$ and $\phi(.)$ and $\Phi(.)$ are the density and distribution functions of standard Normal distribution, respectively. The value of the Bayes factor is strongly affected by the choice of the value of c, which should have small influence on the resulting inference.

Various alternatives have been proposed in order to overcome these limitations of the Bayes factor. They are presented in the next section.

6.3.1   Bayesian testing alternatives

A first alternative is the posterior Bayes factor (*PBF*; Aitkin, 1991) defined by

$$PBF(H_0; H_1) = \frac{\overline{p(\boldsymbol{x}|H_0)}}{p(\boldsymbol{x}|H_1)},$$

where $\overline{p(\boldsymbol{x}|H_j)}$ is the posterior mean of the marginal likelihood function $p(\boldsymbol{x}|H_j)$, which can be computed by

$$\overline{p(\boldsymbol{x}|H_j)} = \int_{\Theta_j} p(\boldsymbol{x}|H_j)\pi(\boldsymbol{\theta}|\boldsymbol{x})d\boldsymbol{\theta} = \frac{\int_{\Theta_j} p(\boldsymbol{x}|H_j)^2\pi(\boldsymbol{\theta})d\boldsymbol{\theta}}{\int_{\Theta_j} p(\boldsymbol{x}|H_j)\pi(\boldsymbol{\theta})d\boldsymbol{\theta}},$$

for $j = 0, 1$.

The posterior Bayes factor does not present the problem of the Bayes factors shown in Example 6.4 since the constant $(2c)^{-1}$ would be canceled in the ratio of the integrals of $p(\boldsymbol{x}|H_j)^2$ and $p(\boldsymbol{x}|H_j)$. The data is, however, used twice. This is at least a debatable procedure, since it is not consistent with basic Bayesian rules.

The problem lying at the heart of posterior Bayes factors is the use of the entire dataset to correct the impropriety of the prior distribution. The partial Bayes factor proposed by O'Hagan (1995) tries to mitigate that by replacing the improper prior by the posterior distribution based on part of the data, called training sample and denoted by $\boldsymbol{x}_t$, and obtaining only the marginal distribution for the other part of the data, called validation sample and denoted by $\boldsymbol{x}_v = \{\boldsymbol{x}\}\backslash\{\boldsymbol{x}_t\}$.

The Bayes factor can be rewritten as

$$BF(H_0; H_1; \boldsymbol{x}) = \frac{m_0(\boldsymbol{x})}{m_1(\boldsymbol{x})}, \text{ where } m_j(\boldsymbol{x}) = \int_{\Theta_j} p(\boldsymbol{x}|H_j)\pi(\boldsymbol{\theta})d\boldsymbol{\theta}.$$

It follows that the partial Bayes factor $(PaBF)$ is then defined as

$$PaBF(H_0; H_1; \boldsymbol{x}) = \frac{q_0(\boldsymbol{x}_v|\boldsymbol{x}_t)}{q_1(\boldsymbol{x}_v|\boldsymbol{x}_t)} = \frac{\int_{\Theta_0} p(\boldsymbol{x}_v|H_0, \boldsymbol{x}_t)\,\pi(\boldsymbol{\theta}|\boldsymbol{x}_t)\,d\boldsymbol{\theta}}{\int_{\Theta_1} p(\boldsymbol{x}_v|H_1, \boldsymbol{x}_t)\,\pi(\boldsymbol{\theta}|\boldsymbol{x}_t)\,d\boldsymbol{\theta}},$$

where

$$q_j(\boldsymbol{x}_v|\boldsymbol{x}_t) = \frac{\int_{\Theta_j} p(\boldsymbol{x}|H_j)\pi(\boldsymbol{\theta})d\boldsymbol{\theta}}{\int_{\Theta_j} p(\boldsymbol{x}_t|H_j)\pi(\boldsymbol{\theta})d\boldsymbol{\theta}} = \frac{m_j(\boldsymbol{x})}{m_j(\boldsymbol{x}_t)}.$$

It can thus be shown that $PaBF(H_0; H_1; \boldsymbol{x}) = BF(\boldsymbol{x})/BF(\boldsymbol{x}_t)$.

The partial Bayes factor corresponds to a Bayes factor if $\boldsymbol{x}_t = \{\emptyset\}$ and $\boldsymbol{x}_v = \boldsymbol{x}$. Moreover, the posterior Bayes factor corresponds to a partial Bayes factor with training and validation samples given by $\boldsymbol{x}$. Note that when the data set is larger, the size of the training sample can also be larger. In any case, whatever the choice of the training and validation samples, data values are used only once.

The fractional Bayes factor (FBF), also proposed by O'Hagan (1995), avoids the arbitrariness of choosing a particular $\boldsymbol{x}_t$. It is defined as

$$FBF(H_0; H_1; b) = \frac{q_0(b, \boldsymbol{x})}{q_1(b, \boldsymbol{x})},$$

where

$$q_j(b, \boldsymbol{x}) = \frac{\int_{\boldsymbol{\Theta}_j} p(\boldsymbol{x}|H_j)\, \pi(\boldsymbol{\theta})\, d\boldsymbol{\theta}}{\int_{\boldsymbol{\Theta}_j} p(\boldsymbol{x}_t|H_j)^b\, \pi(\boldsymbol{\theta})\, d\boldsymbol{\theta}}$$

and $b = n/m$, where m is the length of $\boldsymbol{x}_t$.

The intrinsic Bayes factor (IBF; Berger and Pericchi, 1996) corresponds to the partial Bayes factor with $\boldsymbol{x}_t$ as a minimal training sample according to the following definition.

Definition 6.4 *Given an improper prior distribution $\pi(.)$, the training sample $\boldsymbol{x}_t$ is proper if $0 < m_i(\boldsymbol{x}_t) < \infty$ for i=1,2. The training sample is minimal if it is proper and has no proper subset.*

Minimal training samples typically consist on p observations when p is the dimension of $\boldsymbol{\Theta}$. When data set is large, the presence of more than one minimal training sample is common, and the intrinsic Bayes factor will depend on the choice of the minimal training sample. In order to increase IBF stability, Berger and Pericchi (1996) proposed to use the average (AIBF) of intrinsic Bayes factors over all minimal training samples.

A Bayesian alternative to significance tests was proposed by G. E. P. Box & G. C. Tiao in the 60's, with further developments by C. Pereira and co-authors in the 90's. Let $p^* = \sup\{p(\boldsymbol{\theta}|\boldsymbol{x}) : \boldsymbol{\theta} \in \boldsymbol{\Theta}_0\}$ be the highest posterior density value obtained under $H_0 : \boldsymbol{\theta} \in \boldsymbol{\Theta}_0$ and $\boldsymbol{C} = \{\boldsymbol{\theta} \in \boldsymbol{\Theta} : p(\boldsymbol{\theta}|\boldsymbol{x}) > p^*\}$ be the set of parameter values that are more probable than those in H_0. If this set contains a substantial amount of probability, this indicates that H_0 is not supported by the information available in the posterior distribution. The posterior probability of $\boldsymbol{C}$ is given by

$$P(\boldsymbol{\theta} \in \boldsymbol{C}|\boldsymbol{x}) = \int_C \pi(\boldsymbol{\theta}|\boldsymbol{x})d\boldsymbol{\theta}.$$

An informal testing procedure consists of rejecting H_0 for small values of the above probability and accepting H_0 for large values of the above probability.

Example 6.5 *Consider $X_1, \ldots, X_n$ a random sample of a $N(\mu, \sigma^2)$ with σ^2 known and assume an improper uniform prior distribution for μ, $p(\mu) = c$. Suppose the interest is to test $H_0 : \mu = \mu_0$ vs $H_0 : \mu \neq \mu_0$.*

The posteriori distribution of μ is $N(\bar{x}, \sigma^2/n)$ and the likelihood function is given by

$$l(\mu) = \exp\left\{-\frac{n}{2\sigma^2}(\mu - \bar{x})^2\right\}.$$

The $m_i(\boldsymbol{x})$ functions are given by $m_0(\boldsymbol{x}) = l(\mu_0)$ and $m_1(\boldsymbol{x}) = c\sqrt{2\pi\sigma^2/n}$.

The Bayes factor is computed as

$$BF = \frac{m_0(\boldsymbol{x})}{m_1(\boldsymbol{x})} = \frac{l(\mu_0)}{c\sqrt{2\pi\sigma^2/n}}.$$

Note that BF depends on the arbitrary constant c.

The posterior Bayes factor depends on the posterior means $\overline{p(\boldsymbol{x}|H_0)} = E[l(\mu)|\boldsymbol{x}, H_0] = l(\mu_0)$ and $\overline{p(\boldsymbol{x}|H_1)} = E[l(\mu)|\boldsymbol{x}, H_1] = \sqrt{2}$, and is given by

$$PBF = \frac{\overline{p(\boldsymbol{x}|H_0)}}{\overline{p(\boldsymbol{x}|H_1)}} = \frac{l(\mu_0)}{\sqrt{2}}.$$

Letting m be the size of the training sample $\boldsymbol{x}_t$ and $\bar{x}_t$ be its mean,

$$m_0(\boldsymbol{x}_t) = l_t(\mu_0) = \exp\left\{-\frac{m}{2\sigma^2}(\mu_0 - \bar{x}_t)^2\right\} \text{ and } m_1(\boldsymbol{x}_t) = c\sqrt{2\pi\sigma^2/m},$$

and

$$q_0(\boldsymbol{x}_v|\boldsymbol{x}_t) = \frac{m_0(\boldsymbol{x})}{m_0(\boldsymbol{x}_t)} = \frac{\exp\left\{-\frac{n}{2\sigma^2}(\mu_0 - \bar{x})^2\right\}}{\exp\left\{-\frac{m}{2\sigma^2}(\mu_0 - \bar{x}_t)^2\right\}}$$

and

$$q_1(\boldsymbol{x}_v|\boldsymbol{x}_t) = \frac{m_1(\boldsymbol{x})}{m_1(\boldsymbol{x}_t)} = \sqrt{\frac{m}{n}}.$$

So the partial Bayes factor is given by

$$PaBF = \frac{q_0(\boldsymbol{x}_v|\boldsymbol{x}_t)}{q_1(\boldsymbol{x}_v|\boldsymbol{x}_t)} = \frac{\exp\left\{\frac{m}{2\sigma^2}(\mu_0 - \bar{x}_t)^2\right\}/\sqrt{m}}{\exp\left\{\frac{n}{2\sigma^2}(\mu_0 - \bar{x})^2\right\}/\sqrt{n}}.$$

Note that $\boldsymbol{x}_t$ is always proper, provided it is not the null sample. So it is minimal if $m = 1$, the dimension of the parameter space. Thus, $\boldsymbol{x}_t = \{x_i\}$ is minimal for all $i \in \{1, \ldots, n\}$.

The intrinsic Bayes factors and average intrinsic Bayes factors are respectively given by

$$IBF_i = \frac{\exp\left\{\frac{1}{2\sigma^2}(\mu_0 - x_i)^2\right\}}{\exp\left\{\frac{n}{2\sigma^2}(\mu_0 - \bar{x})^2\right\}/\sqrt{n}},$$

and

$$AIBF = \frac{\sum_{i=1}^{n}\exp\left\{\frac{1}{2\sigma^2}(\mu_0 - x_i)^2\right\}/n}{\exp\left\{\frac{n}{2\sigma^2}(\mu_0 - \bar{x})^2\right\}/\sqrt{n}}.$$

For computing the fractional Bayes factor, we need to calculate first $q_0(b, \boldsymbol{x}) = l(\mu_0)/(\mu_0)^b$ and

$$q_1(b, \boldsymbol{x}) = \frac{\int_{\Theta_1} l(\mu)\,\pi(\mu|H_0)\,d\mu}{\int_{\Theta_1}[l(\mu)]^b\,\pi(\mu|H_0)d\mu} = \frac{c\sqrt{2\pi\sigma^2/n}}{c\sqrt{2\pi\sigma^2/bn}} = \sqrt{b}.$$

Then,

$$FBF(H_0; H_1) = \frac{q_0(b, \boldsymbol{x})}{q_1(b, \boldsymbol{x})} = \frac{1}{\sqrt{b} \, [l(\mu_0)]^{b-1}}.$$

For computing the Bayesian significance test, note that $p^* = \sup\{p(\mu|\boldsymbol{x}) : \mu = \mu_0\} = p(\mu_0|\boldsymbol{x})$ *and* $C = \{\mu \in \mathbb{R}\backslash\{\mu_0\} : p(\mu|\boldsymbol{x}) > p(\mu_0|\boldsymbol{x})\} = (\bar{x} - v, \bar{x} + v)$ *where* $v = |\bar{x} - \mu_0|$.

The probabilistic evidence against H_0 *may be assessed by the posterior probability*

$$P(\mu \in C|\boldsymbol{x}) = \int_{\bar{x}-v}^{\bar{x}+v} \pi(\mu|\boldsymbol{x})d\mu = \Phi\left(\frac{\sqrt{n}\,(\bar{x}+v)}{\sigma}\right) - \Phi\left(-\frac{\sqrt{n}\,(\bar{x}-v)}{\sigma}\right).$$

So, the null hypothesis H_0 *should be rejected for large values of the above probability.*

Example 6.6 *(Example 6.5 continued) Considering the observed values* $\boldsymbol{x} = (5.68, \; -2.92, \; 0.25, \; 3.80, \; -0.96, \; 6.16, \; 4.42, \; -5.60, \; 3.16, \; -1.85, \; -1.78, \; -1.60, \; -0.66, \; 0.21, \; -1.57)$ *generated from an* $N(0, 7)$ *distribution, assume one wishes to test* $H_0 : \mu = 0$. *Then, application of the results of Example 6.5 leads to* $PBF = 0.569$, $AIBF = 11.272$, $FBF = 1 / (\sqrt{b} \, 0.953^b)$, *which depends on* b, *and the posterior probability of the set of values more probable than* H_0 *is 0.530.*

Figure 6.3 shows the $PaBF$ *values for* $\boldsymbol{x}_t = (x_1, \ldots, x_r)$ *with* $r = 1, \ldots, n$. *The PaBFs are always greater than 1. Thus, there is no evidence against* H_0. *Figure 6.4 shows the FBF values for different values of* b. *The smaller value of the FBF is obtained for* $b \cong 2.2$, *but it is still larger than 1. Figure 6.5 shows the posterior probability density function of* μ *with indication of region more probable than* H_0 *a posteriori.*

It will be seen in Chapter 8 that it is common to test if a few of the components of $\boldsymbol{\theta}$ are null. Writing $\boldsymbol{\theta} = (\boldsymbol{\theta}_1 \, , \boldsymbol{\theta}_2)$, where $\boldsymbol{\theta}_1$ is $(p-q)$-dimensional and $\boldsymbol{\theta}_2$ is q-dimensional, one may wish to test whether $\boldsymbol{\theta}_2 = \boldsymbol{0}$. In this case, $P(\boldsymbol{\theta} \mid H_0)$ will be concentrated on a $(p-q)$-dimensional subspace of R^p passing through the point $\boldsymbol{\theta}_2 = \boldsymbol{0}$. This point will be returned to in Section 6.5.

6.4   Hypothesis testing and confidence intervals

You may have noticed the strong connection between hypothesis testing and confidence intervals. This connection is clearer with the frequentist framework, but it is also relevant under the Bayesian approach as will shortly be seen.

Starting with the classical approach, consider a level α test for the hypothesis that $\boldsymbol{\theta} = \boldsymbol{\theta}_0$. Assume that, based on a given sample size, the test yields

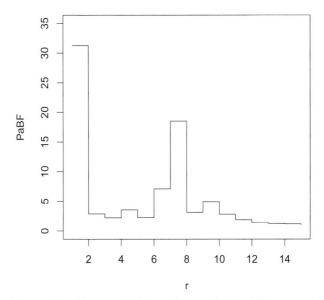

Figure 6.3: *The partial Bayes factor plot for different r values.*

an acceptance region $A = A(\boldsymbol{\theta}_0)$ where it is important to explicitly denote the dependence of the region on $\boldsymbol{\theta}_0$. So,

$$P(\mathbf{X} \in A(\boldsymbol{\theta}_0) \mid \boldsymbol{\theta}_0) = 1 - \alpha.$$

Varying the value of $\boldsymbol{\theta}_0$ in $\boldsymbol{\Theta}$ leads to a family of regions, all depending on $\boldsymbol{\theta}$ and with probability $1 - \alpha$. This automatically implies that, after observing the value $\mathbf{x}$ for $\mathbf{X}$, the region

$$\{\boldsymbol{\theta} : \mathbf{x} \in A(\boldsymbol{\theta})\}$$

will have confidence $1 - \alpha$.

Example 6.7 *Let $\boldsymbol{X} = (X_1, \ldots, X_n)$ be a random sample from the $N(\theta, \sigma^2)$ distribution with known σ^2. The UMP unbiased test of level α to test H_0: $\theta = \theta_0$ versus H_1: $\theta \neq \theta_0$ has acceptance region $\{\boldsymbol{X} : \theta_0 - z_{\alpha/2}\sigma/\sqrt{n} \leq \bar{X} \leq \theta_0 + z_{\alpha/2}\sigma/\sqrt{n}\}$. This interval can be rewritten as $\{\boldsymbol{X} : \bar{X} - z_{\alpha/2}\sigma/\sqrt{n} \leq \theta_0 \leq \bar{X} + z_{\alpha/2}\sigma/\sqrt{n}\}$. Replacing now θ_0 by θ gives the $1 - \alpha$ confidence interval for θ*

$$\left\{\theta : \bar{x} - \frac{\sigma}{\sqrt{n}} z_{\alpha/2} \leq \theta \leq \bar{x} + \frac{\sigma}{\sqrt{n}} z_{\alpha/2}\right\}.$$

Of course, this is exactly the same interval obtained in Section 4.4.

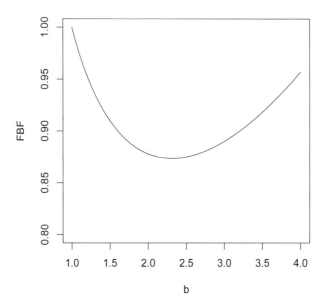

Figure 6.4: *The fractional Bayes factor plot for different values of b.*

This relation is more heavily explored in the next section when asymptotic tests are introduced. In fact, this relation can be used also in the reverse direction with hypothesis tests obtained from confidence regions. To see this, suppose that $\{\boldsymbol{\theta} : G(\mathbf{x}, \boldsymbol{\theta}) \in \boldsymbol{C}\}$ is a $100(1 - \alpha)\%$ confidence region for $\boldsymbol{\theta}$. Then, for every value $\boldsymbol{\theta}_0$ of $\boldsymbol{\theta}$, it is true that $P(G(\mathbf{X}, \boldsymbol{\theta}_0) \in \boldsymbol{C} \mid \boldsymbol{\theta}_0) = 1 - \alpha$. A level α test for the hypothesis $\boldsymbol{\theta} = \boldsymbol{\theta}_0$ can be defined by the critical region $\{\mathbf{X} : G(\mathbf{X}, \boldsymbol{\theta}_0) \notin \boldsymbol{C}\}$.

This brings us naturally to an alternative definition of Bayesian hypothesis testing. The method consists of constructing a credibility region for $\boldsymbol{\theta}$ with probability $1 - \alpha$ and accepting the hypothesis $\boldsymbol{\theta} = \boldsymbol{\theta}_0$ if the above region contains the value $\boldsymbol{\theta}_0$. This method was proposed and extensively used by Lindley (1965).

Example 6.8 *Taking again a random sample* $\mathbf{X} = (X_1, \ldots, X_n)$ *from the* $N(\theta, \sigma^2)$ *distribution with known* σ^2 *and prior* $\theta \sim N(\mu, \tau^2)$ *gives the posterior* $\theta \mid x \sim N(\mu_1, \tau_1^2)$. *The* $100(1 - \alpha)\%$ *credibility interval for* θ *is given by* $[\mu_1 - \tau_1 z_{\alpha/2}, \mu_1 + \tau_1 z_{\alpha/2}]$. *The hypothesis* $\theta = \theta_0$ *can be accepted if* θ_0 *belongs to the interval above. In the non-informative case,* $\mu_1 \to \bar{x}$ *and* $\tau_1 \to \sigma/\sqrt{n}$, *and the hypothesis is accepted if* θ_0 *belongs to the interval* $[\bar{x} - z_{\alpha/2}\sigma/\sqrt{n}, \bar{x} + z_{\alpha/2}\sigma/\sqrt{n}]$, *coinciding with the classical test.*

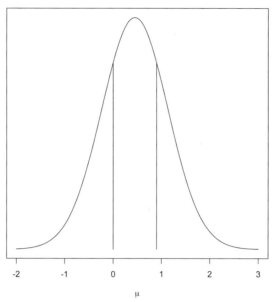

Figure 6.5 *The posterior probability density function of μ. The two vertical lines indicate the limits of region with higher posterior probability than H_0.*

This procedure provides a unifying approach for obtaining tests of hypotheses from interval estimation, irrespective of whether a Bayesian or a frequentist approach is used. One should aim to construct regions (intervals, in the scalar case) with the smallest possible volume (length, in the scalar case). The value of α is typically low but there is no prescription about the value to be adopted in any given problem.

In the previous section, Bayesian tests of a simple hypothesis in the form $\theta = \theta_0$ were defined by attributing a lump prior probability p to this hypothesis, and the remaining $1 - p$ was distributed according to some probability distribution of $\theta|H_1$. This specification is criticized for giving a sometimes unjustified special, different status to the value θ_0.

This discussion is in fact wider and goes beyond the boundaries of the Bayesian thinking. It has to do with the capacity of single hypothesis to represent adequately the situation under study. Some authors suggest that these hypotheses are always at most useful approximations of more realistic hypotheses where θ actually belongs to a neighborhood of θ. We will not pursue this discussion here further other than acknowledging that the spectrum of thoughts on this matter goes from total rejection to total acceptance of this formulation. Good references for this discussion are Berger and Delampady (1987) and Lindley (1993).

6.5 Asymptotic tests

Asymptotic tests are those based on asymptotic approximation for the distribution of the test quantity, irrespective of whether it is under the Bayesian or frequentist paradigm. This is a very broad definition including results based on the central limit theorem. The use of asymptotic theory to develop useful test statistics was carried out by Bartlett, Wald and Wilks in the 40's among others. We shall concentrate here on four usual tests based on the maximum likelihood estimator (Wald, 1943), on the score function (Rao, 1948), on the maximum likelihood ratio (Wilks, 1938), and on a recent test based on the gradient function (Terrell, 2002). In most cases, we are led to a limiting χ^2 distribution.

In many cases, it is not possible to analytically obtain the exact distribution of the MLR statistic $\lambda(\mathbf{X})$, and asymptotic methods are frequently used. Suppose that $\boldsymbol{\theta} \in \boldsymbol{\Theta} \subset R^p$ and one wishes to test the hypothesis $H_0: \boldsymbol{\theta} = \boldsymbol{\theta}_0$. In this case, a Taylor series expansion of the function $L(\boldsymbol{\theta}_0; \mathbf{X}) = \log p(\mathbf{X} \mid \boldsymbol{\theta}_0)$ around $\hat{\boldsymbol{\theta}}$ gives

$$
\begin{aligned}
L(\boldsymbol{\theta}_0; \mathbf{X}) &\simeq L(\hat{\boldsymbol{\theta}}; \mathbf{X}) + [\mathbf{U}(\mathbf{X}; \hat{\boldsymbol{\theta}})]'(\boldsymbol{\theta}_0 - \hat{\boldsymbol{\theta}}) - \frac{1}{2}(\boldsymbol{\theta}_0 - \hat{\boldsymbol{\theta}})'\mathbf{J}(\hat{\boldsymbol{\theta}})(\boldsymbol{\theta}_0 - \hat{\boldsymbol{\theta}}) \\
&= L(\hat{\boldsymbol{\theta}}; \mathbf{X}) - \frac{1}{2}(\boldsymbol{\theta}_0 - \hat{\boldsymbol{\theta}})'\mathbf{J}(\hat{\boldsymbol{\theta}})(\boldsymbol{\theta}_0 - \hat{\boldsymbol{\theta}}),
\end{aligned}
$$

since $\mathbf{U}(\mathbf{X}; \hat{\boldsymbol{\theta}}) = \mathbf{0}$. The higher order terms are neglected since, under H_0, $\boldsymbol{\theta}_0$ and $\hat{\boldsymbol{\theta}}$ are close for large n. Therefore,

$$
\begin{aligned}
-2 \log \lambda(\mathbf{X}) &= -2 \log \left[\frac{p(\mathbf{X} \mid \boldsymbol{\theta}_0)}{p(\mathbf{X} \mid \hat{\boldsymbol{\theta}})} \right] \\
&= -2 \left[L(\boldsymbol{\theta}_0; \mathbf{X}) - L(\hat{\boldsymbol{\theta}}; \mathbf{X}) \right] \\
&\simeq (\boldsymbol{\theta}_0 - \hat{\boldsymbol{\theta}})'\mathbf{J}(\hat{\boldsymbol{\theta}})(\boldsymbol{\theta}_0 - \hat{\boldsymbol{\theta}}).
\end{aligned}
$$

Since the maximum likelihood estimator is asymptotically Normal and $\mathbf{J}(\hat{\boldsymbol{\theta}})/n$ converges almost surely to its expectation $\mathbf{I}(\boldsymbol{\theta}_0)$ under H_0, the quadratic form to the right-hand side of the equation has a χ_p^2 asymptotic distribution. Therefore, the asymptotic distribution of $W_{LR} = -2 \log \lambda(\mathbf{X})$ is a χ_p^2, and the MLR test with asymptotic level α accepts H_0 if $W_{LR} < \chi_{\alpha,p}^2$.

More generally, if the null hypothesis is of the form $H_0: \boldsymbol{\theta} \in \boldsymbol{\Theta}_0$ with dim $\Theta_0 = p - q > 0$, then a revised test statistic W_{LR} is written as

$$W_{LR} = -2\log \lambda(\boldsymbol{X}) = -2\log \left[\frac{p(\boldsymbol{X}|\hat{\boldsymbol{\theta}}_0)}{p(\boldsymbol{X}|\hat{\boldsymbol{\theta}})} \right],$$

where $\hat{\boldsymbol{\theta}}_0$ denotes the maximum likelihood estimator of $\boldsymbol{\theta}$ restricted to Θ_0. The asymptotic distribution of W_{LR} is a χ_q^2, and the MLR test with asymptotic level α accepts H_0 if $W_{LR} < \chi_{\alpha,q}^2$.

These results are also useful in the construction of confidence intervals. The idea is once again to explore the relation between confidence intervals and hypothesis tests. Suppose that H_0: $\boldsymbol{\theta} = (\theta_1, \ldots, \theta_{p-q}, \theta_{p-q+1,0}, \ldots, \theta_{p,0})$. In other words, the last q components of $\boldsymbol{\theta}$ are fixed. Let the maximum likelihood ratio now be denoted by $\lambda(\boldsymbol{X}; \theta_{p-q+1,0}, \ldots, \theta_{p,0})$. A $100(1 - \alpha)\%$ confidence region for $(\theta_{p-q+1}, \ldots, \theta_p)$ is given by

$$\left\{ (\theta_{p-q+1}, \ldots, \theta_p) : -2\log \lambda(\mathbf{x}; \theta_{p-q+1}, \ldots, \theta_p) < \chi_{\alpha,q}^2 \right\}.$$

Similarly, from the Bayesian point of view, the asymptotic posterior distribution of $-2\log \lambda(\mathbf{x}; \theta_{p-q+1}, \ldots, \theta_p)$ is a χ_q^2. Tests and confidence regions can be constructed as described earlier. Note that we have chosen to test for the last q components for simplicity. The same results hold for a test on any q components of $\boldsymbol{\theta}$. One simply has to reorder the $\boldsymbol{\theta}$ components accordingly.

The asymptotic distributions of the maximum likelihood estimator and of the score function lead to two classes of classical tests. Assume that $\boldsymbol{\theta} \in \Theta \subset R^p$ and one wishes to test H_0: $\boldsymbol{\theta} = \boldsymbol{\theta}_0$. Defining then

$$W_E(\boldsymbol{\theta}_0) = (\hat{\boldsymbol{\theta}} - \boldsymbol{\theta}_0)'\mathbf{I}(\hat{\boldsymbol{\theta}})(\hat{\boldsymbol{\theta}} - \boldsymbol{\theta}_0)$$

and

$$W_U(\boldsymbol{\theta}_0) = [U(\mathbf{X}; \boldsymbol{\theta}_0)]'\mathbf{I}^{-1}(\boldsymbol{\theta}_0)\mathbf{U}(\mathbf{X}; \boldsymbol{\theta}_0),$$

we have that both statistics have a χ_p^2 asymptotic distribution under H_0. So, the tests of asymptotic level α reject H_0 if $W_E(\boldsymbol{\theta}_0) > \chi_{\alpha,p}^2$ and $W_U(\boldsymbol{\theta}_0) > \chi_{\alpha,p}^2$, respectively. Given the almost sure convergence of $J(\hat{\boldsymbol{\theta}})$ and $J(\boldsymbol{\theta}_0)$ to $I(\boldsymbol{\theta}_0)$, these replacements can be made in the definitions of W_E and W_U and the same results are obtained. Equivalent Bayesian results can be obtained that the asymptotic posterior distributions of $W_E(\boldsymbol{\theta})$ and $W_U(\boldsymbol{\theta})$ are χ_p^2. These results were partially presented in Section 5.3. Hypothesis testing then can be made as described in the previous section.

The tests based on W_E and W_U can also be applied to situations where $H_0 : \boldsymbol{\theta} \in \Theta_0$, with dim $\Theta_0 = p - q > 0$. In this case, the value of $\boldsymbol{\theta}_0$ is replaced in their expressions by $\hat{\boldsymbol{\theta}}_0$, the maximum likelihood estimator of $\boldsymbol{\theta}$ under H_0. Their asymptotic distribution becomes a χ_q^2, just as in the maximum likelihood ratio test.

Note that three general tests have been defined and they all have asymptotic χ_ν^2 distribution under H_0 where the number ν of degrees of freedom is the difference between the dimensions of $\boldsymbol{\Theta}$ and $\boldsymbol{\Theta}_0$. The test based on the maximum likelihood ratio depends on maximization under both hypotheses whereas the score test requires maximization under the null hypothesis and the test based on the maximum likelihood estimator requires maximization only under the alternative hypothesis. In most but not all cases, the null hypothesis provides a simplification to the model. It is then simpler to estimate parameters under the null hypotheses which computationally favors the score test. When estimation under the null hypothesis is harder, it is simpler to perform the test based on the maximum likelihood estimator. A good reference for comparison between and interpretation of these tests is Buse (1982).

Example 6.9 *Assume a specific model for observations $Y_1, \ldots, Y_n$ is of the form $Y_i \sim N(f_i(\boldsymbol{\theta}), \sigma^2)$ where $\boldsymbol{\theta} = (\theta_0, \theta_1, \theta_2)$ and $f_i(\boldsymbol{\theta}) = \theta_0 + \theta_1 \exp(\theta_2 z_i)$, $i = 1, \ldots, n$. Let the null hypothesis be in the form $H_0: \theta_2 = \theta_{2,0}$. Under the null hypothesis the model becomes a simple linear regression with known regressor variable $x_i = \exp(\theta_{2,0} z_i)$, $i = 1, \ldots, n$. It is then simpler to estimate the model under the null hypothesis. If, however, the null hypothesis is of the form $H_0: \partial f_i / \partial z_i = 1/3$, then estimation under the null hypothesis becomes a non-linear problem with restriction whereas estimation under the alternative is only a non-linear problem. In those cases, it is easier to use the test based on the maximum likelihood estimator.*

Another interesting question concerns the appropriateness of the asymptotic approximation. It can be shown that the approximation of $\lambda(\mathbf{X})$ to the χ_p^2 distribution is of order n^{-1}. Bartlett (1947) showed that $P(\lambda(\mathbf{X}) \leq x) = P(Z \leq x) + O(n^{-1})$, where $Z \sim \chi_p^2$, and obtained a corrected MLR statistic $\lambda^*(\mathbf{X}) = \lambda(\mathbf{X})[1 + b(\boldsymbol{\theta}_0)/n]^{-1}$ such that $E[\lambda^*(\mathbf{X})] = p + O(n^{-2})$ for many multivariate problems. Lawley (1956) showed that all moments of λ^* agree with those of a χ_p^2 distribution to order n^{-2}. Cordeiro (1987) proved that this approximation order is also valid for the distribution function of λ^*. This result was extended further to any test statistic with asymptotic χ_p^2 distribution by Cordeiro & Ferrari (1991).

A particular case of special interest is when n items are observed and classified independently into one of p possible groups. The parameter of interest is the vector with the group probabilities $\boldsymbol{\theta} = (\theta_1, \ldots, \theta_{p-1})$ where the probability for the pth group is obtained from the unit sum restriction. Suppose one wishes to test H_0: $\boldsymbol{\theta} = \boldsymbol{\theta}_0 = (\theta_{1,0}, \ldots, \theta_{p-1,0})$. Then it can be shown (see Exercise 6.21) that the test statistics W_E and W_U are given by

$$W_E(\boldsymbol{\theta}_0) = \sum_{i=1}^{p} \frac{(N_i - n\theta_{i,0})^2}{N_i} \text{ and } W_U(\boldsymbol{\theta}_0) = \sum_{i=1}^{p} \frac{(N_i - n\theta_{i,0})^2}{n\theta_{i,0}}.$$

Both have an asymptotic χ_{p-1}^2 distribution under H_0. The only difference between them is the replacement of N_i by $n\theta_{i,0}$. But, under H_0, $N_i/n \to \theta_{i,0}$ almost surely when $n \to \infty$, by the strong law of large numbers. This is an indication of an asymptotic equivalence between the two statistics. The tests reject H_0 when the values of the statistics W_E and W_U are larger than the $1 - \alpha$ quantile of the χ_{p-1}^2 distribution.

These tests try to measure how well a given hypothesis fits to the data. For this reason, they are known as goodness-of-fit tests and are heavily used in statistics whenever a situation can be represented in p mutually exclusive categories. Of course the assessment of the fit of a model is much more general and leads to a variety of other tests. Analogously, the Bayesian asymptotic result is that when the above quantities are written as functions of $\boldsymbol{\theta}$ (instead of $\boldsymbol{\theta}_0$), they will have asymptotic χ_{p-1}^2 posterior distribution.

When H_0 is no longer a simple hypothesis but has instead dimension $p - q - 1 > 0$, the goodness-of-fit statistics are given by

$$W_E(\hat{\boldsymbol{\theta}}_0) = \sum_{i=1}^{p} \frac{(N_i - n\hat{\theta}_{i,0})^2}{N_i} \text{ and } W_U(\hat{\boldsymbol{\theta}}_0) = \sum_{i=1}^{p} \frac{(N_i - n\hat{\theta}_{i,0})^2}{n\hat{\theta}_{i,0}},$$

where $\hat{\theta}_{i,0}$ is the MLE of θ_i, $i = 1, \ldots, p-1$ under H_0. Both test statistics have asymptotic χ_q^2 distribution under H_0 and reject H_0 if and only if the value of the statistic is larger than the $1 - \alpha$ quantile of the χ_q^2 distribution.

An important application of goodness-of-fit tests is the test of the fit of a given distribution to a dataset. In this case, a random sample $\mathbf{X} = (X_1, \ldots, X_n)$ from an unknown distribution is observed and one wishes to test whether this distribution is of a given known form. Partitioning the line into intervals I_i, $i = 1, \ldots, p$, the number of observations in each interval can be counted. These counts have jointly multinomial distribution with probabilities $\theta_{i,0} = P(X \in I_i | H_0)$, and the test of the fit is given as above. If the null hypothesis specifies a class of distributions with $q > 0$ unknown parameters, then one should first estimate the unknown parameters, by maximum likelihood, say, under H_0. Then, the statistics W_E and W_U can be evaluated with the estimates $\hat{\theta}_{i,0} = \hat{P}(X \in I_i | H_0)$. The test statistics will now have an asymptotic χ_{p-q-1}^2 distribution and the level α test rejects H_0 if the value of the test statistic is larger than the $1 - \alpha$ quantile of the χ_{p-q-1}^2 distribution.

Example 6.10 *Assume a random sample of size n, $X = (X_1, \ldots, X_n)$, and one wishes to test*

$$H_0 : X_1, \ldots, X_n \sim N(\mu, \sigma^2),$$

with a null hypothesis under a two dimensional parameter vector.

Under H_0, the maximum likelihood estimators of μ and σ^2 are given by

$$\hat{\mu} = \frac{1}{n} \sum_{i=1}^{n} X_i \quad and \quad \hat{\sigma}^2 = \frac{1}{n} \sum_{i=1}^{n} (X_i - \hat{\mu})^2.$$

For any fixed value of p and defined partition of $\mathbb{R}$ by $I = \{I_1, \ldots, I_p\}$, we have

$$\hat{\theta}_i = \int_{I_i} \frac{1}{\sqrt{2\pi\hat{\sigma}^2}} \exp\left[\frac{(x - \hat{\mu})^2}{2\hat{\sigma}^2}\right] dx, \quad i = 1, \ldots, p.$$

So the statistics of the tests based on the maximum likelihood estimator and on the score functions are given by

$$W_E(\hat{\boldsymbol{\theta}}_0) = \sum_{i=1}^{p} \frac{(N_i - n\hat{\theta}_{i,0})^2}{N_i} \quad and \quad W_U(\hat{\boldsymbol{\theta}}_0) = \sum_{i=1}^{p} \frac{(N_i - n\hat{\theta}_{i,0})^2}{n\hat{\theta}_{i,0}},$$

and in each test the null hypothesis is rejected if the statistics are larger than the $1 - \alpha$ quantile of the χ^2_{p-3} distribution.

Another important application of these tests to contingency tables is left as an exercise.

All these tests have the property that their power function converges to 1 when $n \to \infty$ for any parameter value in the alternative. This result can be obtained from the Taylor series expansion of the log likelihood around the point in the alternative. It follows essentially from the consistency of MLE. Tests with this property are said to be consistent.

Recently, Terrell (2002) proposed the so-called gradient test, as an alternative to the tests based on the maximum likelihood estimator, on the maximum likelihood ratio and on the score function. Suppose that interest is to test $H_0 : \boldsymbol{\theta}_2 = \boldsymbol{\theta}_{2,0}$, where $\boldsymbol{\theta} = (\boldsymbol{\theta}_1, \boldsymbol{\theta}_2)$, $\boldsymbol{\theta}_1 = (\theta_1, \ldots, \theta_{p-q})$ and $\boldsymbol{\theta}_2 = (\theta_{p-q+1}, \ldots, \theta_p)$, the gradient statistic under H_0 is given by

$$W_G(\tilde{\boldsymbol{\theta}}, \hat{\boldsymbol{\theta}}) = [U_2(\mathbf{X}; \tilde{\boldsymbol{\theta}})]'(\hat{\boldsymbol{\theta}}_2 - \boldsymbol{\theta}_{2,0}),$$

where $U_2(\mathbf{X}; \cdot)$ correspond to the $\boldsymbol{\theta}_2$ score function, $\hat{\boldsymbol{\theta}}$ is the unrestricted maximum likelihood estimator of $\boldsymbol{\theta}$ and $\tilde{\boldsymbol{\theta}}$ is the maximum likelihood estimator of $\boldsymbol{\theta}$ under H_0. Under H_0, W_G also has an asymptotic χ^2_q distribution with q degrees of freedom and the null hypothesis is rejected with level α if $W_G > \chi^2_{\alpha,q}$. So the gradient statistic has the same asymptotic distribution of the other tests. The advantage of the gradient statistics is that it is simpler to be computed, since it does not require calculation of any information matrix, neither expected nor observed.

The four asymptotic tests above can be compared in the case $H_0 : \theta = \theta_0$ vs $H_1 : \theta \neq \theta_0$. The test statistics are given by

$$
\begin{aligned}
W_{LR} &= 2[L(\hat{\theta}) - L(\theta_0)], \\
W_E &= (\hat{\theta} - \theta_0)^2 I(\hat{\theta}), \\
W_U &= [U(\theta_0)]^2 I(\theta_0)^{-1}, \\
W_G &= (\hat{\theta} - \theta_0)U(\theta_0).
\end{aligned}
$$

The test statistics can be seen as a balance of the distance between $\hat{\theta}$ and θ_0 weighed by the likelihood function in different ways. Moreover, the likelihood ratio statistic depends on the likelihood value calculated under the unrestricted hypothesis, while the W_G depends only on the score function under the null hypothesis.

Figure 6.6 shows an interpretative diagram of the hypothesis tests for comparison of the roles played by each of the four asymptotic tests presented above. The maximum likelihood ratio statistic is twice the distance between the log-likelihoods, evaluated at its maximum value $\hat{\theta}$ and at the value under H_0. For fixed $\hat{\theta}$, θ_0 and significance level α, the greater the curvature of the log-likelihood, the greater the value of the statistic will be and the easier it is to reject the null hypothesis. Large deviations of $\hat{\theta}$ from θ_0 are taken as evidence that the data do not confirm the null hypothesis. But to know when the deviation is large or small the test based on the maximum likelihood estimator uses the Fisher information $I(\hat{\theta})$, which can be understood as the expected value of the log-likelihood function curvature in $\hat{\theta}$. The score function $U(\mathbf{X}; \theta_0)$ is the variation rate of the tangent line of the log-likelihood function in θ_0, which tends to zero as θ_0 approaches $\hat{\theta}$. For fixed values of θ_0 and $\hat{\theta}$ the value of $U(\mathbf{X}; \theta_0)$ is directly proportional to the log-likelihood curvature in θ_0. The score test uses the inverse of Fisher information to balance the square of $U(\mathbf{X}; \theta_0)$, while the gradient statistic uses $U(\theta_0)$ to balance the distance between θ_0 and $\hat{\theta}$.

Exercises

§6.2

1. Assume that $X_1, \ldots, X_n$ are iid with density $p(x|\theta) = \theta x^{\theta-1} I_x([0, 1])$, and $\theta > 0$ is unknown. Determine the UMP test of level 0.05 for H_0: $\theta \leq 1$ against H_1: $\theta > 1$.

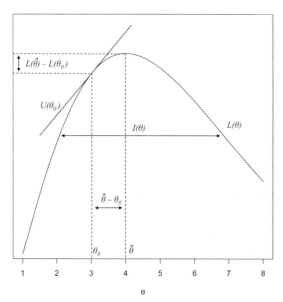

Figure 6.6 *Interpretative diagram of the asymptotic hypothesis tests. A sample of n values was generated from a probability distribution function with parameter $\theta = 3$. Assume one wishes to test the hypotheses $H_0 : \theta = 3$ vs $H_1 : \theta \neq 3$, and $\hat\theta = 4$ was observed.*

2. Let $X \sim Bin(n, \theta)$ and suppose one wishes to test H_0: $\theta = 1/2$ versus H_1: $\theta \neq 1/2$.

(a) Show that the maximum likelihood ratio test statistic is $|2X - n|$.

(b) Find the critical region for a test of level as close as possible to 0.05 when $n = 25$.

(c) Repeat the previous item, using instead the Normal approximation.

3. Suppose that k independent tests about the same hypothesis H: $\theta = \theta_0$ have been performed based on independent statistics $T_1, \ldots, T_k$ with continuous distributions under H_0. Let $\alpha(T_1), \ldots, \alpha(T_k)$ be their respective p-values.

(a) Show that $\alpha(T_1), \ldots, \alpha(T_k)$ form a random sample from the uniform distribution on $(0, 1)$, under H_0.

(b) Define $F = -2 \sum_{i=1}^{k} \log \alpha(T_i)$ as the test statistic for a combined test of H. Which values of F should lead to the rejection of the hypothesis H?

(c) Show that $F \sim \chi^2_{2k}$, under H and specify the critical region of a level α test.

4. Let $\mathbf{X} = (X_1, \ldots, X_n)$ and $\mathbf{Y} = (Y_1, \ldots, Y_n)$ be independent samples from the exponential distributions with means θ_1 and θ_2, respectively, and suppose we wish to test the hypothesis H_0 of equality of the distributions.

(a) Show that the maximum likelihood ratio test rejects H_0 if

$$\left| \frac{\bar{X}}{\bar{X} + \bar{Y}} - \frac{1}{2} \right| > c.$$

(b) Show that the value of c is given by

$$\frac{1}{1 + \underline{F}_\beta(2n, 2n)} - \frac{1}{2}$$

for the level α test and β is such that $P(\underline{F}_\beta(2n, 2n) < F(2n, 2n) < \underline{F}_\beta(2n, 2n) + 2) = 1 - \alpha$.

Hint: Show that under H_0, $\bar{X}/\bar{Y} \sim F(2n, 2n)$.

5. Show that the power of the t test is a strictly increasing function of $|\theta - \theta_0|$. What is the expression of the p-value for this test?

6. Show that the uniform distribution over the interval $[0, \theta]$ has monotone likelihood ratio; i.e., show that the ratio of sampling densities is a monotonically non-increasing function of $T(\mathbf{X}) = \max_i X_i$.

7. Consider a random sample $X_1, \ldots, X_n$ from the $N(\theta, \sigma^2)$ distribution with both parameters unknown and define the hypotheses $H_0 : \theta = \theta_0$ and $H_1 : \theta \neq \theta_0$. Show that working with the full parameter space Θ instead of parameter space Θ_1 under H_1 amounts to a zero probability change in the maximum likelihood ratio statistic.

8. Let $X_1, \ldots, X_p$ be independent random variables with respective $Pois(\mu_i)$ distributions, $i = 1, \ldots, p$, and we wish to test the hypothesis H_0 of the equality of the distributions.

(a) Obtain the maximum likelihood ratio test for H_0.

(b) Show that $(X_1, \ldots, X_p) | X_1 + \cdots + X_p = S$ has Multinomial distribution with parameters S and $(\theta_1, \ldots, \theta_p)$ where $\theta_i = \mu_i/(\mu_1 + \ldots + \mu_p)$, $i = 1, \ldots, p$.

(c) Justify the following test (commonly used in such situations): reject H_0 if $\sum_{i=1}^{p} (X_i - \bar{X})^2 / \bar{X} \geq \chi^2_{p-1, \alpha}$.

(d) Are the test in (a) and the test in (c) similar?

9. Prove that the tests in Section 6.2.3 are unbiased, similar and invariant under linear transformations.

§6.3

10. Let $X \sim Exp(\theta)$ and assume one wishes to test H_0: $\theta = \theta_0$ versus H_1: $\theta = \theta_1$, where $\theta_1 < \theta_0$.

  (a) Prove that the level α likelihood ratio test accepts H_0 if $X < -\theta_0^{-1} \log \alpha$.

  (b) Obtain the p-value associated with $X = 3$, when $\theta_0 = 1$.

  (c) Assuming that $\theta_1 = 1/2$, calculate the Bayes factor for $X = 3$ and $\theta_0 = 1$.

  (d) Supposing $p(H_0) = p(H_1)$, calculate the posterior probability of H_0.

  (e) Compare the results of the classical test obtained in item (a) with those from the Bayesian test.

11. Suppose that $X \sim Cauchy(\theta, 1)$ and that one wishes to test H_0: $\theta = 0$ versus H_1: $\theta \neq 0$. To do that, set a mass probability $p > 0$ to H_0 and the remaining probability over H_1 distributed according to a density $p(\theta)$.

  (a) Prove that $p(H_0 \mid x) \to p$ when $|x| \to \infty$.

  (b) What conclusions can be drawn from this result?

12. The data set below represent the number of vehicles that travel through a section of a highway during a 15 minutes interval in the afternoons of six consecutive days of two consecutive weeks.

Table 6.2: *Number of Vehicles.*

Week	Day of the Week					
	Mon	Tue	Wed	Thu	Fri	Sat
1st	50	65	52	63	84	102
2nd	56	49	60	45	112	90

Assume that the number of vehicles follows a Poisson distribution. Assume further that the average number of vehicles traveling from Monday through Thursday is λ and for Friday and Saturday is μ.

  (a) Obtain 95% credibility intervals for λ and μ based on non-informative priors and comment on the results.

  (b) Test the hypothesis that $2\lambda = \mu$.

13. Suppose that independent random variables X_1 and X_2 where $P(X_i = 0) = 1 - \theta_i$ and $P(X_i = 1) = \theta_i$, $i = 1, 2$, are observed and one wishes to test H_0: $\theta_1 = \theta_2 = \theta$ versus H_1: $\theta_1 \neq \theta_2$. Assume also that all prior distributions are uniform; that is, under H_0, θ is uniform over the unit interval, and under H_1, (θ_1, θ_2) is uniform over the unit square.

(a) Show that the maximum likelihood estimator of θ under H_0 is $(X_1 + X_2)/2$ and the MLE's of θ_i under H_1 are X_i, $i = 1, 2$.

(b) Obtain the posterior distributions under the two hypotheses, that is, the distributions of $\theta \mid \mathbf{x}, H_0$ and $(\theta_1, \theta_2) \mid \mathbf{x}, H_1$.

(c) Obtain the generalized maximum likelihood estimators under H_0 and H_1 and compare them with the maximum likelihood estimators obtained in item (a).

(d) Obtain the predictive distributions of (X_1, X_2) under H_0 and under H_1.

(e) Show that $BF(H_0, H_1) = 4/3$, if $x_1 = x_2$, and $4/6$ if $x_1 \neq x_2$ and interpret the result.

14. Assume $\theta \sim N(\mu, \tau^2)$ and observe $X \sim N(\theta, \sigma^2)$, σ^2 known. Suppose one wishes to verify if a quantity θ is smaller than a prespecified value θ_0.

(a) Obtain the prior probability of H_0: $\theta < \theta_0$.

(b) Obtain the posterior probability of H_0.

(c) Prove that the probability of H_0 increases after observing $X = x$ iff $x < \theta_0(1 - 1/\sqrt{2})$ in the case $\sigma^2 = \tau^2$ and $\mu = 0$.

§6.4

15. Construct a level α test for the hypothesis $\theta = \theta_0$ from confidence intervals, assuming that independent $X_i \sim Pois(\theta t_i)$ are observed with known t_i, for $i = 1, \dots, n$. The hypothesis can be accepted if $\sum_{i=1}^{n} x_i = 10$, $\sum_{i=1}^{n} t_i = 5$, $\theta_0 = 1$ and $\alpha = 0.05$? And if $\alpha = 0.01$?

16. Let $X_1 \sim N(\theta_1, 1)$ and $X_2 \sim N(\theta_2, 1)$ be independent and define $\rho = \theta_1/\theta_2$.

(a) Show that the test that rejects the hypothesis $\rho = \rho_0$ when $|X_1 - \rho_0 X_2| > (1 + \rho_0)^{1/2} z_{\alpha/2}$ has level α.

(b) Obtain a $100(1 - \alpha)\%$ confidence region for ρ from the test described in (a). Draw a graph of the region and interpret the result.

§6.5

17. Let $\mathbf{X} = (X_1, \dots, X_n)$ be a random sample from the $Exp(\theta)$ distribution and suppose one wishes to test $H: \theta \leq 1$.

(a) Show that the likelihood ratio test rejects H when $\sum_{i=1}^{n} X_i < c$.

(b) What is the value of c for the test with level α?

(c) Show that the power of this test is a strictly monotonic function of θ.

(d) Draw a graph of the power for $\alpha = 0.05$ and $n = 15$.

(e) Construct a test based on asymptotic results and compare it with the test based on exact results.

18. Let $\mathbf{X} = (X_1, \ldots, X_{10})$ be a random sample from an unknown distribution. After observing $\mathbf{X} = (1, 0.7, 0.2, -1.3, -0.5, 1.52, -0.85, 0.25, \ 0.47, -0.67)$, test whether one can assume that the data comes from a standard Normal distribution. Check also whether one can assume that the data comes from a Normal distribution. Obtain the p-values of the two tests and compare them.

19. Show that the t test of $H_0\colon \theta = \theta_0$ versus $H_1\colon \theta \neq \theta_0$ based on a random sample of size n from the $N(\theta, \sigma^2)$, θ and σ^2 unknown, is consistent.

20. Suppose that $X_1, \ldots, X_n$ are iid with $N(0, \sigma^2)$ and one wishes to test $H_0\colon \sigma = \sigma_0$ versus $H_1\colon \sigma \neq \sigma_0$.

(a) Show that the maximum likelihood ratio test with level α accepts H_0 if $\sum_{i=1}^{n} (X_i - \overline{X})^2 / \sigma_0^2 \in [c_1, c_2]$ where c_1 and c_2 are such that $F_{n-1}(c_2) - F_{n-1}(c_1) = 1 - \alpha$ and F_ν is the distribution function of the χ_ν^2 distribution.

(b) What condition must be satisfied by c_1 and c_2 for an unbiased test?

(c) Show that the Normal approximation gives $c_1 = n - \sqrt{2n} z_{\alpha/2}$ and $c_2 = n + \sqrt{2n} z_{\alpha/2}$ where z_γ is the $1 - \gamma$ quantile of the $N(0,1)$ distribution.

(d) Show that the equal tail test, where $F_{n-1}(c_2) = 1 - \alpha/2$ and $F_{n-1}(c_1) = \alpha/2$ is asymptotically unbiased, i.e., that the test is unbiased when $n \to \infty$.

21. Prove that in the case of multinomial observations, the statistics W_E and W_U are respectively given by

$$W_E(\boldsymbol{\theta}_0) = \sum_{i=1}^{p} \frac{(N_i - n\theta_{i,0})^2}{N_i} \quad \text{and} \quad W_U(\boldsymbol{\theta}_0) = \sum_{i=1}^{p} \frac{(N_i - n\theta_{i,0})^2}{n\theta_{i,0}}.$$

22. A contingency table is a table of multiple classification of observational units into cells. In the simplest case of double entry, the cells are defined by the intersection of two factors: A with levels $A_1, \ldots, A_p$ and B with levels $B_1, \ldots, B_r$. Define the probabilities $\theta_{ij} = P(A = A_i, B = B_j)$, $\forall (i,j)$ and assume that n independent observations are made, with each one of them classified into a single cell (i, j) and the counts N_{ij} associated with the cells registered.

(a) Show that $\{N_{ij}\}$ has multinomial distribution with parameters n and $\{\theta_{ij}\}$. Define the parameter space Θ and obtain the maximum likelihood estimators of $\{\theta_{ij}\}$, $\forall (i, j)$.

(b) The factors A and B are said to be independent if $P(A = A_i, B = B_j) = P(A = A_i)P(B = B_j)$. Define the parametric space Θ_0 under independence between A and B and calculate its dimension.
Hint: Define $\theta_{i+} = \sum_{j=1}^{r} \theta_{ij}$ and $\theta_{+j} = \sum_{i=1}^{p} \theta_{ij}$.

(c) Calculate the maximum likelihood estimators of $\{\theta_{ij}\}$ under Θ_0.

(d) Obtain the goodness-of-fit and maximum likelihood ratio tests of level α, specifying the critical regions and the distributions involved.

Chapter 7

Prediction

All that has been done up to now concerns estimation, that is, understanding a phenomenon through probabilistic assertions that relate directly or indirectly to unobserved quantities of interest. Thus, these assertions can never be refuted. The quantities we deal with do not exist from a practical point of view. Their usefulness is only associated with the valuable help in describing in the best possible way the random process under study. An improvement was obtained when we criticized the adopted models by putting them under hypotheses tests but we are still restricted to the observed data. A real test of a theory or model is obtained when its assertions are applied to future experiences and observations. In this chapter, we will deal only with prediction.

Statistical prediction has a well-defined meaning and is an integral part of the inferential procedure as will be seen. It deals with making probabilistic statements about quantities to be observed in the future. Note that at the moment inference is made, the problem is similar to that already studied on parametric estimation. So, much of the material described in this chapter is a mere adaptation of the material from previous chapters. The fundamental difference here is that all statements here can be confronted against reality and are thus subject to approval or dismissal without dispute.

Even though notions of decision theory were introduced when we dealt with estimation, we think the appropriate moment for decision making is when we make predictions. The decision maker has a clear view of the consequences and respective losses associated with his/her acts when he faces his/her positions about quantities that will become known.

7.1   Bayesian prediction

The typical situation of the prediction problem is that in which a quantity $\mathbf{X}$ related to an unobserved quantity $\boldsymbol{\theta}$ through $P_1(\mathbf{x} \mid \boldsymbol{\theta})$ is observed, and we are interested in producing statements about another random quantity $\mathbf{Y}$ that is related to $\mathbf{X}$ and $\boldsymbol{\theta}$ through $P_2(\mathbf{y} \mid \boldsymbol{\theta}, \mathbf{x})$. So, after observing $\mathbf{X} = \mathbf{x}$ we have updated information to make inference about $\mathbf{Y}$, and this information is contained in the distribution of $\mathbf{Y} \mid \mathbf{X}$. Therefore, the distribution of $\mathbf{Y} \mid \mathbf{X} = \mathbf{x}$ is

$$
\begin{aligned}
p(\mathbf{y} \mid \mathbf{x}) &= \int p(\mathbf{y}, \boldsymbol{\theta} \mid \mathbf{x}) \, d\boldsymbol{\theta} \\
&= \int p(\mathbf{y} \mid \boldsymbol{\theta}, \mathbf{x}) \, p(\boldsymbol{\theta} \mid \mathbf{x}) \, d\boldsymbol{\theta} \\
&= \int p(\mathbf{y} \mid \boldsymbol{\theta}) \, p(\boldsymbol{\theta} \mid \mathbf{x}) \, d\boldsymbol{\theta},
\end{aligned}
$$

with the last equality valid in the common case of conditional independence between $\mathbf{X}$ and $\mathbf{Y}$ given $\boldsymbol{\theta}$. That happens, for instance, when we sample future and past observations from the same population. In the trivial case in which it is possible to directly specify the distribution of $\mathbf{Y} \mid \mathbf{x}$, the above calculations involving the removal of unobserved quantities are not needed.

In the case of a random sample $\mathbf{X} = (X_1, \ldots, X_n)$ from $p(\cdot \mid \boldsymbol{\theta})$ and a single future observation Y that also comes from the same population, that is, Y has density $p(\cdot \mid \boldsymbol{\theta})$, then

$$
p(y \mid \mathbf{x}) = \int p(y \mid \boldsymbol{\theta}) p(\boldsymbol{\theta} \mid \mathbf{x}) \, d\boldsymbol{\theta} \text{ where } p(\boldsymbol{\theta} \mid \mathbf{x}) \propto p(\boldsymbol{\theta}) \prod_{i=1}^{n} p(x_i \mid \boldsymbol{\theta}),
$$

and $p(y \mid \boldsymbol{\theta})$ and $p(x_i \mid \boldsymbol{\theta})$ have the same form. One can then write

$$
p(y \mid \mathbf{x}) = E_{\boldsymbol{\theta} \mid \mathbf{x}} \left[p(y \mid \boldsymbol{\theta}) \right].
$$

In the case of prediction in samples from the one-parameter exponential family with conjugate prior, we have that

$$
\begin{aligned}
p(x \mid \theta) &= a(x) \exp\{u(x)\phi(\theta) + b(\theta)\} \\
p(\theta) &= k(\alpha, \beta) \exp\{\alpha\phi(\theta) + \beta b(\theta)\}
\end{aligned}
$$

and so

$$p(\theta \mid \mathbf{x}) = k\left(\alpha + \sum_{i=1}^{n} u(x_i), \beta + n\right)$$

$$\times \exp\left\{\left[\alpha + \sum_{i=1}^{n} u(x_i)\right]\phi(\theta) + [\beta + n]b(\theta)\right\}$$

$$\text{and } p(\mathbf{x}) = \prod_{i=1}^{n} a(x_i)\frac{k(\alpha, \beta)}{k\left(\alpha + \sum_{i=1}^{n} u(x_i), \beta + n\right)}.$$

The expression of $p(y \mid \mathbf{x})$ is obtained similarly to $p(\mathbf{x})$, using the posterior density $p(\theta \mid \mathbf{x})$ instead of $p(\theta)$. So,

$$p(y \mid \mathbf{x}) = a(y)\ \frac{k\left(\alpha + \sum_{i=1}^{n} u(x_i), \beta + n\right)}{k\left(\alpha + \sum_{i=1}^{n} u(x_i) + u(y), \beta + n + 1\right)}.$$

Example 7.1 *Let $X_1, \ldots, X_n$ be a sample from the exponential distribution with parameter θ. The conjugate distribution is $G(\gamma, \delta)$. The densities can be written in the form*

$$p(x \mid \theta) = \exp\{-x\theta + \log\theta\},\ x > 0,$$

$$p(\theta) = \frac{\delta^\gamma}{\Gamma(\gamma)}\exp\{-\delta\theta + (\gamma - 1)\log\theta\},\ \theta > 0.$$

Identifying with the above notation, we have

$$\alpha = -\delta,$$
$$\beta = \gamma - 1,$$
$$k(\alpha, \beta) = (-\alpha)^{\beta+1}/\Gamma(\beta + 1),$$
$$a(x) = 1,\ and$$
$$u(x) = -x,$$

from which we obtain

$$p(y \mid \mathbf{x}) = \frac{[-\alpha - \sum_{i=1}^{n} u(x_i)]^{\beta+n+1}/\Gamma(\beta + n + 1)}{[-\alpha - \sum_{i=1}^{n} u(x_i) - u(y)]^{\beta+n+2}/\Gamma(\beta + n + 2)},\ y > 0$$

$$= \frac{\beta + n + 1}{-\alpha - \sum_{i=1}^{n} u(x_i) - u(y)}\left(1 + \frac{u(y)}{\alpha + \sum_{i=1}^{n} u(x_i)}\right)^{-(\beta+n+1)},$$

for $y > 0$. Rewriting as a function of γ and δ leads to

$$p(y \mid \mathbf{x}) = \frac{\gamma + n}{\delta + \sum_{i=1}^{n} x_i + y}\left(1 + \frac{y}{\delta + \sum_{i=1}^{n} x_i}\right)^{-(\gamma+n)},\ y > 0.$$

Note that this density is strictly decreasing with y, just as the sampling density.

One can proceed to inference with the predictive distribution, as previously seen in Chapter 4. In particular, we can make inference by point prediction. To do this, we put the problem into the framework of decision theory whose elements are

1. States of nature – here, represented by the possible values of the quantity $\mathbf{Y}$ to be observed in the future and that we wish to predict;

2. Space of possible actions – containing possible actions to be taken; here, taking an action is to choose a value for $\mathbf{Y}$, its point predictor $\boldsymbol{\delta}$;

3. Loss function – to each possible value of $\mathbf{Y}$ and to each predictor $\boldsymbol{\delta}$ we have a loss $L(\mathbf{Y}, \boldsymbol{\delta})$.

Put in this form, the problem is mathematically identical to that from Section 4.1, that is, we choose the predictor $\boldsymbol{\delta}$ so as to minimize the expected loss

$$\int L(\mathbf{y}, \boldsymbol{\delta}) \, p(\mathbf{y} \mid \mathbf{x}) \, d\mathbf{y}.$$

This decision theoretic approach to prediction is pursued by Aitchinson and Dunsmore (1975) and Geisser (1993). The difference with respect to parameter estimation is not in the equation but in the interpretations of its elements. We can objectively quantify the loss we will incur by predicting $\mathbf{Y}$ by $\boldsymbol{\delta}$ because $\mathbf{Y}$ is observable. This quantification is less clear in the case of estimation and only makes sense when related to observed quantities associated with the parameters. The example of John and his doctor in Section 4.1 illustrates this point. It was only possible to construct the loss table after referring to observed quantities in the future such as death and definition of the disease state of the patient.

Usually we do not take decisions against values of theoretical objects used to understand a phenomenon (parameters) but only after evaluating the consequences these values will have upon observables. This point is the basis of the predictivist approach to inference which itself is not free from arguments. Its advantage, however, is that it allows judgments that are not ambiguous with a clear and unquestionable meaning. A prediction can always be confronted against reality while estimation never is.

So, a point predictor can be chosen according to the loss function we incur. Using results from Section 4.1, we have that

1. The predictor associated with absolute loss is the median of the predictive distribution.

2. The predictor associated with quadratic loss is the mean of the predictive distribution.

3. The predictor associated with the $0 - 1$ loss is the mode of the predictive distribution.

Example 7.2 *(Example 7.1 continued) The predictors described above are given by*

a) The solution med of the equation

$$\frac{1}{2} = \int_0^{med} p(y \mid \boldsymbol{x}) dy,$$

given by $med = (\delta + \sum_{i=1}^n x_i)(2^{1/\gamma+n} - 1);$

b) The predictive mean given by

$$
\begin{aligned}
E(Y \mid \boldsymbol{x}) &= E_{\theta|x}[E(Y \mid \theta)] \\
&= E_{\theta|x}[1/\theta] \\
&= \frac{\delta + \sum_{i=1}^n x_i}{\gamma + n - 1};
\end{aligned}
$$

c) The mode of the predictive distribution of $Y \mid \boldsymbol{x}$, *given by 0.*

Observe that for large n, the predictive mean is approximately equal to $\bar{x}$ *and for the non-informative prior* $(\gamma, \delta \to 0)$ *the predictive mean is*

$$E(Y \mid \boldsymbol{x}) = \frac{\sum_{i=1}^n x_i}{n - 1}.$$

In analogy to the estimation problem, $100(1 - \alpha)\%$ predictive regions may be obtained for $\mathbf{Y}$. All that is needed is to find a region $\boldsymbol{C}$ such that $P(\mathbf{Y} \in \boldsymbol{C}|\mathbf{x}) \geq 1 - \alpha$. In the case of a scalar Y, the region C may be reduced to an interval $[a_1, a_2]$ satisfying $P(a_1 < Y < a_2 \mid \mathbf{x}) \geq 1 - \alpha$. Comments of Section 4.6 are still valid here: for a given value of α one wishes to find the interval with the smallest possible length. This reasoning leads naturally to the choice of regions where their predictive density is higher. This is formalized by the concept of regions of highest predictive density (HPRD, in short), $\boldsymbol{C}$ given by

$$C = \{\mathbf{y} : p(\mathbf{y} \mid \mathbf{x}) \geq k(\alpha)\},$$

where $k(\alpha)$ is the highest constant guaranteeing that $P(\mathbf{Y} \in C \mid \mathbf{x}) \geq 1 - \alpha$.

Example 7.3 *(Example 7.2 continued) As the predictive density is strictly decreasing, the HPRD interval for Y must be in the form* $[0, a]$ *where a is such that*

$$\int_0^a p(y \mid \boldsymbol{x}) \, dy = 1 - \alpha.$$

Solving for a gives $\left(\delta + \sum_{i=1}^{n} x_i\right)\left((1-\alpha)^{-1/\gamma+n} - 1\right)$ *and the HPRD for* Y *is given by* $\left[0, \left(\delta + \sum_{i=1}^{n} x_i\right)\left[(1-\alpha)^{-1/\gamma+n} - 1\right]\right]$. *In the case of a non-informative prior the HPRD interval for* Y *becomes* $\left[0, \left(\sum_{i=1}^{n} x_i\right)\left[(1-\alpha)^{-1/n} - 1\right]\right]$.

In the case of a large sample, the posterior distribution of $\boldsymbol{\theta} \mid \mathbf{x}$ gets concentrated around $\hat{\boldsymbol{\theta}}$, the maximum likelihood estimator of $\boldsymbol{\theta}$. So,

$$p(\mathbf{y} \mid \mathbf{x}) = \int p(\mathbf{y} \mid \boldsymbol{\theta}) \, p(\boldsymbol{\theta} \mid \mathbf{x}) \, d\boldsymbol{\theta} \doteq p(\mathbf{y} \mid \hat{\boldsymbol{\theta}}).$$

This approximation neglects the variability of the parameter and leaves only the sampling variability of the quantity to be predicted. In the general case, both forms of variability are important and should be taken into account.

7.2   Classical prediction

There are no clear rules as to how to proceed to make prediction of future observations in classical inference. One of the most used procedures is to substitute the value of the parameter appearing in the sampling distribution of future observations by some estimate based on past data. This is the plug-in estimator for the parameter, introduced in Section 4.4.

Specifically, assuming that $\mathbf{Y}$ with sampling distribution $p(\mathbf{y} \mid \boldsymbol{\theta})$ must be predicted based on observations $\mathbf{X}$ from a sample of $p(\mathbf{x} \mid \boldsymbol{\theta})$, one uses the distribution $p(\mathbf{y} \mid \hat{\boldsymbol{\theta}})$ where $\hat{\boldsymbol{\theta}}$ is an estimator of $\boldsymbol{\theta}$ based on $\mathbf{X}$. The most common choice is the maximum likelihood estimator, which takes us back to the discussion in the final paragraph of the last section. The drawback of this procedure is not to take into account the variability associated with the estimation of $\boldsymbol{\theta}$.

As an example, assume that $E(\mathbf{Y} \mid \boldsymbol{\theta}) = \mathbf{b}(\boldsymbol{\theta})$ and that $V(\mathbf{Y} \mid \boldsymbol{\theta}) = \mathbf{B}(\boldsymbol{\theta})$. It is common practice to take $\hat{\mathbf{Y}}(\mathbf{x}) = E[\mathbf{Y}|\hat{\boldsymbol{\theta}}] = \mathbf{b}(\hat{\boldsymbol{\theta}})$ as a point predictor of $\mathbf{Y}$ and $\mathbf{B}(\hat{\boldsymbol{\theta}})$ as a measure of the variability associated with that prediction. This procedure underestimates the variability of the prediction by not taking into account the variability of the estimation of $\boldsymbol{\theta}$. After all, we know $\hat{\boldsymbol{\theta}} \neq \boldsymbol{\theta}$, with certainty. So, there is an error associated with equating the above two quantities.

Example 7.4 *(Example 7.1 continued) The maximum likelihood estimator of* θ *is* $1/\overline{X}$ *and the mean of* Y *is* $1/\theta$. *After making the substitutions we get the point predictor* $\overline{X}$ *for* Y. *Note that the estimator of the variance of* Y, $1/\overline{X}^2$ *does not consider the variability of* $\hat{\theta}$ *with respect to* θ.

The above prediction has an associated error given by

$$e = \hat{Y}(x) - Y,$$

with the prediction estimator $\hat{Y}(x)$ being unbiased for Y, since the expected value of e is equal to zero, that is,

$$\begin{aligned} E(e) &= E[\hat{Y}(x)] - E(Y) \\ &= E[E(Y|\hat{\theta})] - E(Y) \\ &= E(Y) - E(Y) \\ &= 0. \end{aligned}$$

The variance of the error is given in the scalar case by

$$\begin{aligned} V(e) &= E(e^2) = E[(\hat{Y}(x) - Y)^2] \\ &= E\{[\hat{Y}(x) - E(Y) + E(Y) - Y]^2\} \\ &= E\{[\hat{Y}(x) - E(Y)]^2\} + E\{[E(Y) - Y]^2\} \\ &\qquad + 2E\{[\hat{Y}(x) - E(Y)][E(Y) - Y]\} \\ &= V[\hat{Y}(x)] + V(Y) + 2E\{[\hat{Y}(x) - E(Y)][E(Y) - Y]\}. \end{aligned}$$

Then, if X and Y are independent, the variance of the error is given by $V[\hat{Y}(x)] + V(Y)$. Similar calculations follow for multivariate prediction for Y, with the variance of the error given by $V[\hat{Y}(x)] + V(Y)$, when X and Y are independent.

Example 7.5 *Let $X = (X_1, \ldots, X_n)$ be a random sample from a Normal distribution with unknown parameter vector $\theta = (\mu, \sigma^2)$, where μ and σ^2 denote the mean and variance, respectively. The maximum likelihood estimator of μ is equal to $\bar{X}$ and the point prediction of Y is given by $\hat{Y}(x) = \bar{x}$. Then, $E[\hat{Y}(X)] = E(\bar{X}) = \mu$, $V[\hat{Y}(X)] = V(\bar{X}) = \sigma^2/n$ and $V(e) = V[\hat{Y}(X)] + V(Y) = \sigma^2(1 + 1/n)$, with $\lim_{n \to \infty} V(e) = \sigma^2$.*

From Example 7.5, we note that $V[\hat{Y}(X)]$ is a function of the variance of the estimator $\hat{\theta} = (\hat{\mu}, \hat{\sigma}^2)$. Then, from the invariance property of the maximum likelihood estimators, an estimate of $V[\hat{Y}(x)]$ is given by $\hat{\sigma}^2(1 + 1/n)$, which is different and more accurate than $\hat{V}[Y|\theta] = \hat{\sigma}^2$, since the variability associated with the estimatior is considered.

Another procedure that allows the variance of the estimator to be taken into account is the use of resampling methods as presented in Section 5.6.3. For simplicity, let $\hat{\theta}_0$ denote the maximum likelihood estimator of the original random sample x of size n. Based on a parametric bootstrap approach, let us generate B random samples with size n of $p(.|\hat{\theta}_0)$. Let us denote them by $x_1, \ldots, x_B$, with the respective maximum likelihood estimators given by $\hat{\theta}_1, \ldots, \hat{\theta}_B$. Then, for each sample, a point prediction for Y is obtained as

$\hat{Y}(x_j) = b(\hat{\theta}_j)$ for $j = 1, \ldots, B$. The covariance matrix of $\hat{Y}(x)$ can be estimated as the sample covariance of $[\hat{Y}(x_1), \ldots, \hat{Y}(x_m)]$. Interested readers can refer to Valliant et al. (2000) for more details on prediction using bootstrap.

One approach avoiding the problem consists of obtaining a pivotal quantity $G(Y, X)$ whose distribution does not depend on θ. This approach was used in Section 4.6 in interval estimation. As earlier, the function G must depend on the sample X in an optimal form (for instance, through minimal sufficient statistics for θ). Once the pivot is obtained, one can make probabilistic statements about it. In particular, for a given value of α ($\alpha \in (0, 1)$), one can obtain that $P(G(Y, X) \in C) \geq 1 - \alpha$. Then, it becomes possible to construct a $100(1 - \alpha)\%$ predictive region for Y given by

$$\{Y : G(Y, x) \in C\}.$$

The problem with this approach is that it is not always possible to find such function G. The example we are dealing with in this chapter is one of the exceptions.

Example 7.6 *(Example 7.4 continued) We now want to find a function $G(Y, X)$ whose distribution does not depend on θ. Preferably, this function would depend on X through $\overline{X}$ which is a minimal sufficient statistic for θ. Fortunately, in this case, this is possible since*

$$Y \sim Exp(\theta) = G(1, \theta) = \frac{\chi_2^2}{2\theta} \text{ and } \sum_{i=1}^{n} X_i \sim G(n, \theta) = \frac{\chi_{2n}^2}{2\theta}$$

and they are independent. So, $Y/\overline{X} \sim F(2, 2n)$, which does not depend on θ. Using the properties of the F distribution, we have that $E(Y/\overline{X}) = 2n/(2n-2)$ from where we can take as a point predictor for Y

$$\overline{X} \frac{2n}{2n - 2} = \frac{\sum_{i=1}^{n} X_i}{n - 1},$$

which differs from the predictor obtained above but coincides with the Bayesian predictor with noninformative prior. The $100(1 - \alpha)\%$ confidence interval for Y can be constructed with the percentiles of the $F(2, 2n)$ distribution as $P(\underline{F}_{\beta/2}(2, 2n - 2) < Y/\overline{X} < \overline{F}_{\gamma/2}(2, 2n - 2)) = 1 - \alpha$, for $\beta + \gamma = \alpha$. Since the $F(2, 2n)$ distribution has a monotonically decreasing density, the minimal length interval is obtained by taking $\beta = 0$ and $\gamma = \alpha$, thus implying a lower limit equal to 0. The $100(1 - \alpha)\%$ prediction interval for Y is then given by $(0, \overline{X} F_\alpha(2, 2n - 2))$.

7.3   Prediction in the Normal model

7.3.1   Bayesian approach

The structure of Normal models allows many easy calculations especially in the case of linearity. The application of these rules to prediction reduces the calculation considerably. Basically consider an observation $Y \mid \theta \sim N(\theta, \sigma^2)$ where σ^2 is known. One can rewrite this model as $Y = \theta + \epsilon$ where $\epsilon \sim N(0, \sigma^2)$ can be regarded as an observation error. Observe that the distribution of ϵ does not depend on θ, and therefore, ϵ and θ are independent. Assume now that the updated distribution of θ (possibly after the observation of a sample) is $N(\mu, \tau^2)$. Y is therefore the sum of independent Normal quantities and has $N(\mu, \sigma^2 + \tau^2)$ distribution. This is the predictive distribution of Y. Point and HPRD interval predictions can be made as described in Section 7.1.

Example 7.7 *Assume the updated distribution of θ is its posterior distribution relative to a sample X from a $N(\theta, \sigma^2)$ distribution. This leads to $Y \mid x \sim N(\mu_1, \sigma^2 + \tau_1^2)$ where μ_1 and τ_1^2 are given by Theorem 2.1. The point predictor of Y in this case will be μ_1, and the $100(1 - \alpha)\%$ HPRD interval for Y is of the form $(\mu_1 - z_{\alpha/2}\sqrt{\sigma^2 + \tau_1^2}, \mu_1 + z_{\alpha/2}\sqrt{\sigma^2 + \tau_1^2})$.*

In the case of non-informative prior, $\mu_1 = \bar{x}$ and $\tau_1^2 = \sigma^2/n$, leading to the predictive distribution

$$Y \mid x \sim N\left(\bar{x}, \sigma^2\left(1 + \frac{1}{n}\right)\right).$$

The point predictor of Y in this case will be $\bar{x}$ and the $100(1 - \alpha)\%$ HPRD interval for Y is of the form $(\bar{x} - z_{\alpha/2}\sigma\sqrt{1 + n^{-1}}, \bar{x} + z_{\alpha/2}\sigma\sqrt{1 + n^{-1}})$.

The above example can be generalized in various forms. If $\mathbf{Y}$ is a vector with sampling distribution $N(\boldsymbol{\theta}, \boldsymbol{\Sigma})$ and the updated distribution of $\boldsymbol{\theta}$ is $N(\boldsymbol{\mu}, \boldsymbol{\tau})$, we have by the same reasoning that the predictive distribution of $\mathbf{Y}$ is $N(\boldsymbol{\mu}, \boldsymbol{\Sigma} + \boldsymbol{\tau})$. Assume now that the sampling mean of $\mathbf{Y}$ is given by the linear relation $\mathbf{X}\boldsymbol{\theta}$, where $\mathbf{X}$ is a matrix of known constants, and that $\boldsymbol{\theta}$ remains with the same distribution. If the matrix $\mathbf{X}$ is square, the dimension of $\boldsymbol{\theta}$ and the hyperparameters of its distribution remain unaltered. This restriction is unnecessary and we can consider any matrix $\mathbf{X}$ with the correspondent change in the dimension and distribution of $\boldsymbol{\theta}$. We shall see in the next chapter that the cases of interest involve a dimension of $\boldsymbol{\theta}$ lower than the dimension of $\mathbf{Y}$ implying a reduction in the dimensionality of the problem. Schematically,

$$\mathbf{Y} = \mathbf{X}\boldsymbol{\theta} + \boldsymbol{\epsilon},$$

where $\boldsymbol{\theta}$ and $\boldsymbol{\epsilon}$ are independent. So, the predictive distribution of $\mathbf{Y}$ is given by the sum of two independent Normal quantities, the first relative to $\boldsymbol{\theta}$ given by a $N(\mathbf{X}\boldsymbol{\mu}, \mathbf{X}\boldsymbol{\tau}\mathbf{X}')$ distribution and so

$$\mathbf{Y} \mid \mathbf{X} \sim N(\mathbf{X}\boldsymbol{\mu}, \mathbf{X}\boldsymbol{\tau}\mathbf{X}' + \boldsymbol{\Sigma}).$$

An example is the case of prediction of a series of future observations for a population from which a sample had previously been observed.

Up to now, the sampling variance was assumed known. Usually it is not, and the usual approach in this case is to try to specify a conjugate distribution for the variance. Although a conjugate analysis when the sampling covariance matrix is totally unknown is possible, we will consider here only the case where the covariance matrix is totally known but for an unknown multiplicative scalar σ^2. Without loss of generality we will assume that $V(\mathbf{Y} \mid \boldsymbol{\theta}, \sigma^2) = \sigma^2 \mathbf{I}_p$ where p is the dimension of $\mathbf{Y}$. For a conjugate analysis, it is necessary that the updated covariance matrix of $\boldsymbol{\theta}$ be proportional to σ^2 as seen in Chapter 3. So, $\boldsymbol{\theta} \mid \sigma^2 \sim N(\boldsymbol{\mu}, \sigma^2\boldsymbol{\tau})$ and $\mathbf{Y} \mid \mathbf{X}, \sigma^2 \sim N(\mathbf{X}\boldsymbol{\mu}, \sigma^2(\mathbf{X}\boldsymbol{\tau}\mathbf{X}^T + \mathbf{I}_p))$. Assuming as before that the updated distribution of σ^2 is $n_0\sigma_0^2\phi \sim \chi_{n_0}^2$ with $\phi = \sigma^{-2}$,

$$
\begin{aligned}
p(\mathbf{y}|\mathbf{x}) &= \int p(\mathbf{y} \mid \mathbf{x}, \phi)\, p(\phi)\, d\phi \\
&\propto \int \phi^{p/2} \exp\{-\phi Q(\mathbf{y})/2\}\, \phi^{(n_0/2)-1} \exp\{-\phi n_0 \sigma_0^2/2\}\, d\phi \\
&= \int \phi^{(n_0+p/2)-1} \exp\{-\phi[n_0\sigma_0^2 + Q(\mathbf{y})]/2\}\, d\phi \\
&\propto [n_0\sigma_0^2 + Q(\mathbf{y})]^{-(n_0+p)/2},
\end{aligned}
$$

where $Q(\mathbf{y}) = (\mathbf{y} - \mathbf{X}\boldsymbol{\mu})^T (\mathbf{X}\boldsymbol{\tau}\mathbf{X}^T + \mathbf{I}_p)^{-1}(\mathbf{y} - \mathbf{X}\boldsymbol{\mu})$.

It is easy to obtain then that

$$\mathbf{Y} \mid \mathbf{X} \sim t_\nu(\mathbf{X}\boldsymbol{\mu}, \sigma_0^2(\mathbf{X}\boldsymbol{\tau}\mathbf{X}' + \mathbf{I}_p)).$$

Note that the only changes with respect to the known variance case are the substitutions of the Normal by the Student-t distribution with n_0 degrees of freedom and of σ^2 by its updated estimator σ_0^2.

The main properties of the multivariate Student-t distribution that are relevant here (assuming that a p-dimensional $\mathbf{U} \sim t_\nu(\mathbf{m}, \mathbf{C})$) are

1. The marginal distribution of any q-dimensional subvector $\mathbf{U}_1$ from $\mathbf{U}$ $(q < p)$ is Student $t_\nu(\mathbf{m}_1, \mathbf{C}_1)$, with paramters obtained from the components of the vector $\mathbf{m}$ and the matrix $\mathbf{C}$ correspondent to the components of the vector $\mathbf{U}_1$. In particular, the jth component of $\mathbf{U}$, $U_j \sim t_\nu(m_j, C_{jj})$ distribution;

2. $\mathbf{LU} \sim t_\nu(\mathbf{Lm}, \mathbf{LCL'})$, for any $r \times p$ matrix $\mathbf{L}$.

Example 7.8 *(Example 7.7 continued) The same results are valid with the modifications cited above. As seen in Section 3.3, the posterior distribution of (θ, ϕ) is given by the Normal-χ^2 with parameters $(\mu_1, c_1, n_1, \sigma_1)$. A future observation Y will have predictive distribution*

$$Y \mid \boldsymbol{x} \sim t_{n_1}(\mu_1, \sigma_1^2(1 + c_1^{-1})).$$

In the case of a non-informative prior, the predictive distribution is given by a $t_{n-1}(\bar{x}, s^2(1 + n^{-1}))$ distribution where s^2 is the unbiased estimator of σ^2. The HPRD intervals coincide with the intervals obtained for known σ^2 except for the substitutions of the percentiles of the $N(0, 1)$ for those of the $t_{n-1}(0, 1)$ distribution and of σ by s.

7.3.2 Classical approach

For the frequentist inference, one should work with only the sampling distributions of Y and $\mathbf{X}$ in search of a pivot whose distribution does not depend on the parameters. Again, we write $Y - \theta \sim N(0, \sigma^2)$ and all that is left is to find suitable estimators for θ and σ^2. Assuming initially that σ^2 is known we have the estimator $\hat{\theta} = \bar{X}$. As Y and $\bar{X}$ are both independent Normal with the same mean, $Y - \bar{X} \sim N(0, \sigma^2(1 + n^{-1}))$ is the pivot with distribution identical to that obtained with a noninformative prior. Confidence intervals for Y will also coincide. The difference in the approaches is theoretical: the Bayesian result is conditional on $\mathbf{X}$ which in fact is the way it will be used in classical inference too.

In the case where σ^2 is unknown, another pivot must be found because the distribution of $Y - \bar{X}$ depends on σ^2. In the Normal case, this is easy because S^2 is independent from $Y - \bar{X}$ and $(n - 1)S^2/\sigma^2 \sim \chi_{n-1}^2$. So,

$$\frac{Y - \bar{X}}{S\sqrt{1 + 1/n}} \sim t_{n-1}(0, 1)$$

is the required pivot. It follows that the $(1 - \alpha)100\%$ confidence interval for Y is given by

$$[\bar{x} - t_{\alpha/2, n-1} \ s \ \sqrt{1 + 1/n} \ , \ \bar{x} + t_{\alpha/2, n-1} \ s \ \sqrt{1 + 1/n}].$$

Again, we get the same results obtained in Example 7.8 with a non-informative prior.

The results of the example above were obtained only for the simplest case of a single future observation following the observation of a sample from the same population. They can be extended to more general situations. Some of these extensions will be seen in the next chapter.

In the purely predictive case without unknown parameters, we have $\mathbf{Y}$ and $\mathbf{X}$ with joint distribution

$$\begin{pmatrix} \mathbf{Y} \\ \mathbf{X} \end{pmatrix} \sim N \left[\begin{pmatrix} \boldsymbol{\mu_Y} \\ \boldsymbol{\mu_X} \end{pmatrix}, \begin{pmatrix} \boldsymbol{\Sigma}_Y & \boldsymbol{\Sigma}_{YX} \\ \boldsymbol{\Sigma}_{XY} & \boldsymbol{\Sigma}_X \end{pmatrix} \right]$$

and the prediction of $\mathbf{Y}$ given the observation of $\mathbf{X}$ is based on the distribution of $\mathbf{Y} \mid \mathbf{X} = \mathbf{x}$ given by

$$N \left(\boldsymbol{\mu_Y} + \boldsymbol{\Sigma}_{XY} \boldsymbol{\Sigma}_X^{-1} (\mathbf{x} - \boldsymbol{\mu_X}), \ \boldsymbol{\Sigma}_Y - \boldsymbol{\Sigma}_{XY} \boldsymbol{\Sigma}_X^{-1} \boldsymbol{\Sigma}_{YX} \right).$$

If Y and X are scalar quantities

$$Y \mid X = x \sim N \left(\mu_Y + \frac{\sigma_{XY}}{\sigma_X^2} (x - \mu_X), \ \sigma_Y^2 - \frac{\sigma_{XY}^2}{\sigma_X^2} \right),$$

where $\sigma_X^2 = V(X)$, $\sigma_Y^2 = V(Y)$ and $\sigma_{XY} = Cov(X, Y)$. We can see from the results above that the knowledge of the value of X reduces the variance of Y. The larger the correlation between X and Y in absolute value the greater the reduction. More important than that is the fact that the optimal predictor (under any reasonable criteria) is a linear function of $\mathbf{X}$. This result is explored in the next section.

7.4   Linear prediction

The problem of linear prediction takes place in the absence of complete distributional information. Assume the presence of two random vectors $\mathbf{X}$ and $\mathbf{Y}$ from which it is known only that

$$E \begin{pmatrix} \mathbf{Y} \\ \mathbf{X} \end{pmatrix} = \begin{pmatrix} \boldsymbol{\mu_Y} \\ \boldsymbol{\mu_X} \end{pmatrix}$$

and

$$Var \begin{pmatrix} \mathbf{Y} \\ \mathbf{X} \end{pmatrix} = \begin{pmatrix} \boldsymbol{\Sigma}_Y & \boldsymbol{\Sigma}_{YX} \\ \boldsymbol{\Sigma}_{XY} & \boldsymbol{\Sigma}_X \end{pmatrix}.$$

The complete distribution of $\mathbf{X}$ and $\mathbf{Y}$ may not be known for a number of reasons. Our problem here is to establish a form of predicting $\mathbf{Y}$ based on $\mathbf{X}$. A few restrictions must be imposed to solve it. The most natural one is to restrict the class of predictors to linear functions of $\mathbf{X}$. This restriction is not only justified by rendering the problem tractable. Linear solutions are obtained as first order approximations. The linear predictor thus serves as a preliminary predictor even in the case where the joint distribution is completely specified.

Choice of the predictor among all predictors of the form $\hat{\mathbf{Y}}(\mathbf{X}) = \mathbf{a} + \mathbf{B}\,\mathbf{X}$ requires the definition of an optimality criterion. It is reasonable that this criterion be based on the quadratic risk (or expected loss) given by

$$\mathbf{R}(\mathbf{Y}, \hat{\mathbf{Y}}) = E\{[\mathbf{Y} - \hat{\mathbf{Y}}(\mathbf{X})]'[\mathbf{Y} - \hat{\mathbf{Y}}(\mathbf{X})]\}$$

and the optimality obtained by the minimization of the trace of the matrix $\mathbf{R}$. The search of the optimal predictor reduces to the search of optimal constants $\mathbf{a}$ and $\mathbf{B}$. The predictor so obtained is called the linear predictor and the associated risk, the linear risk. Observe that the predictor that (globally) minimizes the risk is the conditional expectation $E(\mathbf{Y} \mid \mathbf{X})$ and the associated risk is the conditional variance. For that reason, the linear predictor is also called the linear expectation and its risk, the linear variance. In the specific case of the Normal distribution, the conditional expectation is a linear function of $\mathbf{X}$. So, the linear predictor minimizes the quadratic expected loss among all possible predictors (linear or not).

In the special case where Y and X are scalar, the risk is given by

$$R(Y, \hat{Y}) = E[Y - \hat{Y}(X)]^2 = E[Y - (a + bX)]^2.$$

In this case, trace $(R) = R$, and to find the values of a and b that minimize R, we calculate

$$\frac{\partial R}{\partial a} = E[2(Y - a - bX)(-1)] = 2[a + b\mu_X - \mu_Y]$$

and

$$\frac{\partial R}{\partial b} = E[2(Y - a - bX)(-X)] = 2[a\mu_X + bE(X^2) - E(XY)].$$

Equating the derivatives to zero gives the system of equations

$$\hat{a} = \mu_Y - \hat{b}\mu_X$$

and

$$\hat{b} = \frac{E(XY) - \hat{a}\mu_X}{E(X^2)}.$$

Replacing $\hat{a}$ in the second equation, it follows that

$$
\begin{aligned}
\hat{b} &= \frac{E(XY) - (\mu_Y - \hat{b}\mu_X)\mu_X}{E(X^2)} \\
&= \frac{E(XY) - \mu_X\mu_Y + \hat{b}\mu_X^2}{E(X^2)} \\
&= \frac{\sigma_{XY} + \hat{b}\mu_X^2}{E(X^2)},
\end{aligned}
$$

which has the solution $\hat{b} = \sigma_{XY}/\sigma_X^2$ where $\sigma_{XY} = Cov(X, Y)$. So the optimal linear predictor is

$$
\hat{Y} = \mu_Y + \frac{\sigma_{XY}}{\sigma_X^2}(X - \mu_X)
$$

and its risk is given by

$$
\begin{aligned}
R(Y, \hat{Y}) &= E[Y - \mu_Y - \frac{\sigma_{XY}}{\sigma_X^2}(X - \mu_X)]^2 \\
&= E(Y - \mu_Y)^2 + \left(\frac{\sigma_{XY}}{\sigma_X^2}\right)^2 E(X - \mu_X)^2 \\
&\qquad\qquad -2\frac{\sigma_{XY}}{\sigma_X^2}E[(Y - \mu_Y)(X - \mu_X)] \\
&= \sigma_Y^2 + \left(\frac{\sigma_{XY}}{\sigma_X^2}\right)^2 \sigma_X^2 - 2\frac{\sigma_{XY}}{\sigma_X^2}\sigma_{XY} \\
&= \sigma_Y^2 - \frac{\sigma_{XY}^2}{\sigma_X^2}.
\end{aligned}
$$

These are the expressions of the predictive mean and variance when the observations are normally distributed.

Example 7.9 *Consider the following model* $Y = \alpha X + \beta Z$, *with* $X \sim N(0, \sigma^2)$ *and* $Z \sim N(0, 1)$, *where* α, β *and* σ^2 *are known parameters. It is easy to show that the probability distributions of* Y *and* $Y|X$, *from the linear properties of the Normal distribution, are given by*

$$
Y \sim N(0, \alpha^2\sigma^2 + \beta^2)
$$

and

$$
Y|X \sim N(\alpha X, \beta^2).
$$

Then, the covariance of X *and* Y *is given by*

$$
\begin{aligned}
\sigma_{XY} &= E\{[X - E(X)][Y - E(Y)]\} \\
&= E(XY) = E[E(XY|X)] \\
&= \alpha E(X^2) = \alpha\sigma^2.
\end{aligned}
$$

The point prediction of Y is given by

$$\hat{Y} = \mu_Y + \frac{\sigma_{XY}}{\sigma_X^2}(x - \mu_X) = 0 + \frac{\alpha\sigma^2}{\sigma^2}(x - 0) = \alpha x$$

and its risk is given by

$$R(\hat{Y}, Y) = \sigma_Y^2 - \frac{\sigma_{XY}^2}{\sigma_X^2} = \alpha^2\sigma^2 + \beta^2 - \frac{\alpha\sigma^2}{\sigma^2} = \alpha^2\sigma^2 + \beta^2 - \alpha.$$

In the multivariate case, it can be shown that the linear predictor is

$$\mu_Y + \Sigma_{XY}\Sigma_X^{-1}(\mathbf{x} - \mu_X)$$

and its risk is given by
$$\Sigma_Y - \Sigma_{XY}\Sigma_X^{-1}\Sigma_{YX},$$

generalizing the result obtained when Y and X are scalar. As mentioned above, the linear predictor is given by the expression of the conditional expectation in the Normal case. This implies that the linear predictor is as close to the global optimum (in terms of minimization of the quadratic risk) as the Normal distribution is close to the joint distribution of $\mathbf{X}$ and $\mathbf{Y}$. The prediction and risk for $\mathbf{Y}$ are obtained directly from the $\mathbf{Y}$ model if $\mathbf{X}$ and $\mathbf{Y}$ are independent variables.

The method of linear prediction was developed in the 70's by Hartigan and Goldstein among others. It does not depend on the point of view adopted for inference. However, it generally appears in the Bayesian literature where it receives the name of linear Bayes methodology.

Exercices

§7.1

1. Let $\mathbf{X} = (X_1, \ldots, X_n)$ be a random sample from a $U[0, \theta]$ distribution and $\theta \sim Pa(\alpha, \beta)$.

   (a) Obtain the predictive distribution for a new observation from the same population based on all the information.

   (b) Assuming that max $x_i > \beta$, what is the probability of $Y > \max x_i$?

   (c) Repeat (a) and (b) for $\alpha \to 0$ and $\beta \to 0$.

2. Show that the problem of choice of a point predictor for Y may be reformulated as that of finding the predictor δ that minimizes

$$\int V(\theta, \delta)\, p(\theta \mid \mathbf{x})\, d\theta$$

where

$$V(\theta, \delta) = E_{Y|\theta}[L(Y, \delta)].$$

3. Consider the problem of choosing the point predictor of $Y \in R^p$ based on the quadratic loss

$$L(\mathbf{Y}, \boldsymbol{\delta}) = (\mathbf{Y} - \boldsymbol{\delta})'\mathbf{M}(\mathbf{Y} - \boldsymbol{\delta}),$$

where $\mathbf{M}$ is a known positive definite matrix. Show that the point predictor is given by $\delta = E(\mathbf{Y} \mid \mathbf{x})$ and that the expected loss is given by the trace of the matrix $\mathbf{M}V(\mathbf{Y} \mid \mathbf{x})$.

§7.2

4. Assume that one wishes to make a prediction for $Y \sim Bin(m, \theta)$ based on observation of $X \sim Bin(n, \theta)$. How can this be done from a frequentist point of view?

5. A geologist wishes to study the incidence of seismic movements in a given region. He then selects m independent but geologically similar observation points and counts the number of movements in a specific time interval. The observational model is $X_i \sim Pois(\theta)$, where X_i, $i = 1, \ldots, m$ is the number of occurrences in the ith observation point and θ is the average rate of seismic movements. From his previous experience, the researcher assumes that $E(\theta) = 2$ movements per time interval and that $V(\theta) = 0.36$ and uses these values to specify a conjugate prior.

   (a) Assuming that $x = (2, 3, 0, 0, 1, 0, 2, 0, 3, 0, 1, 2)$ was observed, what is the posterior distribution?

   (b) He wishes to predict the expected number of seismic movements and their precision in an $(m + 1)$th site based on the observation he had made. Establish the necessary hypotheses and perform the calculations using the data above.

§7.3

6. Consider the observation of independent samples $\mathbf{Y} = (Y_1, \ldots, Y_m)$ and $\mathbf{X} = (X_1, \ldots, X_n)$ from an $N(\theta, \sigma^2)$ population. Indicate how to obtain point and confidence interval predictors for the components of $\mathbf{Y}$ using

   (a) a conjugate prior for (θ, σ^2);
   (b) a noninformative prior for (θ, σ^2);

(c) frequentist inference.

§7.4

7. Prove that, in the multivariate case, the linear predictor is given by

$$\mu_Y + \Sigma_{XY}\Sigma_X^{-1}(\mathbf{x} - \mu_X)$$

and its risk is given by

$$\Sigma_Y - \Sigma_{XY}\Sigma_X^{-1}\Sigma_{YX}.$$

Chapter 8

Introduction to linear models

This chapter considers one of the most important problems in Statistics, the modeling of a random variable affected by other variables. The scope is introductory since it deserves at least one whole book to be considered in depth. A complete approach to the subject can be found in books such as Draper and Smith (1966) and Broemeling (1985).

The class of Normal linear models is characterized in the first section. In this same section the class of generalized linear models will also be introduced, as an extension of the Normal linear models to the exponential family. In the following sections, classical and Bayesian inference is developed for these models. Other broad classes of models, that are natural extensions of the Normal linear models, are described in Sections 8.4 (hierarchical linear models), 8.5 (dynamic linear models) and 8.6 (linear models with shrinkage).

8.1 The linear model

In this Section, the problem of observing a random variable Y with values affected by other variables will be discussed. For example, the income of a firm is affected by its capital and by the number of people it employs; the production of a machine is influenced by the maintenance scheme implemented and how trained its operator is; the arterial blood pressure depends upon the age of the patient, and so on.

In all these cases, the variability of Y is explained by some other variables. As a first approximation only the linear relationship will be used to describe how these variables influence Y. Let $X_1, \ldots, X_p$ denote the set of p explanatory variables. It follows that, at least to an initial approximation,

$$E(Y) = \beta_1 X_1 + \ldots + \beta_p X_p.$$

If a model with an intercept is wished, we need only to specify $X_1 = 1$.

The model introduced above is called a linear model or linear regression model. The case with $p = 1$, where only one explanatory variable is involved, is named a simple linear regression. Note that the expectation of Y is calculated conditionally on the values of the variables $X_1, \ldots, X_p$. Throughout this chapter it will be assumed that the values of the explanatory variables are known.

Assume that a sample $Y_1, \ldots, Y_n$ is observed with corresponding explanatory variables $(X_{1i}, \ldots, X_{pi})$, for $i = 1, \ldots, n$. The observation errors from the sample $Y_1, \ldots, Y_n$ can be specified as

$$e_i = Y_i - \beta_1 X_{1i} + \cdots + \beta_p X_{pi}, \ i = 1, \ldots, n$$

and the parameters $\boldsymbol{\beta} = (\beta_1, \ldots, \beta_p)$ can easily be estimated. In this case, the least squares estimator is given by the value of $\boldsymbol{\beta}$ that minimizes $\sum_{i=1}^{n} e_i^2$. Note that no hypothesis about the distribution of Y or of the errors was needed. If the hypotheses of normality and independence of the distribution of the Y_i's are assumed, and in addition, a common variance $Var(Y_i) = \sigma^2$ is assumed for all i, then it is easy to obtain the likelihood function as

$$l(\boldsymbol{\beta}, \sigma^2; y_1, \ldots, y_n) \propto \sigma^{-n} \exp \left\{ -\frac{1}{2\sigma^2} \sum_{i=1}^{n} e_i^2 \right\}.$$

Then, maximization of the likelihood and minimization of $\sum_{i=1}^{n} e_i^2$ are equivalent; that is, the maximum likelihood and the least squares estimators of $\boldsymbol{\beta}$ coincide.

It is useful, for further development, to adopt a matrix notation. Let us define

$$\mathbf{Y} = \begin{pmatrix} Y_1 \\ \vdots \\ Y_n \end{pmatrix} \quad \text{and} \quad \mathbf{X} = \begin{pmatrix} \mathbf{x}_1' \\ \vdots \\ \mathbf{x}_n' \end{pmatrix} = \begin{pmatrix} x_{11} & \cdots & x_{p1} \\ \vdots & & \vdots \\ x_{1n} & \cdots & x_{pn} \end{pmatrix}$$

following that $\mathbf{Y} \mid \boldsymbol{\beta}, \sigma^2 \sim N(\mathbf{X}\boldsymbol{\beta}, \sigma^2 \mathbf{I}_n)$ where $\mathbf{I}_p$ is the $p \times p$ identity matrix. The likelihood function can be rewritten as $\sigma^{-n} \exp \left\{ -S(\boldsymbol{\beta})/2\sigma^2 \right\}$, where

$$S(\boldsymbol{\beta}) = \sum_{i=1}^{n} (y_i - \mathbf{x}_i' \boldsymbol{\beta})^2 = (\mathbf{y} - \mathbf{X}\boldsymbol{\beta})'(\mathbf{y} - \mathbf{X}\boldsymbol{\beta}).$$

Moreover, the model presented previously can be rewritten through the following equations

$$Y_i \ \sim \ N(\mu_i, \sigma^2), \text{ for } i = 1, \ldots, n, \text{ independent,}$$
$$\mu_i \ = \ \lambda_i, \text{ for } i = 1, \ldots, n,$$
$$\lambda_i \ = \ \mathbf{x}_i' \boldsymbol{\beta}, \text{ for } i = 1, \ldots, n,$$

where the (apparently unnecessary) second equation states the relationship between the mean of the observations and the explanatory structure in the model. Stating the main objective in this form it is clear that there is no reason to be restricted to the Normal distribution and the class of linear relationship. One of the most relevant extensions in the model presented above constitutes the class of the generalized linear models, which allows modeling of observational distribution by a member of the exponential family and relating the mean of the distribution with the explanatory relationship through a differentiable function, not necessarily identity. Therefore the Y_i's have density given by

$$
\begin{aligned}
p(y_i \mid \theta) &= a(y_i) \exp\{u(y_i)\phi(\theta) + b(\theta)\}, \\
g(\mu_i) &= \lambda_i \text{ where } g \text{ is differentiable and } \mu_i = E(Y_i \mid \theta), \\
\lambda_i &= \mathbf{x}_i'\boldsymbol{\beta},
\end{aligned}
$$

for $i = 1, \ldots, n$, independent. This class is broad enough to include many of the most used models in Statistics. A complete description of these models, including many inferential aspects, can be found in McCullagh and Nelder (1989).

Example 8.1 *(Example 5.2 continued) Let Y_i follows a Bernoulli distribution with success probability $\mu_i \in (0, 1)$ and the mean of Y_i is given by $\mu_i = E(Y_i | x_i)$, for $i = 1, \ldots, n$. Let the connection with the explanatory variable be given by $logit(\mu_i) = \mathbf{x}_i'\boldsymbol{\beta}$. This link function associates values in the unit interval to the real line.*

With these assumptions, we can rewrite the model above as

$$
p(y_i \mid \mu_i) = \exp\{y_i \log(\mu_i/(1 - \mu_i)) + \log(1 - \mu_i)\}, \text{for } i = 1, \ldots, n,
$$

$$
logit(\mu_i) = \log\left(\frac{\mu_i}{1 - \mu_i}\right) = \lambda_i, \text{ for } i = 1, \ldots, n,
$$

$$
\lambda_i = \mathbf{x}_i'\boldsymbol{\beta}, \text{ for } i = 1, \ldots, n.
$$

The model presented in Example 8.1 is the so-called logistic model, one of the most popular linear models. The next example illustrates its use in an important practical area.

Example 8.2 *The logistic model can be used for determining the probability of default of customers according to a dichotomous explanatory variable. Let Y_i be equal to 1 if the i-th customer defaults and 0 otherwise, $X_{1i} = 1$ (in order to include the intercept) and X_{2i} be an economic factor for the i-th customer, for $i = 1, 2, \ldots, n$, where n is the sample size. The economic factor can be any variable that is relevant to the banking behaviour of a customer, such as his* *monthly income. So the chances of default and not default according to the values of X_2 are given by*

$$\mu(X_2 = 1) = \frac{\exp(\beta_1 + \beta_2)}{1 + \exp(\beta_1 + \beta_2)} \ and \ \mu(X_2 = 0) = \frac{\exp(\beta_1)}{1 + \exp(\beta_1)}.$$

8.2 Classical estimation of linear models

Classical estimators for β and σ^2 can be obtained from the likelihood function exhibited earlier, that is, $\sigma^{-n} \exp\{-S(\beta)/2\sigma^2\}$. Beginning with the estimation of β we see that the least square and maximum likelihood estimators do coincide and are given by the value of β that minimizes the quadratic form $S(\beta)$. Differentiating this expression with respect to the elements of the parameter vector β, it follows that

$$\frac{\partial S(\beta)}{\partial \beta} = 2(\mathbf{X}'\mathbf{X}\beta - \mathbf{X}'\mathbf{y})$$

(see Exercise 8.1). The solution of $\partial S(\beta)/\partial \beta = \mathbf{0}$ provides the point of minimum $\hat{\beta}$ of $S(\beta)$, since the matrix of second derivatives is positive definite. Besides, it must satisfy

$$\mathbf{X}'\mathbf{X}\hat{\beta} = \mathbf{X}'\mathbf{y},$$

which comprises a system of p equations, which are known as the Normal equations. If $\mathbf{X}'\mathbf{X}$ is a full rank matrix, or if the columns of $\mathbf{X}$ are linearly independent, then $\mathbf{X}'\mathbf{X}$ has an inverse and the maximum likelihood estimator of β is given by

$$\hat{\beta} = (\mathbf{X}'\mathbf{X})^{-1}\mathbf{X}'\mathbf{y}.$$

From here on it will be assumed that the matrix $\mathbf{X}'\mathbf{X}$ has full rank. This restriction means that the number p of regression coefficients must be smaller than the number n of observation points. Even when $p < n$, $\mathbf{X}$ may not be of full rank if there are some redundancies in its specification or in the model specification. These redundancies can be useful for model understanding but can be eliminated without any loss from the following inference procedures. The case where p approaches or is larger than n is more problematic and will be addressed in Section 8.6.

The quadratic form can be expressed as

$$S(\beta) = (\beta - \hat{\beta})'\mathbf{X}'\mathbf{X}(\beta - \hat{\beta}) + S_e,$$

where

$$\begin{aligned} S_e &= \mathbf{y}'\mathbf{y} - \hat{\beta}'\mathbf{X}'\mathbf{X}\hat{\beta} \\ &= (\mathbf{y} - \mathbf{X}\hat{\beta})'(\mathbf{y} - \mathbf{X}\hat{\beta}). \end{aligned}$$

The maximum likelihood estimator of σ^2 is obtained as the solution of the equation

$$\frac{\partial \log l(\hat{\beta}, \hat{\sigma}^2; \mathbf{y})}{\partial \sigma^2} = -\frac{n}{2}\frac{1}{\hat{\sigma}^2} + \frac{S(\hat{\beta})}{2\hat{\sigma}^4} = 0$$

and will be denoted by $\hat{\sigma}^2$. Since $S(\hat{\beta}) = S_e$ by the above development, the maximum likelihood estimator of σ^2 is given by S_e/n.

The sampling distribution of these estimators can easily be obtained. Since $\hat{\beta}$ is a linear function of $\mathbf{Y}$, its sampling distribution is a multivariate Normal with mean and variance given by

$$E(\hat{\beta} \mid \beta, \sigma^2) = (\mathbf{X}'\mathbf{X})^{-1}\mathbf{X}'E(\mathbf{Y} \mid \beta, \sigma^2) = (\mathbf{X}'\mathbf{X})^{-1}\mathbf{X}'\,\mathbf{X}\beta = \beta$$

and

$$Var(\hat{\beta} \mid \beta, \sigma^2) = (\mathbf{X}'\mathbf{X})^{-1}\mathbf{X}'V(\mathbf{Y} \mid \beta, \sigma^2)\mathbf{X}(\mathbf{X}'\mathbf{X})^{-1} = \sigma^2(\mathbf{X}'\mathbf{X})^{-1}.$$

So, $\hat{\beta}$ is an unbiased estimator of β. It also has minimum variance since the score function is linear in $\hat{\beta}$. The quadratic form given by $[S(\beta) - S_e]/\sigma^2$ has a χ_p^2 sampling distribution. On the other hand, it can be shown that S_e is independent of $\hat{\beta}$ and so of $S(\beta) - S_e$. Since $S(\beta)/\sigma^2 \sim \chi_n^2$, $S_e/\sigma^2 \sim \chi_{n-p}^2$ and, therefore, $s^2 = S_e/(n-p)$ is an unbiased estimator of σ^2. So, it follows that $(\hat{\beta}-\beta)/s$ has a multivariate Student-t sampling distribution with $n-p$ degrees of freedom, location parameter $\mathbf{0}$ and scale parameter matrix $(\mathbf{X}'\mathbf{X})^{-1}$.

The above statistic can be used as pivot to obtain confidence intervals for β or its components. In particular, it is easy to obtain the sampling distribution of $(\hat{\beta}_j - \beta_j)/s$ which is $t_{n-p}(0, c_{jj})$ where c_{jj} is the (j,j)th element of the matrix $(\mathbf{X}'\mathbf{X})^{-1}$, $j = 1, \ldots, p$. The $100(1-\alpha)\%$ confidence interval for β_j is given by

$$[\hat{\beta}_j - t_{\alpha/2, n-p}\, s\, c_{jj}^{1/2}, \hat{\beta}_j + t_{\alpha/2, n-p}\, s\, c_{jj}^{1/2}].$$

Alternatively, from the independence between $S(\beta) - S_e$ and S_e, it follows that $[S(\beta) - S_e]/ps^2 \sim F(p, n-p)$ and confidence regions for β can also be obtained.

Example 8.3 *(Simple linear regression) Consider the model* $y_i = \beta_0 + \beta_1(x_i - \bar{x}) + e_i$, $i = 1, \ldots, n$ *with only one explanatory variable and where* $\bar{x}$ *is the mean of the observed values* x_i's. *Then* $\mathbf{X} = (\mathbf{1}_n, \mathbf{x} - \bar{x}\mathbf{1}_n)$,

$$\mathbf{X}'\mathbf{X} = \begin{pmatrix} n & 0 \\ 0 & \sum_{i=1}^{n}(x_i - \bar{x})^2 \end{pmatrix} \quad and \quad \hat{\beta} = \begin{pmatrix} \hat{\beta}_0 \\ \hat{\beta}_1 \end{pmatrix} \quad with$$

$$\hat{\beta}_0 = \bar{y}, \ \hat{\beta}_1 = \frac{\sum_{i=1}^{n}(y_i - \bar{y})(x_i - \bar{x})}{\sum_{i=1}^{n}(x_i - \bar{x})^2} \quad and \quad s^2 = \frac{\sum_{i=1}^{n}(y_i - \hat{\beta}_0 - \hat{\beta}_1(x_i - \bar{x}))^2}{n-2},$$

where $\mathbf{1}_n$ is an n-dimensional vector of 1's and $\boldsymbol{x}' = (x_1, \ldots, x_n)$. Due to the centering of the values of the x_i's, the covariance matrix of the sampling distribution of $\hat{\boldsymbol{\beta}}$ is diagonal. As this is a multivariate Normal, this is equivalent to saying that $\hat{\beta}_0$ and $\hat{\beta}_1$ are independent.

Each of these estimators will provide separate information for each parameter. Confidence intervals can be built up based on the independent sampling distributions

$$\frac{\hat{\beta}_0 - \beta_0}{s} \sim t_{n-2}(0, 1/n) \ \ and \ \ \frac{\hat{\beta}_1 - \beta_1}{s} \sim t_{n-2}\left(0, \left[\sum_{i=1}^{n}(x_i - \bar{x})^2\right]^{-1}\right).$$

Example 8.4 *(Analysis of variance with one classification factor)* Let $y_{ji} = \beta_j + e_{ji}$, $i = 1, \ldots, n_j$, $j = 1, \ldots, k$ be the model, that is, the n_j observations in group j have the same mean β_j, for $j = 1, \ldots, k$. This model is frequently referred to as one-way ANOVA (an acronym for ANalysis Of VAriance). The total number of observations n is given by $\sum_{j=1}^{k} n_j$. Other parameterizations are possible, with the most usual given by $\beta_j = \mu + \alpha_j$ where μ is the global mean and α_j the deviation of the group mean with respect to the global mean. This parametrization is not used here due to the redundancy $\sum_{j=1}^{k} \alpha_j = 0$, which we are trying to avoid, but can easily be retrieved from the β_j's by taking $\mu = \sum_j \beta_j$ and $\alpha_j = \beta_j - \mu$, for $j = 1, \ldots, k$.

The model is completely characterized with parameter vector defined by $\boldsymbol{\beta} = (\beta_1, \ldots, \beta_k)'$, the observation vector

$$\boldsymbol{y} = \begin{pmatrix} y_{11} \\ \vdots \\ y_{1n_1} \\ \vdots \\ y_{k1} \\ \vdots \\ y_{kn_k} \end{pmatrix} \ \ and \ the \ design \ matrix \ \boldsymbol{X} = \begin{pmatrix} 1 & 0 & \cdots & \cdots & 0 \\ \vdots & \vdots & & & \vdots \\ 1 & 0 & \cdots & \cdots & 0 \\ & & \ddots & & \\ 0 & \cdots & \cdots & 0 & 1 \\ \vdots & & & \vdots & \vdots \\ 0 & \cdots & \cdots & 0 & 1 \end{pmatrix}.$$

As in the previous example, the matrix $(\boldsymbol{X}'\boldsymbol{X})^{-1}$ has a diagonal form with (j,j)-th element given by n_j^{-1} and $\hat{\beta}_j = \bar{Y}_j$ where $\bar{Y}_j$ is the observation mean in group j. Besides that,

$$S_e = \sum_{j=1}^{k} \sum_{i=1}^{n_j} (Y_{ji} - \bar{Y}_j)^2,$$

and $s^2 = S_e/(n-k)$ is the unbiased estimator of σ^2. Therefore, $(\hat{\beta}_j - \beta_j)/s \sim t_{n-k}(0, n_j^{-1})$, $j = 1, \ldots, k$.

Hypothesis tests based on the likelihood ratio can be obtained. One of the most useful is the model validation test, that is, the test of the hypothesis H_0: $\beta_2 = \cdots = \beta_p = 0$ in the model $Y_i = \beta_1 + \beta_2 x_{2i} + \cdots + \beta_p x_{pi} + e_i$, $i = 1, \ldots, n$. Under H_0, none of the explanatory variables have influence over the value of Y. As we have seen in Section 4.2, the maximum likelihood estimator of (β_1, σ^2) is $(\bar{Y}, \hat{\sigma}_0^2)$ where $\hat{\sigma}_0^2 = \Sigma(Y_i - \bar{Y})^2/n$. The maximized likelihood under H_0 is given by $\hat{\sigma}_0^{-n} e^{-n/2}$. So, the maximum likelihood ratio is given by

$$\left(\frac{n\hat{\sigma}_0^2}{S_e}\right)^{n/2} = \left(1 + \frac{n\hat{\sigma}_0^2 - S_e}{S_e}\right)^{n/2} = \left(1 + \frac{p-1}{n-p}F\right)^{n/2},$$

where $F = (n\hat{\sigma}_0^2 - S_e)/[(p-1)s^2]$.

The maximum likelihood ratio test of significance level α rejects H_0 if $F > c$ where c is implicitly given by the equation $\alpha = Pr(F > c \mid H_0)$. Using results about quadratic forms of Normal random variables, it is possible to show that $n\hat{\sigma}_0^2 - S_e$ is independent of S_e. As $n\hat{\sigma}_0^2/\sigma^2 \sim \chi_{n-1}^2$, then it follows that $(n\hat{\sigma}_0^2 - S_e)/\sigma^2 \sim \chi_{p-1}^2$ and so $F \sim F(p-1, n-p)$, under H_0. Therefore, $c = \bar{F}_\alpha(p-1, n-p)$. Confidence regions for $(\beta_2, \ldots, \beta_p)$ follows from the above test statistic.

Example 8.5 *(Example 8.3 continued) The model validation test in this case corresponds to the test of H_0: $\beta_1 = 0$. The expression $n\hat{\sigma}_0^2$ is given by*

$$
\begin{aligned}
\sum_{i=1}^{n}(Y_i - \bar{Y})^2 &= \sum_{i=1}^{n}\{[Y_i - \bar{Y} - \hat{\beta}_1(x_i - \bar{x})] + \hat{\beta}_1(x_i - \bar{x})\}^2 \\
&= \sum_{i=1}^{n}[Y_i - \bar{Y} - \hat{\beta}_1(x_i - \bar{x})]^2 + \sum_{i=1}^{n}\hat{\beta}_1^2(x_i - \bar{x})^2 \\
&\quad +2\sum_{i=1}^{n}[Y_i - \bar{Y} - \hat{\beta}_1(x_i - \bar{x})]\hat{\beta}_1(x_i - \bar{x}) \\
&= S_e + \hat{\beta}_1\sum_{i=1}^{n}(Y_i - \bar{Y})(x_i - \bar{x}) \\
&= S_e + \hat{\beta}_1^2\sum_{i=1}^{n}(x_i - \bar{x})^2.
\end{aligned}
$$

So, $F = \hat{\beta}_1^2/[s^2/\sum_{i=1}^{n}(x_i - \bar{x})^2] \sim F(1, n-2) \sim t_{n-2}^2$ and the model validation test rejects the null hypothesis H_0 if $\bar{F} > F_\alpha(1, n-2)$ or equivalently if

$$|\hat{\beta}_1| > t_{\alpha/2, n-2}\, s / \sqrt{\sum_{i=1}^{n}(x_i - \bar{x})^2}.$$

This result can be extended to multiple regression in the following sense. Consider the model $Y_i = \beta_1 + \beta_2(x_{2i} - \bar{x}_2) + \ldots + \beta_p(x_{pi} - \bar{x}_p) + e_i$, $i = 1, \ldots, n$ where all the explanatory variables were conveniently centered. The likelihood ratio test for $H_j: \beta_j = 0$, $j = 2, \ldots, p$ with level α rejects H_j if

$$|\hat{\beta}_j| > t_{\alpha/2, n-p} \, s \, / \sqrt{\sum_{j=1}^{n}(x_{ji} - \bar{x}_j)^2}.$$

Example 8.6 *(Example 8.4 continued). Now the model validation test is concerned with the following hypothesis, $H_0: \beta_1 = \ldots = \beta_k$. The expression of $n\hat{\sigma}_0^2$ is*

$$\sum_{j=1}^{k}\sum_{i=1}^{n_i}(Y_{ji} - \bar{\bar{Y}})^2 = \sum_{j=1}^{k}\sum_{i=1}^{n_j}[(Y_{ji} - \bar{Y}_j) + (\bar{Y}_j - \bar{\bar{Y}})]^2,$$

where $\bar{\bar{Y}} = \frac{1}{n}\sum_{j=1}^{k} n_j \bar{Y}_j$.
   It follows that

$$\sum_{j=1}^{k}\sum_{i=1}^{n_i}(Y_{ji} - \bar{\bar{Y}})^2 = \sum_{j=1}^{k}\sum_{i=1}^{n_j}(Y_{ji} - \bar{Y}_j)^2 + \sum_{j=1}^{k} n_j(\bar{Y}_j - \bar{\bar{Y}})^2$$

$$+ 2\sum_{j=1}^{k}\sum_{i=1}^{n_j}(Y_{ji} - \bar{Y}_j)(\bar{Y}_j - \bar{\bar{Y}})$$

$$= S_e + \sum_{j=1}^{k} n_j(\bar{Y}_j - \bar{\bar{Y}})^2,$$

because the third term vanishes.
   We have already seen that squared deviations around the mean for normally distributed variables are independent of the sample mean. Then, S_e is independent of the means $\bar{Y}_1, \ldots, \bar{Y}_k$ and so is also independent from the second term in the above expression. This term is also in the form of squared deviations of k Normal observations with respect to its mean and so has the sampling distribution given by $\sigma^2 \chi^2_{k-1}$. Therefore,

$$F = \frac{\sum_{j=1}^{k} n_j(\bar{Y}_j - \bar{\bar{Y}})^2/(k-1)}{s^2} \sim F(k-1, n-k),$$

confirming the result described above for the general linear model. F is the MLR test statistic for H_0 and the test of significance level α rejects H_0 if $F > \bar{F}_\alpha(k-1, n-k)$.

   Finally, we will address the relevant question of how to make predictions using the classical linear model. The problem here is how to make inference about a m dimensional vector of future observations Y^* with explanatory variables gathered in an $m \times p$ matrix $\mathbf{x}^*$. Once again, the trick is to find a convenient pivot, which in this case is given by $(\mathbf{Y}^* - \mathbf{x}^*\hat{\beta})/s$. To verify this result, it suffices to observe that $\mathbf{Y}^*$, $\hat{\beta}$ and s^2 are independent and so $\mathbf{Y}^* - \mathbf{x}^*\hat{\beta}$ is independent of s^2. The numerator is distributed as Normal with mean and variance given by

$$
\begin{aligned}
E[\mathbf{Y}^* - \mathbf{x}^*\hat{\beta}|\beta, \sigma^2] &= \mathbf{x}^*\beta - \mathbf{x}^*\beta = 0 \\
V[\mathbf{Y}^* - \mathbf{x}^*\hat{\beta}|\beta, \sigma^2] &= \sigma^2 + \mathbf{x}^*\sigma^2(\mathbf{X}'\mathbf{X})^{-1}\mathbf{x}^{*'} \\
&= \sigma^2\mathbf{C}^* \text{ where } \mathbf{C}^* = \mathbf{I}_m + \mathbf{x}^*(\mathbf{X}'\mathbf{X})^{-1}\mathbf{x}^{*'}.
\end{aligned}
$$

Since $(n-p)s^2/\sigma^2 \sim \chi^2_{n-p}$, the pivotal quantity $(\mathbf{Y}^* - \mathbf{x}^*\hat{\beta})/s$ has $t_{n-p}(\mathbf{0}, \mathbf{C}^*)$ distribution, independently of the parameters in the model. So, confidence intervals for Y^* with confidence level $100(1-\alpha)\%$ can be built. These results are easily extended to the prediction of a vector of future observations following the same steps as previously.

Example 8.7 *(Example 8.3 continued) For a simple linear regression, we have*

$$
\hat{y}^* = \mathbf{x}^*\hat{\beta} = \begin{pmatrix} \hat{\beta}_0 + \hat{\beta}_1(x_1^* - \bar{x}^*) \\ \vdots \\ \hat{\beta}_0 + \hat{\beta}_1(x_m^* - \bar{x}^*) \end{pmatrix},
$$

where $\bar{x}^* = \sum_{i=1}^m x_i^*/m$. The $100(1-\alpha)\%$ confidence interval for Y^* can be obtained by using the pivot $(Y^* - \mathbf{x}^*\hat{\beta})/s$, which is distributed as $t_{n-2}(\mathbf{0}, \mathbf{C}^*)$, that is,

$$
\left[\hat{y}_i^* - t_{\alpha/2, n-2} \ s \ \sqrt{c_{ii}^*} \ , \ \hat{y}_i^* + t_{\alpha/2, n-2} \ s \ \sqrt{c_{ii}^*} \right],
$$

where c_{ii}^* is the (i, i)th element of the matrix $\mathbf{C}^*$, for $i = 1, \ldots, m$.

Example 8.8 *(Example 8.4 continued) For analysis of variance with one factor of classification, considering the Example 8.4, we have*

$$
\hat{y}^* = \mathbf{x}^*\hat{\beta} = \begin{pmatrix} \sum_{j=1}^k \hat{\beta}_j x_{1j}^* \\ \vdots \\ \sum_{j=1}^k \hat{\beta}_j x_{mj}^* \end{pmatrix}.
$$

A $100(1-\alpha)\%$ confidence interval for Y_i^* is given by

$$
\left[\hat{y}_i^* - t_{\alpha/2, n-k} \ s \ \sqrt{c_{ii}^*} \ , \ \hat{y}_i^* + t_{\alpha/2, n-k} \ s \ \sqrt{c_{ii}^*} \right],
$$

for $i = 1, \ldots, m$.

8.3   Bayesian estimation of linear models

In Bayesian inference, a prior distribution for the parameters must be specified in addition to the likelihood function. First, the results involving proper priors (in fact, natural conjugate, as will be shown) will be presented. The analysis with non-informative prior will be presented afterwards and some comparisons with the classical inference will be made. The examples presented in the previous section will be revisited.

The prior distribution adopted for the parameters is a multivariate generalization of the Normal-χ^2 presented in Section 3.3. Assume that the parametric vector $\boldsymbol{\beta}$ has a conditional prior distribution $N(\boldsymbol{\mu}_0, \phi^{-1}\mathbf{C}_0^{-1})$ where $\phi = \sigma^{-2}$ and that $n_0\sigma_0^2\phi \sim \chi_{n_0}^2$. In this way, the prior distribution is fully specified with density given by

$$
\begin{aligned}
p(\boldsymbol{\beta}, \phi) &= (2\pi)^{-p/2}|\phi\mathbf{C}_0|^{1/2} \exp\left\{-\frac{\phi}{2}(\boldsymbol{\beta} - \boldsymbol{\mu}_0)'\mathbf{C}_0(\boldsymbol{\beta} - \boldsymbol{\mu}_0)\right\} \\
&\quad \times \frac{(n_0\sigma_0^2/2)^{n_0/2}}{\Gamma(n_0/2)}\phi^{(n_0/2)-1} \exp\left\{-\frac{n_0\sigma_0^2}{2}\phi\right\} \\
&\propto \phi^{[(n_0+p)/2]-1} \exp\left\{-\frac{\phi}{2}[n_0\sigma_0^2 + (\boldsymbol{\beta} - \boldsymbol{\mu}_0)'\mathbf{C}_0(\boldsymbol{\beta} - \boldsymbol{\mu}_0)]\right\}.
\end{aligned}
$$

Then, as in the univariate case, the conditional distribution of $\phi \mid \boldsymbol{\beta}$ can be obtained from the joint prior distribution of $\boldsymbol{\beta}$ and ϕ collecting only the terms involving ϕ. It is given by $[n_0\sigma_0^2 + (\boldsymbol{\beta} - \boldsymbol{\mu}_0)'\mathbf{C}_0(\boldsymbol{\beta} - \boldsymbol{\mu}_0)]\phi \mid \boldsymbol{\beta} \sim \chi_{n_0+p}^2$. The marginal distribution of $\boldsymbol{\beta}$ can be obtained dividing $p(\boldsymbol{\beta}, \phi)$ by $p(\phi|\boldsymbol{\beta})$ or, as done earlier, integrating the joint distribution with respect to ϕ. Its density is given by

$$
p(\boldsymbol{\beta}) \propto [n_0\sigma_0^2 + (\boldsymbol{\beta} - \boldsymbol{\mu}_0)'\mathbf{C}_0(\boldsymbol{\beta} - \boldsymbol{\mu}_0)]^{-(n_0+p)/2},
$$

which corresponds to the density of a $t_{n_0}(\boldsymbol{\mu}_0, \sigma_0^2\mathbf{C}_0^{-1})$ distribution, as seen in Chapter 4. The normalizing constant is

$$
\frac{\Gamma[(n_0 + k)/2]}{\Gamma(n_0/2)n_0^{k/2}} (n_0\sigma_0^2)^{n/2} \mid \mathbf{C}_0 \mid^{1/2}.
$$

On the other hand, the likelihood of $\boldsymbol{\beta}$ and ϕ is given by

$$
\phi^{n/2} \exp\left\{-\frac{\phi}{2}[S_e + (\boldsymbol{\beta} - \hat{\boldsymbol{\beta}})'\mathbf{X}'\mathbf{X}(\boldsymbol{\beta} - \hat{\boldsymbol{\beta}})]\right\}
$$

and has the same form of the prior density. Therefore, the posterior is given by

$$
\begin{aligned}
p(\boldsymbol{\beta}, \phi|\mathbf{y}) &\propto \phi^{((n+n_0+p)/2)-1} \\
&\times \exp\left\{-\frac{\phi}{2}[n_0\sigma_0^2 + S_e + (\boldsymbol{\beta}-\boldsymbol{\mu}_0)'\mathbf{C}_0(\boldsymbol{\beta}-\boldsymbol{\mu}_0) + (\boldsymbol{\beta} - \hat{\boldsymbol{\beta}})'\mathbf{X}'\mathbf{X}(\boldsymbol{\beta}-\hat{\boldsymbol{\beta}})]\right\}.
\end{aligned}
$$

It can be shown that

$$(\boldsymbol{\beta} - \boldsymbol{\mu}_0)'\mathbf{C}_0(\boldsymbol{\beta} - \boldsymbol{\mu}_0) + (\boldsymbol{\beta} - \hat{\boldsymbol{\beta}})'\mathbf{X}'\mathbf{X}(\boldsymbol{\beta} - \hat{\boldsymbol{\beta}})$$
$$= (\boldsymbol{\beta} - \boldsymbol{\mu}_1)'\mathbf{C}_1(\boldsymbol{\beta} - \boldsymbol{\mu}_1) + \boldsymbol{\mu}_0'\mathbf{C}_0\boldsymbol{\mu}_0 + \hat{\boldsymbol{\beta}}'\mathbf{X}'\mathbf{X}\hat{\boldsymbol{\beta}} + \boldsymbol{\mu}_1'\mathbf{C}_1\boldsymbol{\mu}_1,$$

where $\boldsymbol{\mu}_1 = \mathbf{C}_1^{-1}(\mathbf{C}_0\boldsymbol{\mu}_0 + \mathbf{X}'\mathbf{y})$ and $\mathbf{C}_1 = \mathbf{C}_0 + \mathbf{X}'\mathbf{X}$. Note that even if $\mathbf{X}$ does not have full rank, $\mathbf{C}_1$ will have and can always be inverted. Inference can then proceed with no difficulties. This finding is the key to solve problems in classical inference associated with rank deficiency of $\mathbf{X}$. This point will be returned to in Section 8.6

To complete inference procedures, we still have to deal with the terms

$$S_e + \boldsymbol{\mu}_0'\mathbf{C}_0\boldsymbol{\mu}_0 + \hat{\boldsymbol{\beta}}'\mathbf{X}'\mathbf{X}\hat{\boldsymbol{\beta}} + \boldsymbol{\mu}_1'\mathbf{C}_1\boldsymbol{\mu}_1$$
$$= \mathbf{y}'\mathbf{y} - \hat{\boldsymbol{\beta}}'\mathbf{X}'\mathbf{X}\hat{\boldsymbol{\beta}} + \boldsymbol{\mu}_0'\mathbf{C}_0\boldsymbol{\mu}_0 + \hat{\boldsymbol{\beta}}'\mathbf{X}'\mathbf{X}\hat{\boldsymbol{\beta}} + \boldsymbol{\mu}_1'(\mathbf{C}_0\boldsymbol{\mu}_0 + \mathbf{X}'\mathbf{y})$$
$$= (\mathbf{y} - \mathbf{X}\boldsymbol{\mu}_1)'\mathbf{y} + (\boldsymbol{\mu}_0 - \boldsymbol{\mu}_1)'\mathbf{C}_0\boldsymbol{\mu}_0.$$

Then, the posterior density of $\boldsymbol{\beta}$ and ϕ can be written as

$$p(\boldsymbol{\beta}, \phi \mid \mathbf{y}) \propto \phi^{p/2} \exp\left\{ -\frac{\phi}{2}(\boldsymbol{\beta} - \boldsymbol{\mu}_1)'\mathbf{C}_1(\boldsymbol{\beta} - \boldsymbol{\mu}_1) \right\}$$

$$\times \phi^{(n_1/2)-1} \exp\left\{ -\frac{\phi}{2}n_1\sigma_1^2 \right\},$$

where $n_1 = n + n_0$ and $n_1\sigma_1^2 = n_0\sigma_0^2 + (\mathbf{y} - \mathbf{X}\boldsymbol{\mu}_1)'\mathbf{y} + (\boldsymbol{\mu}_0 - \boldsymbol{\mu}_1)'\mathbf{C}_0\boldsymbol{\mu}_0$. This density has the same form of the prior and, so, is natural conjugate to the Normal linear models. In particular, $\boldsymbol{\beta} \mid \mathbf{y} \sim t_{n_1}(\boldsymbol{\mu}_1, \sigma_1^2\mathbf{C}_1^{-1})$ and $\beta_j \mid \mathbf{y} \sim t_{n_1}(\mu_{1j}, \sigma_1^2(\mathbf{C}_1^{-1})_{jj})$, $j = 1, \ldots, p$. The posterior mean and variance-covariance matrix of $\boldsymbol{\beta}$ are given, respectively, by

$$\boldsymbol{\mu}_1 \quad \text{and} \quad \frac{n_1}{n_1 - 2}\sigma_1^2\mathbf{C}_1^{-1}, \, n_1 > 2.$$

The posterior distribution of ϕ is $n_1\sigma_1^2\phi \mid \mathbf{y} \sim \chi_{n_1}^2$ with mean σ_1^{-2}. The point estimators of $\boldsymbol{\beta}$ and ϕ are given by $\boldsymbol{\mu}_1$ and σ_1^{-2}, respectively. Confidence intervals for β_j and ϕ are obtained with the percentiles of t_{n_1} and $\chi_{n_1}^2$ distributions, respectively. It is also possible to make inference about the joint distribution of the $\boldsymbol{\beta}$ based on the fact that $(\boldsymbol{\beta} - \boldsymbol{\mu}_1)'\mathbf{C}_1(\boldsymbol{\beta} - \boldsymbol{\mu}_1)/\sigma_1^2 \mid \mathbf{y} \sim F(p, n_1 - p)$.

The non-informative prior can be used to represent, in some sense, the absence of initial information. Using the same approach as in Section 3.4, as $\boldsymbol{\beta}$ is essentially a (multivariate) location parameter and ϕ a scale parameter, it follows that the joint non-informative prior distribution is given by $p(\boldsymbol{\beta}, \phi) \propto$

ϕ^{-1}. This prior is a particular degenerated case of the natural conjugate prior. Making the convenient substitutions, the posterior density is

$$p(\boldsymbol{\beta}, \phi \mid \mathbf{y}) \quad \propto \quad \phi^{(n/2)-1} \exp\left\{-\frac{\phi}{2}[S_e + (\boldsymbol{\beta} - \hat{\boldsymbol{\beta}})'\mathbf{X}'\mathbf{X}(\boldsymbol{\beta} - \hat{\boldsymbol{\beta}})]\right\}$$

$$\propto \phi^{p/2} \exp\left\{-\frac{\phi}{2}(\boldsymbol{\beta} - \hat{\boldsymbol{\beta}})'\mathbf{X}'\mathbf{X}(\boldsymbol{\beta} - \hat{\boldsymbol{\beta}})\right\} \phi^{((n-p)/2)-1} \exp\left\{-\frac{\phi}{2}(n-p)s^2\right\}.$$

Therefore, the posterior will remain in the same class with only changes to the values of the hyperparameters of the relevant distributions. So, $\boldsymbol{\beta} \mid \mathbf{y} \sim t_{n-p}(\hat{\boldsymbol{\beta}}, s^2(\mathbf{X}'\mathbf{X})^{-1})$ and $(n-p)s^2\phi \mid \mathbf{y} \sim \chi^2_{n-p}$, and the quadratic form in $\boldsymbol{\beta}$ will be reduced to $(\boldsymbol{\beta} - \hat{\boldsymbol{\beta}})'\mathbf{X}'\mathbf{X}(\boldsymbol{\beta} - \hat{\boldsymbol{\beta}})/s^2$ with posterior distribution $F(p, n - p)$. These distributions provide the parallel Bayesian results to those obtained in the previous section using the classical approach.

Example 8.9 *(Example 8.3 continued) Suppose that in the prior distribution, β_0 and β_1 are conditionally independent given ϕ with distributions $N(\mu_0, (c_0\phi)^{-1})$ and $N(\mu_1, (c_1\phi)^{-1})$, respectively, and $n_0\sigma_0\phi \sim \chi^2_{n_0}$. (The case where β_0 and β_1 are not conditionally independent is left as an exercise.) Since $\mathbf{X}'\mathbf{X}$ is a diagonal matrix, the quadratic form $(\boldsymbol{\beta} - \hat{\boldsymbol{\beta}})'\mathbf{X}'\mathbf{X}(\boldsymbol{\beta} - \hat{\boldsymbol{\beta}})$ is reduced to $S(\beta_0) + S(\beta_1)$ where*

$$S(\beta_0) = n(\beta_0 - \bar{y})^2 \text{ and } S(\beta_1) = \sum_{i=1}^{n}(x_i - \bar{x})^2(\beta_1 - \hat{\beta}_1)^2,$$

and the likelihood function is given by

$$l(\beta_0, \beta_1, \phi; \boldsymbol{y}) \propto \phi^{n/2} \exp\left\{-\frac{\phi}{2}[S_e + S(\beta_0) + S(\beta_1)]\right\}.$$

Combining with the prior, the posterior distribution follows

$$p(\beta_0, \beta_1, \phi \mid \boldsymbol{y}) \quad \propto \quad \phi^{(n+n_0+2)/2-1}$$

$$\times \exp\left\{-\frac{\phi}{2}[n_0\sigma_0^2 + S_e + c_0\beta_0 - \mu_0)^2 + S(\beta_0) + c_1(\beta_1 - \mu_1)^2 + S(\beta_1)]\right\}$$

from where it is easy to see that, conditional on ϕ, β_0 and β_1 remain independent with respective $N(\mu_j^*, (c_j^*\phi)^{-1})$ $(j = 0, 1)$ distributions, where $c_0^* = c_0 + n$, $c_1^* = c_1 + \sum_{i=1}^{n}(x_i - \bar{x})^2$,

$$\mu_0^* = \frac{c_0\mu_0 + n\bar{y}}{c_0 + n} \text{ and } \mu_1^* = \frac{c_1\mu_1 + \sum_{i=1}^{n}(x_i - \bar{x})^2\hat{\beta}_1}{c_1 + \sum_{i=1}^{n}(x_i - \bar{x})^2}.$$

The posterior distribution of ϕ is $n_1\sigma_1^2\phi \mid \boldsymbol{y} \sim \chi^2_{n_1}$ with $n_1 = n + n_0$ and

$$n_1\sigma_1^2 = \sum_{i=1}^{n} y_i^2 - n\mu_0^*\hat{\beta}_0 - \mu_1^*\hat{\beta}_1 \sum_{i=1}^{n}(x_i - \bar{x})^2 + c_0(\beta_0 - \mu_0)^2 + c_1(\beta_1 - \mu_1)^2.$$

The marginal posterior distributions of β_0 and β_1 are Student-t distributions with n_1 degrees of freedom and parameters μ_j^* and σ_1^2/c_j^*, $j = 0, 1$, respectively.

In the case of non-informative prior $p(\beta_0, \beta_1, \phi) \propto \phi^{-1}$, the posterior is given by

$$
\begin{aligned}
p(\beta_0, \beta_1, \phi \mid \boldsymbol{y}) \quad &\propto \quad \phi^{-1}\,\phi^{n/2} \exp\left\{-\frac{\phi}{2}[(n-2)s^2 + S(\beta_0) + S(\beta_1)]\right\} \\
&\propto \quad \phi^{1/2} \exp\left\{-\frac{\phi}{2}S(\beta_0)\right\} \phi^{1/2} \exp\left\{-\frac{\phi}{2}S(\beta_1)\right\} \\
&\qquad \times \phi^{((n-2)/2)-1} \exp\left\{-\frac{\phi}{2}(n-2)s^2\right\}.
\end{aligned}
$$

Examining the above expression, it is easy to identify that $\beta_0 \mid \phi, \boldsymbol{y} \sim N(\bar{y}, (n\phi)^{-1})$, $\beta_1 \mid \phi, \boldsymbol{y} \sim N(\hat{\beta}_1, [\phi \sum_{i=1}^{n}(x_i - \bar{x})^2]^{-1})$ and $(n-2)s^2\phi \mid \boldsymbol{y} \sim \chi_{n-2}^2$. Then, $\beta_0 \mid \boldsymbol{y} \sim t_{n-2}(\bar{y}, s^2/n)$ and $\beta_1 \mid \boldsymbol{y} \sim t_{n-2}\left(\hat{\beta}_1, s^2/\sum_{i=1}^{n}(x_i - \bar{x})^2\right)$, and so, the inference coincides numerically with the one obtained following the classical approach.

Example 8.10 *(Example 8.4 continued) Assume that $\beta_j \mid \phi \sim N(\mu_j, (c_j\phi)^{-1})$, $j = 1, \ldots, k$, are conditionally independent and that $n_0\sigma_0^2\phi \sim \chi_{n_0}^2$. An alternative to the conditional independence of the β_j's will be considered in the next section. As in Example 8.3, $\boldsymbol{X}'\boldsymbol{X}$ is a diagonal matrix and the quadratic form $(\boldsymbol{\beta} - \hat{\boldsymbol{\beta}})'\boldsymbol{X}'\boldsymbol{X}(\boldsymbol{\beta} - \hat{\boldsymbol{\beta}})$ reduces to*

$$
\sum_{j=1}^{k} n_j(\beta_j - \bar{y}_j)^2.
$$

The likelihood is given by

$$
l(\beta_1, \ldots, \beta_k, \phi; \boldsymbol{y}) \propto \phi^{n/2} \exp\left\{-\frac{\phi}{2}\left[(n-k)s^2 + \sum_{j=1}^{k} n_j(\beta_j - \bar{y}_j)^2\right]\right\}.
$$

Combining with the prior distribution, the following posterior is obtained

$$
\begin{aligned}
p(\beta_1, \ldots, \beta_k, \phi \mid \boldsymbol{y}) \quad &\propto \quad \phi^{((n+n_0+k)/2)-1} \\
&\times \exp\left\{-\frac{\phi}{2}\left[n_0\sigma_0^2 + (n-k)s^2 + \sum_{j=1}^{k} c_j(\beta_j - \mu_j)^2 + \sum_{j=1}^{k} n_j(\beta_j - \bar{y}_j)^2\right]\right\} \\
&\propto \left\{\prod_{j=1}^{k} \phi^{1/2} \exp\left[-\frac{\phi}{2}c_j^*(\beta_j - \mu_j^*)^2\right]\right\} \phi^{(n_1/2)-1} \exp\left\{-\frac{\phi}{2}n_1\sigma_1^2\right\},
\end{aligned}
$$

where $c_j^* = c_j + n_j$, $\mu_j^* = (c_j\mu_j + n_j\bar{y}_j)/c_j^*$, $j = 1, \ldots, k$, $n_1 = n + n_0$ and

$$n_1\sigma_1^2 = n_0\sigma_0^2 + (n-k)s^2 + \sum_{j=1}^{k} \frac{n_j c_j}{n_j + c_j}(\bar{y}_j - \mu_j)^2.$$

It is easy to obtain that $\beta_j \mid \phi, \boldsymbol{y} \sim N[\mu_j^*, (c_j^*\phi)^{-1}]$, thus retaining the prior independence and $n_1\sigma_1^2\phi \mid \boldsymbol{y} \sim \chi_{n_1}^2$, and therefore, the marginal posterior distributions of β_j are $t_{n_1}(\mu_j^*, \sigma_1^2/c_j^*)$, $j = 1, \ldots, k$. The point estimators of β_j and ϕ are given by μ_j^* and σ_1^{-2}, $j = 1, \ldots, k$, respectively. Confidence intervals for β_j and ϕ (or σ^2) can be obtained using the percentiles of the t_{n_1} and $\chi_{n_1}^2$ distributions, respectively.

In the case of non-informative prior, we will have the posterior

$$p(\beta_1, \ldots, \beta_k, \phi \mid \boldsymbol{y}) \propto \phi^{(n/2)-1} \exp\left\{-\frac{\phi}{2}\left[(n-k)s^2 + \sum_{j=1}^{k} n_j(\beta_j - \bar{y}_j)^2\right]\right\}$$

$$\propto \left\{\prod_{j=1}^{k} \phi^{1/2} \exp\left[-\frac{\phi}{2}n_j(\beta_j - \bar{y}_j)^2\right]\right\} \phi^{((n-k)/2)-1} \exp\left\{-\frac{\phi}{2}(n-k)s^2\right\}.$$

Making the same identifications as in Example 8.3, the marginal posterior distributions $\beta_j \mid \boldsymbol{y} \sim t_{n-k}(\bar{y}_j, s^2/n_j)$, $j = 1, \ldots, k$ and $(n-k)s^2\phi \mid \boldsymbol{y} \sim \chi_{n-k}^2$ are easily obtained. These distributions are similar to the ones obtained in classical inference.

Predictive distributions are needed to perform hypothesis testing and prediction from a Bayesian point of view. As we have already seen, these distributions can be obtained by integrating the sample distribution with respect to the distribution of the parameters. Fortunately, in the Normal case, this task is simplified by the linear structure of the distribution. Consider the model

$$\mathbf{Y} = \mathbf{X}\boldsymbol{\beta} + \mathbf{e}, \text{ where } \mathbf{e} \sim N(\mathbf{0}, \phi^{-1}\mathbf{I}_n), \text{ with } \phi = 1/\sigma^2,$$

and suppose that the conditional prior distribution of $\boldsymbol{\beta}$ is

$$\boldsymbol{\beta} \mid \phi \sim NM(\boldsymbol{\mu}, \phi^{-1}\mathbf{C}^{-1}).$$

Combining these two results, we get

$$\begin{aligned}\mathbf{Y} \mid \phi &\sim \mathbf{X}N(\boldsymbol{\mu}, \phi^{-1}\mathbf{C}^{-1}) + N(\mathbf{0}, \phi^{-1}\mathbf{I}_n) \\ &\sim N[\mathbf{X}\boldsymbol{\mu}, \phi^{-1}(\mathbf{I}_n + \mathbf{X}\mathbf{C}^{-1}\mathbf{X}')],\end{aligned}$$

since the above Normal distributions are independent. Supposing now that $\nu\sigma_0^2\phi \sim \chi_\nu^2$ leads to

$$\mathbf{Y} \sim t_\nu[\mathbf{X}\boldsymbol{\mu}, \sigma_0^2(\mathbf{I}_n + \mathbf{X}\mathbf{C}^{-1}\mathbf{X}')],$$

and the marginal distribution of Y_i is $t_\nu[\mathbf{x}_i\boldsymbol{\mu}, \sigma_0^2(1 + \mathbf{x}_i\mathbf{C}^{-1}\mathbf{x}_i')]$, $i = 1, \ldots, n$.

Comparison of the hypothesis H_0 and H_1 requires the evaluation of the marginal distributions of $(\mathbf{Y} \mid H_l)$, for $l = 0, 1$. The model validation test may be based on the Bayes factor

$$BF(H_0; H_1) = \frac{p(\mathbf{y} \mid H_0)}{p(\mathbf{y} \mid H_1)},$$

where the denominator is the density of the above distribution given by

$$\frac{\Gamma[(n_0 + n)/2]}{\Gamma(n_0/2)n_0^{n/2}} (n_0\sigma_0^2)^{n/2} \mid \mathbf{I}_n + \mathbf{X}\mathbf{C}_0^{-1}\mathbf{X}' \mid^{-1/2}$$
$$\times [n_0\sigma_0^2 + (\mathbf{y} - \mathbf{X}\boldsymbol{\mu}_0)'(\mathbf{I}_n + \mathbf{X}\mathbf{C}_0^{-1}\mathbf{X}')^{-1}(\mathbf{y} - \mathbf{X}\boldsymbol{\mu}_0)]^{-(n_0+n)/2}.$$

If it is desirable to test the model validation hypothesis, we build H_0 : $\beta_2 = \ldots = \beta_p = 0$. Under H_0, the model simplifies to $\mathbf{Y} = \mathbf{1}_n\beta_1 + \mathbf{e}$ and $\beta_1 \mid \phi \sim N[\mu_{00}, (c_{00}\phi)^{-1}]$, and therefore, $\mathbf{Y} \mid \phi \sim N[\mathbf{1}_n\mu_{00}, \phi^{-1}(\mathbf{I}_n + c_{00}^{-1}\mathbf{1}_n\mathbf{1}_n')]$. Then, the density $p(\mathbf{y} \mid H_0)$ is given by

$$\frac{\Gamma[(n_0 + n)/2]}{\Gamma(n_0/2)n_0^{n/2}} (n_0\sigma_0^2)^{n/2} \mid \mathbf{I}_n + c_{00}^{-1}\mathbf{1}_n\mathbf{1}_n' \mid^{-1/2}$$
$$\times [n_0\sigma_0^2 + (\mathbf{y} - \mathbf{1}_n\mu_{00})'(\mathbf{I}_n + c_{00}^{-1}\mathbf{1}_n\mathbf{1}_n')^{-1}(\mathbf{y} - \mathbf{1}_n\mu_{00})]^{-(n_0+n)/2}.$$

The Bayes factor is, then, given by

$$\left\{ \frac{|\mathbf{I}_n + \mathbf{X}\mathbf{C}_0^{-1}\mathbf{X}'|}{|\mathbf{I}_n + c_{00}^{-1}\mathbf{1}_n\mathbf{1}_n'|} \right\}^{\frac{1}{2}}$$

$$\times \left\{ \frac{|n_0\sigma_0^2 + (\mathbf{y}-\mathbf{X}\boldsymbol{\mu}_0)'(\mathbf{I}_n + \mathbf{X}\mathbf{C}_0^{-1}\mathbf{X}')^{-1}(\mathbf{y}-\mathbf{X}\boldsymbol{\mu}_0)|}{|n_0\sigma_0^2 + (\mathbf{y}-\mathbf{1}_n\mu_{00})'(\mathbf{I}_n + c_{00}^{-1}\mathbf{1}_n\mathbf{1}_n')^{-1}(\mathbf{y}-\mathbf{1}_n\mu_{00})|} \right\}^{\frac{n_0+n}{2}}.$$

In the case of the prediction of a m-dimensional vector of future observations $\mathbf{Y}^*$ with explanatory variable matrix $\mathbf{x}^*$, the same result used above can be applied as far as the prediction is based on the predictive distribution of

$\mathbf{Y}^* \mid \mathbf{y}$ with density given by

$$
\begin{aligned}
p(\mathbf{y}^* \mid \mathbf{y}) &= \int \int p(\mathbf{y}^* \mid \boldsymbol{\beta}, \phi, \mathbf{y}) \, p(\boldsymbol{\beta}, \phi \mid \mathbf{y}) \, d\boldsymbol{\beta} \, d\phi \\
&= \int \left\{ \int p(\mathbf{y}^* \mid \boldsymbol{\beta}, \phi) \, p(\boldsymbol{\beta} \mid \phi, \mathbf{y}) d\boldsymbol{\beta} \right\} \, p(\phi \mid \mathbf{y}) \, d\phi
\end{aligned}
$$

and the calculus is similar to that one involved in the evaluation of the Bayes factor. An important difference is that the marginalizations are with respect to the posterior distribution of the parameters while the Bayes factor is obtained using marginalizations with respect to the prior distribution. Then, using the adopted notation it follows that

$$
\mathbf{Y}^* \mid \mathbf{y} \sim t_{n_1}(\mathbf{x}^* \boldsymbol{\mu}_1, \sigma_1^2 (\mathbf{I}_m + \mathbf{x}^* \mathbf{C}_1^{-1} \mathbf{x}^{*\prime})),
$$

and so, point predictions and credibility intervals for $\mathbf{Y}^*$ can be easily obtained. The analysis using non-informative priors leads to $\boldsymbol{\mu}_1 \to \hat{\boldsymbol{\beta}}$, $\mathbf{C}_1 \to \mathbf{X}'\mathbf{X}$, $n_1 \to n$ and $\sigma_1^2 \to s^2$ and the predictive distribution of $\mathbf{Y}^*$ reduces to

$$
\mathbf{Y}^* \mid \mathbf{y} \sim t_{n-p}(\mathbf{x}^* \hat{\boldsymbol{\beta}}, s^2(\mathbf{I}_m + \mathbf{x}^* (\mathbf{X}'\mathbf{X})^{-1} \mathbf{x}^{*\prime})).
$$

This distribution coincides with the predictive distribution of the classical approach providing the same point and interval predictions.

Example 8.11 *(Example 8.3 continued) Consider a simple linear regression and suppose that it is intended to make predictions for $\mathbf{Y}^*$ based on the explanatory variable*

$$
\boldsymbol{x}^* = \begin{pmatrix} 1, & x_1^* - \bar{x}^* \\ \vdots & \vdots \\ 1, & x_m^* - \bar{x}^* \end{pmatrix},
$$

where $\bar{x}^* = \sum_{i=1}^m x_i^* / m$. (The general case with k explanatory variables is left as an exercise for the reader.)

Considering the same informative priors as in the Example 8.9, we have

$$
\hat{\mathbf{Y}}^* = E[\mathbf{Y}^* | \mathbf{y}] = \boldsymbol{x}^* \boldsymbol{\mu}_1,
$$

where $\boldsymbol{\mu}_1 = (\mu_0^*, \mu_1^*)$, with μ_0^* and μ_1^* as given in Example 8.9. Then, we calculate the prediction of Y_i^* as

$$
\begin{aligned}
\hat{Y}_i^* &= \mu_0^* + (x_i^* - \bar{x}^*) \mu_1^* \\
&= \frac{c_0 \mu_0 + n\bar{y}}{c_0 + n} + (x_i^* - \bar{x}^*) \frac{c_1 \mu_1 + \sum_{i=1}^n (x_i - \bar{x})^2 \hat{\beta}_1}{c_1 + \sum_{i=1}^n (x_i - \bar{x})^2},
\end{aligned}
$$

for $i = 1, \ldots, m$.

The $100(1 - \alpha)\%$ credibility intervals for $\boldsymbol{Y}^*$ are obtained directly from the predictive distribution of $\boldsymbol{Y}^* | \boldsymbol{y}$. From Example 8.9, we can write

$$\boldsymbol{C}_1 = \begin{pmatrix} c_0^* & 0 \\ 0 & c_1^* \end{pmatrix} = \begin{pmatrix} c_0 + n & 0 \\ 0 & c_1 + \sum_{i=1}^n (x_i - \bar{x})^2 \end{pmatrix}.$$

Consequently, the quadratic form $\boldsymbol{x}^* \boldsymbol{C}_1^{-1} \boldsymbol{x}^{*\prime}$ is given by

$$\boldsymbol{x}^* \boldsymbol{C}_1^{-1} \boldsymbol{x}^{*\prime} = \begin{pmatrix} \frac{1}{c_0+n} + \frac{(x_1^* - \bar{x}^*)^2}{c_1 + \sum_{i=1}^n (x_i - \bar{x})^2} & \cdots & \frac{1}{c_0+n} + \frac{(x_1^* - \bar{x}^*)(x_m^* - \bar{x}^*)}{c_1 + \sum_{i=1}^n (x_i - \bar{x})^2} \\ \vdots & \ddots & \vdots \\ \frac{1}{c_0+n} + \frac{(x_1^* - \bar{x}^*)(x_m^* - \bar{x}^*)}{c_1 + \sum_{i=1}^n (x_i - \bar{x})^2} & \cdots & \frac{1}{c_0+n} + \frac{(x_m^* - \bar{x}^*)^2}{c_1 + \sum_{i=1}^n (x_i - \bar{x})^2} \end{pmatrix}.$$

Then the $100(1 - \alpha)\%$ credibility intervals for Y_i^*'s, considering $\boldsymbol{C}_1^* = \boldsymbol{I}_m + \boldsymbol{x}^* \boldsymbol{C}_1^{-1} \boldsymbol{x}^{*\prime}$, are given by

$$[\hat{Y}_i^* - t_{\alpha/2, n1} \; \sigma_1 \; \sqrt{c_{1,ii}^*} \; , \; \hat{Y}_i^* + t_{\alpha/2, n1} \; \sigma_1 \; \sqrt{c_{1,ii}^*}],$$

where $c_{1,ii}^* = 1 + 1/c_0 + n + (x_i^* - \bar{x}^*)^2 / [c_1 + \sum_{i=1}^n (x_i - \bar{x})^2]$ is the (i,i)-th element of $\boldsymbol{C}_1^*$ for $i = 1, \ldots, m$.

8.4   Hierarchical linear models

In Section 3.5 we have seen how to combine structural information with strictly subjective information to build up the prior distribution in stages. This strategy is explored within the context of linear models in this section. The linear structure of the model combines very well with the hierarchical modeling in the context of linear Normal models, and this structure can be explored in more detail. Specifically, the hierarchical structure described in Chapter 3 is used with the additional assumption of linearity and normality. This setup preserves the model linearity as a whole and the conjugacy at any stage of the model. This area was organized systematically and underwent substantial research development following a paper by Lindley and Smith (1972).

In order to specify the model and its prior, we can rewrite it as

$$\begin{aligned} \boldsymbol{Y} \mid \boldsymbol{\beta}_1, \phi &\sim N(\boldsymbol{X}_1 \boldsymbol{\beta}_1, \phi^{-1} \boldsymbol{I}_n) \\ \boldsymbol{\beta}_1 \mid \boldsymbol{\beta}_2, \phi &\sim N(\boldsymbol{X}_2 \boldsymbol{\beta}_2, \phi^{-1} \boldsymbol{C}_1^{-1}) \\ \boldsymbol{\beta}_2 \mid \phi &\sim N(\boldsymbol{\mu}, \phi^{-1} \boldsymbol{C}_2^{-1}) \\ n_0 \sigma_0^2 \phi &\sim \chi_{n_0}^2, \end{aligned}$$

where the matrix with explanatory variables is renamed $\boldsymbol{X}_1$ due to the presence of another matrix including the second stage explanatory variables, $\boldsymbol{X}_2$.

This matrix includes the values which explain the variations in the $\boldsymbol{\beta}_1$ and the coefficient of this explanation is given by $\boldsymbol{\beta}_2$. Bayesian inference then proceeds by obtaining the posterior distributions of $\boldsymbol{\beta}_1$, $\boldsymbol{\beta}_2$ and ϕ.

Classical inference can be performed after removal of the last two equations that specify the prior distribution of the higher order parameters $\boldsymbol{\beta}_2$ and ϕ. In this case, the model becomes similar to the specification already seen in Section 4.4. Inference can be performed by obtaining the marginal likelihood for $\boldsymbol{\beta}_2$ and ϕ, obtaining their (maximum likelihood) estimators. Inference about $\boldsymbol{\beta}_1$ is based on its posterior distribution conditional on $\boldsymbol{\beta}_2$ and ϕ being replaced by their plug-in estimators.

The model specified above is the simplest in the class of hierarchical models containing only two stages. More stages can be included in the model specification depending on the structure of the problem. The form of the model is not modified. Only some extra equations are included, each in the form

$$\boldsymbol{\beta}_j \mid \boldsymbol{\beta}_{j+1}, \phi \sim N(\mathbf{X}_{j+1}\boldsymbol{\beta}_{j+1}, \phi^{-1}\mathbf{C}_j^{-1}).$$

Generally, for higher stages, it is more difficult to specify the equations as the level of elaboration involved gets deeper and deeper. The dependence of the variances on ϕ is not by chance. It allows the use of results about conjugacy developed in Sections 3.3 and 8.3. All the derivations that follow are done only for the model with two stages to keep the notation simple, although there is no technical problem with extending the results to models with K stages, $K > 2$.

It is worth mentioning first that the analysis conditional on $\boldsymbol{\beta}_2$ is completely similar to the developments made in the last section. If $\boldsymbol{\beta}_2$ is known, the prior does not depend on its probabilistic specification. So, all that is needed is to apply the results of the last section substituting $\boldsymbol{\mu}_0$ by $\mathbf{X}_2\boldsymbol{\beta}_2$. In particular, it can be obtained that $\boldsymbol{\beta}_1 \mid \mathbf{y} \sim t_{n_1}(\boldsymbol{\mu}_1, \sigma_1^2 \mathbf{C}^{*-1})$ and $n_1 \sigma_1^2 \phi \sim \chi_{n_1}^2$, where $\boldsymbol{\mu}_1 = \mathbf{C}^{*-1}(\mathbf{C}_1 \mathbf{X}_2 \boldsymbol{\beta}_2 + \mathbf{X}_1' \mathbf{y})$, $\mathbf{C}^* = \mathbf{C}_1 + \mathbf{X}_1' \mathbf{X}_1$, $n_1 = n + n_0$ and $n_1 \sigma_1^2 = n_0 \sigma_0^2 + (\mathbf{y} - \mathbf{X}_1 \mathbf{X}_2 \boldsymbol{\beta}_2)' \mathbf{y} + (\mathbf{X}_2 \boldsymbol{\beta}_2 - \boldsymbol{\mu}_1)' \mathbf{C}_0 \mathbf{X}_2 \boldsymbol{\beta}_2$.

Considering again the case with $\boldsymbol{\beta}_2$ unknown, it is worth noting that the distributions of $\boldsymbol{\beta}_1 \mid \phi$ and $\boldsymbol{\beta}_1$ can be obtained via marginalization as done at the end of the last section to obtain the predictive distribution of the observations. Therefore, combining the distributions of $\boldsymbol{\beta}_1 \mid \boldsymbol{\beta}_2, \phi$ and $\boldsymbol{\beta}_2 \mid \phi$, it follows that

$$\begin{aligned}
\boldsymbol{\beta}_1 \mid \phi &\sim \mathbf{X}_2 N(\boldsymbol{\mu}, \phi^{-1}\mathbf{C}_2^{-1}) + N(\mathbf{0}, \phi^{-1}\mathbf{C}_1^{-1}) \\
&\sim N(\mathbf{X}_2\boldsymbol{\mu}, \phi^{-1}\mathbf{C}_1^{*-1}) \text{ where } \mathbf{C}_1^{*-1} = \mathbf{C}_1^{-1} + \mathbf{X}_2\mathbf{C}_2^{-1}\mathbf{X}_2',
\end{aligned}$$

and after integrating ϕ out, we obtain that $\boldsymbol{\beta}_1 \sim t_{n_0}(\mathbf{X}_2\boldsymbol{\mu}, \sigma_0^2 \mathbf{C}_1^{*-1})$.

Example 8.12 *(Example 8.4 continued) In the last section, it was assumed that the means β_j, $j = 1, \ldots, k$ were independently distributed a priori. A reasonable alternative for the case in which similarity among the k groups can be assumed is to suppose that the means are a random sample from a population of means. This population is fictitious and, to fix ideas, taken as homogeneous. To keep the structure presented above, it is assumed that this population is Normal. So, it follows that $\beta_1, \ldots, \beta_k$ is a sample from the $N(\mu, (c_1\phi)^{-1})$, where c_1 measures the precision of the population of means relative to the precision of the likelihood. This specification is complete for classical inference.*

Bayesian inference requires the model to be completed with the specification of the prior distributions of $\mu \mid \phi$ and ϕ. If follows that the prior is completely specified by

$$
\begin{aligned}
1st\ level\ &:\quad \boldsymbol{\beta} \mid \mu, \phi \sim N[\mathbf{1}_k\mu, (c_1\phi)^{-1}\boldsymbol{I}_k] \\
2nd\ level\ &:\quad \mu \mid \phi \sim N[\mu_0, (c_2\phi)^{-1}] \\
&\quad\ \ n_0\sigma_0^2\phi \sim \chi_{n_0}^2,
\end{aligned}
$$

the distribution of the $\boldsymbol{\beta} \mid \phi$ is $N[\mathbf{1}_k\mu_0, \phi^{-1}(c_1^{-1}\boldsymbol{I}_k + c_2^{-1}\mathbf{1}_k\mathbf{1}_k')]$ and the components of $\boldsymbol{\beta}$ are not independent anymore. In particular, the prior correlation between any two distinct components is given by $(1 + c_2/c_1)^{-1}$. This may be helpful in the specification of the constants c_1 and c_2. A larger value for the constant c_2/c_1 leads to a smaller prior correlation and vice versa.

It is interesting to see that the distributions of $\boldsymbol{\beta}_1$ and $\boldsymbol{\beta}_2$ conditional on ϕ are multivariate Normal; therefore, the theory developed in the last section is directly applicable and the predictive and posterior distributions can easily be obtained. So, the predictive distribution for $\mathbf{Y} \mid \phi$ is given by

$$
\mathbf{X}_1 N(\mathbf{X}_2\boldsymbol{\mu}, \phi^{-1}\mathbf{C}_1^{*-1}) + N(\mathbf{0}, \phi^{-1}\mathbf{I}_n)
$$

$$
\sim N(\mathbf{X}_1\mathbf{X}_2\boldsymbol{\mu}, \phi^{-1}(\mathbf{I}_n + \mathbf{X}_1\mathbf{C}_1^{*-1}\mathbf{X}_1')),
$$

and the marginal for $\mathbf{Y}$ is obtained from the above expression substituting the Normal distribution by t_{n_0} distribution and ϕ^{-1} by its posterior estimate σ_0^2. The posterior distribution of $(\boldsymbol{\beta}_1, \phi)$ is a multivariate Normal-χ^2 with parameters $\boldsymbol{\mu}_1$, $\mathbf{C}^*$, n_1 and σ_1^2 given by

$$
\begin{aligned}
\boldsymbol{\mu}_1 &= \mathbf{C}^{*-1}(\mathbf{C}_1^*\mathbf{X}_2\boldsymbol{\mu}_0 + \mathbf{X}_1'\mathbf{y}), \\
\mathbf{C}^* &= \mathbf{C}_1^* + \mathbf{X}_1'\mathbf{X}_1, \\
n_1 &= n + n_0,\ \text{and} \\
n_1\sigma_1^2 &= n_0\sigma_0^2 + (\mathbf{y} - \mathbf{X}_1\boldsymbol{\mu}_1)'\mathbf{y} + (\mathbf{X}_2\boldsymbol{\mu}_0 - \boldsymbol{\mu}_1)'\mathbf{C}_1^{*-1}\mathbf{X}_2\boldsymbol{\mu}_0.
\end{aligned}
$$

Example 8.13 *(Example 8.4 continued) Consider again the analysis of variance model with one classification factor with a hierarchical prior. Applying the above results, it follows that*

$$\boldsymbol{C}_1^{*-1}\boldsymbol{X}_2\boldsymbol{\mu}_0 + \boldsymbol{X}_1'\boldsymbol{y} = (c_1^{-1} + kc_2^{-1})\mu_0\boldsymbol{1}_k + (n_1\bar{y}_1,\dots,n_k\bar{y}_k)'$$

$$\boldsymbol{C}^* = diag\,(c_1 + n_1,\dots,c_1 + n_k) - \frac{c_1^2}{kc_1 + c_2}\boldsymbol{1}_k\boldsymbol{1}_k'.$$

It can still be shown in the case of non-informative prior in the second level $(c_2 \to 0)$ and equal number of observations in each group that the posterior mean of β_j is in the form

$$w_j\bar{y}_j + (1 - w_j)\bar{\bar{y}}, \quad with\ 0 \le w_j \le 1\,,\ j = 1,\dots,k,$$

where $\bar{\bar{y}}$ is the average of group averages. This type of estimator is known as a shrinkage estimator since it brings all the group estimates closer to the global mean, shrinking the distance among the means. Shrinkage estimators and their properties were presented in Example 4.21.

To obtain a posterior distribution of $\boldsymbol{\beta}_2$ we can first obtain the likelihood function of $\boldsymbol{\beta}_2$ and ϕ. Using again the results involving combination of Normal, it follows that

$$\boldsymbol{Y} \mid \boldsymbol{\beta}_2, \phi \ \sim\ \boldsymbol{X}_1 N(\boldsymbol{X}_2\boldsymbol{\beta}_2, \phi^{-1}\boldsymbol{C}_1^{-1}) + N(\boldsymbol{0}, \phi^{-1}\boldsymbol{I}_n)$$

$$\sim\ N(\boldsymbol{X}_1\boldsymbol{X}_2\boldsymbol{\beta}_2, \phi^{-1}\boldsymbol{C}_0^{-1}),$$

where $\boldsymbol{C}_0^{-1} = \boldsymbol{I}_n + \boldsymbol{X}_1\boldsymbol{C}_1^{-1}\boldsymbol{X}_1'$. This equation provides the likelihood $l(\boldsymbol{\beta}_2, \phi; \boldsymbol{Y})$.

Therefore, likelihood and prior are Normal as earlier with a slight difference since the observational variance is not proportional to an identity matrix. The same results are still valid with the respective substitutions, that is, the posterior distribution of $\boldsymbol{\beta}_2$, and ϕ is multivariate Normal-χ^2 with parameters $\boldsymbol{\mu}_2$, $\boldsymbol{C}_2^*$, n_2 and σ_2^2 given by

$$\boldsymbol{\mu}_2 = \boldsymbol{C}_2^{*-1}(\boldsymbol{C}_2^{-1}\boldsymbol{\mu}_0 + \boldsymbol{X}_1'\boldsymbol{X}_2'\boldsymbol{C}_0\boldsymbol{y}),$$

$$\boldsymbol{C}_2^* = \boldsymbol{C}_2 + \boldsymbol{X}_1'\boldsymbol{X}_2'\boldsymbol{C}_0\boldsymbol{X}_1\boldsymbol{X}_2,$$

$$n_1 = n + n_0, \ and$$

$$n_1\sigma_1^2 = n_0\sigma_0^2 + (\boldsymbol{y} - \boldsymbol{X}_1\boldsymbol{X}_2\boldsymbol{\mu}_2)'\boldsymbol{C}_0\boldsymbol{y} + (\boldsymbol{\mu}_0 - \boldsymbol{\mu}_2)'\boldsymbol{C}_2^{*-1}\boldsymbol{\mu}_0.$$

The hierarchical linear models can also be considered from the classical point of view. The above likelihood allows point estimation of $\boldsymbol{\beta}_2$ by $\hat{\boldsymbol{\beta}}_2 = (\boldsymbol{X}_1'\boldsymbol{X}_2'\boldsymbol{C}_0\boldsymbol{X}_1\boldsymbol{X}_2)^{-1}\boldsymbol{X}_1'\boldsymbol{X}_2'\boldsymbol{C}_0\boldsymbol{Y}$. Note that this is also the Bayes estimator of

β_2 in the case of the non-informative prior, obtained by letting $\mathbf{C}_2 \to 0$. Empirical Bayes (EB) point estimation of β_1 is obtained from the expression of μ_1 by pluging-in the above estimator for β_2. Thus, the EB estimator for β_1 is given by $(\mathbf{C}_1 + \mathbf{X}_1'\mathbf{X}_1)^{-1}(\mathbf{C}_1\mathbf{X}_2\hat{\beta}_2 + \mathbf{X}_1'\mathbf{y})$. Finally, $\sigma^2 = \phi^{-1}$ can be estimated by the residual sum of squares as described in Section 4.3.2.

Example 8.14 *(Mixed linear model) An important hierarchical model that is heavily used in applications is the mixed linear model. The name is derived from the combination of fixed effects and random effects. This model was popularized by Laird and Ware (1982) and is expressed in matrix notation as*

$$\begin{aligned} \mathbf{Y} \mid \beta, \gamma, \phi &\sim N(\mathbf{X}\beta + \mathbf{Z}\gamma, \phi^{-1}\mathbf{I}_n) \\ \gamma \mid \phi &\sim N(\mathbf{0}, \phi^{-1}\mathbf{C}_1^{-1}), \end{aligned}$$

where $\mathbf{X}$ contains the values of explanatory variables for the fixed part of the model and $\mathbf{Z}$ contains the values of explanatory variables for the random part of the model. It is not difficult to see that the model can be cast into the general framework of a hierarchical model. The details are left as an exercise.

Model parameters can be estimated in a number of ways. Laird and Ware (1982) describe the empirical Bayes estimation and the estimation using the EM algorithm and also mention the possibility of Bayesian estimation. For this latter method, the model must be completed with a prior distribution for (β, ϕ). Note that the specification of the random component corresponds to a prior distribution.

The use of the EM algorithm for estimation of parameters in hierarchical models was mentioned in Example 8.14. In fact, it can be implemented in any hierarchical model. Let $\theta = (\beta_2, \phi)$ be the parameter vector and β_1 a vector of latent variables. From Section 5.2.2, the EM algorithm can be used for computing the maximum likelihood estimators by considering the *complete* likelihood $l(\theta; \mathbf{y}, \beta_1) = p(\mathbf{y}, \beta_1 | \theta)$.

First, in the E-step, the expected value of the latent variable β_1 is calculated with respect to its conditional distribution $p(\beta_1 | \mathbf{y}, \theta)$. Note that this corresponds to the conditional posterior distribution of β_1, which was already obtained in this section. Then, in the M-step, the value of θ is updated by

$$\theta^{j+1} = \arg\max_{\theta} \; Q(\theta; \theta^j),$$

where $Q(\theta; \theta^j) = L(\theta; \mathbf{y}, \hat{\beta}_1^j) = \log l(\theta; \mathbf{y}, \hat{\beta}_1^j)$.

In the above context, the EM algorithm is defined through two basic steps as follows:

1. Evaluate the expected value of β_1 with the current value of the parameters $\boldsymbol{\theta}^j$ according to $\hat{\boldsymbol{\beta}}_1^j = \boldsymbol{\mu}_{\boldsymbol{\beta}_1}(\boldsymbol{\theta}^j)$.

2. Update the current value of $\boldsymbol{\theta}^j$ by $\boldsymbol{\theta}^{j+1} = \arg\max_{\boldsymbol{\theta}} \; l(\boldsymbol{\theta}|\boldsymbol{y}, \hat{\boldsymbol{\beta}}_1^j)$.

So, the EM algorithm basically iterates between successive applications of the empirical Bayes estimation procedure previously described. The algorithm requires initialization with a value of $\boldsymbol{\theta}^1$ and is stopped when $\max(|\boldsymbol{\theta}^{t+1} - \boldsymbol{\theta}^t|) < \epsilon$, for some pre-specified tolerance ϵ.

8.5   Dynamic linear models

The dynamic linear models are a broad class of models with time varying parameters, useful to model time series data and regression. It is well documented from a Bayesian perspective in the book by West and Harrison (1997) and from a classical perspective in the book by Harvey (1989). Another useful reference is the book by Durbin and Koopman (2001). In the later case, these models are also referred to as structural models. A more encompassing name accepted from both classical and Bayesian perspectives is state space models. In this section some basic aspects of dynamic models will be introduced and some examples in time series as well as regression will be addressed.

8.5.1   Definition and examples

Dynamic linear models are parametric models where the parameter variation and the available data information are described probabilistically. They are characterized by a pair of equations, named observational equation and parameters evolution or system equation. These are respectively given by

$$
\begin{aligned}
Y_t &= \mathbf{x}_t' \boldsymbol{\beta}_t + \epsilon_t, \; \epsilon_t \sim N(0, \sigma_t^2) \\
\boldsymbol{\beta}_t &= \mathbf{G}_t \boldsymbol{\beta}_{t-1} + \boldsymbol{\omega}_t, \quad \boldsymbol{\omega}_t \sim N(\mathbf{0}, \mathbf{W}_t),
\end{aligned}
$$

where Y_t is a time sequence of observations, conditionally independent on the sequence of parameters $\boldsymbol{\beta}_t$, $\boldsymbol{x}_t$ is a vector of explanatory variables as described in Section 8.1, $\boldsymbol{\beta}_t$ is a $p \times 1$ vector of parameters, $\mathbf{G}_t$ is a $p \times p$ matrix describing the parameter evolution and, finally, σ_t^2 and $\mathbf{W}_t$ are the variances of the errors associated with the unidimensional observation and with the p-dimensional vector of parameters, respectively.

Typically, the initial state $\boldsymbol{\beta}_0$, σ_t^2, $\mathbf{W}_t$, and possibly some components of $\mathbf{G}_t$ are unknown and must be estimated. The above model can be analyzed from both frequentist and Bayesian perspectives. In the former, the state parameters $\boldsymbol{\beta}_t$ are viewed as random components that provide the link through time. In the latter, these are simply parameters that are subject to a prior distribution relating them through time. Also, from a Bayesian perspective, the model must be completed with a prior distribution for all unknown parameters. Let us momentarily assume that only $\boldsymbol{\beta}_0$ is unknown and we have prior distribution $\boldsymbol{\beta}_0 \sim N(\boldsymbol{m}_0, \boldsymbol{C}_0)$. We shall include the other unknowns progressively in the estimation as we advance through the section.

Summarizing, a dynamic linear model is completely specified by the quadruple $\{\boldsymbol{x}_t, \boldsymbol{G}_t, \sigma_t^2, \boldsymbol{W}_t\}$. Two special cases are time series models characterized by $\boldsymbol{x}_t = \boldsymbol{x}$ and $\boldsymbol{G}_t = \boldsymbol{G}$, $\forall t$ and dynamic regression models described by $\boldsymbol{G}_t = \boldsymbol{I}_p$.

Example 8.15 *The simplest model in time series is the first order polynomial model, which corresponds to a first order Taylor series approximation of a smooth time function, named the time series trend. This model is completely defined by the quadruple $\{1, 1, \sigma_t^2, W_t\}$. The above equations specialize to*

$$
\begin{aligned}
Y_t &= \beta_t + \epsilon_t, \quad \epsilon_t \sim N(0, \sigma_t^2) \\
\beta_t &= \beta_{t-1} + \omega_t, \quad \omega_t \sim N(0, W_t),
\end{aligned}
$$

where β_t is unidimensional. This model is also known as a local linear model.

Although the model of the example is very simple, it can be applied in many short-term forecasting systems involving a large number of time series such as in stock control or production planning. The observational and parameters variance can also evolve in time, offering a broad scope for modeling.

A slightly more elaborated model, named the linear growth model (LGM, in short), is derived after including an extra parameter $\beta_{2,t}$ to describe the local underlying growth of the process. Then, it follows that

$$
\begin{aligned}
Y_t &= \beta_{1,t} + \epsilon_t \quad \epsilon_t \sim N(0, \sigma_t^2) \\
\beta_{1,t} &= \beta_{1,t-1} + \beta_{2,t-1} + \omega_{1,t} \\
\beta_{2,t} &= \beta_{2,t-1} + \omega_{2,t} \quad \boldsymbol{\omega}_t \sim N(\boldsymbol{0}, \boldsymbol{W}_t).
\end{aligned}
$$

The parameter $\beta_{1,t}$ is interpreted as the local level of the process, and it is easy to verify that $\boldsymbol{x}_t = (1, 0)$ and $\boldsymbol{G}_t = \begin{pmatrix} 1 & 1 \\ 0 & 1 \end{pmatrix}$, $\forall t$ characterizing a time series model.

Example 8.16 *(Example 8.3 continued) Suppose, in this example, that pairs of values (x_t, Y_t) are observed through time and that it is wished to model the existing relationship between x_t and Y_t. Assuming that the linear model is a good approximation for the relationship between these values, a simple linear* *regression model can be set. Its parameters can be estimated via classical or Bayesian methods, as described in Sections 8.2 and 8.3, respectively.*

Since the linear relationship is only a local approximation for the true functional dependence involving x and Y, a model with varying parameters may be more appropriate. This was found to be more adequate in many applications. For example, the omission of some variables can justify the parameter oscillation. The non-linearity of the functional relationship connecting x and Y or some structural changes which occurred in the process under investigation can also be responsible for the parameter instability over time. These situations can be modeled as

$$\begin{aligned} Y_t &= \mathbf{x}_t'\boldsymbol{\beta}_t + \epsilon_t, \\ \boldsymbol{\beta}_t &= \boldsymbol{\beta}_{t-1} + \boldsymbol{\omega}_t, \end{aligned}$$

where $\boldsymbol{x}_t = (1, x_t)'$ and $\boldsymbol{\omega}_t \sim N(\mathbf{0}, \boldsymbol{W}_t)$. Note that, in this case, $\boldsymbol{G}_t = \boldsymbol{I}_2$.

As we can observe, the choice of $\boldsymbol{x}_t$ and $\mathbf{G}_t$ depends on the model and the nature of the data that are being analyzed. To complete the model specification the variances σ_t^2 and $\mathbf{W}_t$ must be set. The observational variance is usually supposed time invariant, as in the previous sections, and $\mathbf{W}_t$ describes the speed of the parameters evolution. In applications, σ_t^2 is often larger than the elements of $\mathbf{W}_t$.

In what follows, the parameter estimation method, including the observational variance, will be described. To make it easier for the conjugate analysis, $\mathbf{W}_t$ is scaled by σ_t^2 and the conditional variances of the $\boldsymbol{\omega}_t$ become $\sigma_t^2 \mathbf{W}_t$. Therefore, the matrix $\mathbf{W}_t$ can be interpreted as a matrix of relative weights with respect to the observational variance. The parameter evolution variance matrix must be assessed subjectively by the user. In order to do that, the notion of discount factor will be useful. Alternatively, it can be estimated by one of the approximating methods described in Chapter 5.

The equations presented earlier can be rewritten as

$$\begin{aligned} Y_t \mid \boldsymbol{\beta}_t &\sim N(\mathbf{x}_t'\boldsymbol{\beta}_t, \sigma_t^2), \\ \boldsymbol{\beta}_t \mid \boldsymbol{\beta}_{t-1} &\sim N(\mathbf{G}_t\boldsymbol{\beta}_{t-1}, \sigma_t^2 \mathbf{W}_t). \end{aligned}$$

Let $D_t = \{D_{t-1}, y_t\}$ with D_0 describing the initial available information, including the values of $\boldsymbol{x}_t$ and $\mathbf{G}_t$, $\forall t$, which are supposed to be known.

It is worth noting that it is assumed that for any time t, the current observation Y_t is independent of the past observations given the knowledge of $\boldsymbol{\beta}_t$. This means that the temporal dynamics are summarized in the state parameters evolution. This linear structure for modeling data observed through

time combines very well with the principles of Bayesian inference by the possibility to describe subjectively the involved probabilities and by its sequential
nature. Therefore, subjective information is coherently combined with past
information to produce convenient inferences.

8.5.2 Evolution and updating equations

The equations described previously enable a joint description of $(Y_t, \boldsymbol{\beta}_t)$ given
the past observed data D_{t-1} via

$$p(y_t, \boldsymbol{\beta}_t | D_{t-1}) = p(y_t | \boldsymbol{\beta}_t, D_{t-1}) p(\boldsymbol{\beta}_t | D_{t-1}).$$

This leads to the predictive distribution after integrating out $\boldsymbol{\beta}_t$.

One of the main characteristics of the dynamic linear model is that, at each
instant of time, all the information available is used to describe the posterior
distribution of the state vector. The theorem that follows shows how to evolve
from the posterior distribution at time $t-1$ to the posterior at t.

Theorem 8.1 *Consider a Normal dynamic linear model with $\sigma_t^2 = \sigma^2$,
$\forall t$. Denote the posterior distribution at $t-1$ by $(\boldsymbol{\beta}_{t-1} | D_{t-1}, \sigma^2) \sim
N(\boldsymbol{m}_{t-1}, \sigma^2 \boldsymbol{C}_{t-1})$ and the marginal posterior distribution of $\phi = \sigma^{-2}$ as
$\phi | D_{t-1} \sim G(n_{t-1}/2, n_{t-1}s_{t-1}/2)$. Then,*

1. *Conditionally on σ^2, it follows*

   (a) *Evolution – the prior distribution at t will be*

   $$\boldsymbol{\beta}_t | \sigma^2, D_{t-1} \sim N(\boldsymbol{a}_t, \sigma^2 \boldsymbol{R}_t),$$

   with $\boldsymbol{a}_t = \boldsymbol{G}_t \boldsymbol{m}_{t-1}$ and $\boldsymbol{R}_t = \boldsymbol{G}_t \boldsymbol{C}_{t-1} \boldsymbol{G}_t' + \boldsymbol{W}_t$.

   (b) *Prediction – the one step ahead predictive distribution will be*

   $$y_t | \sigma^2, D_{t-1} \sim N(f_t, \sigma^2 Q_t),$$

   with $f_t = \mathbf{x}_t' \boldsymbol{a}_t$ and $Q_t = \mathbf{x}_t' \boldsymbol{R}_t \mathbf{x}_t + 1$.

   (c) *Updating – the posterior distribution at t will be*

   $$\boldsymbol{\beta}_t | \sigma^2, D_t \sim N(\boldsymbol{m}_t, \sigma^2 \boldsymbol{C}_t),$$

   *with $\boldsymbol{m}_t = \boldsymbol{a}_t + \boldsymbol{A}_t e_t$ and $\boldsymbol{C}_t = \boldsymbol{R}_t - \boldsymbol{A}_t \boldsymbol{A}_t' Q_t$, where $\boldsymbol{A}_t = \boldsymbol{R}_t \mathbf{x}_t' / Q_t$
   and $e_t = y_t - f_t$.*

2. *The precision ϕ is updated by the relation*

   $$\phi | D_t \sim G(n_t/2, n_t s_t/2),$$

   with $n_t = n_{t-1} + 1$ and $n_t s_t = n_{t-1}s_{t-1} + e_t^2/Q_t$.

3. *Unconditionally on σ^2, we will have*

   (a) $\boldsymbol{\beta}_t|D_{t-1} \sim t_{n_{t-1}}(\boldsymbol{a}_t, s_{t-1}\boldsymbol{R}_t)$;

   (b) $Y_t|D_{t-1} \sim t_{n_{t-1}}(f_t, Q_t^*)$, *with* $Q_t^* = s_{t-1}Q_t$;

   (c) $\boldsymbol{\beta}_t|D_t \sim t_{n_{t-1}}(\boldsymbol{m}_t, s_t\boldsymbol{C}_t)$.

The above iterative procedure starts at $t = 0$ with prior specifications $\boldsymbol{\beta}_0|\phi, D_0 \sim N(\boldsymbol{m}_0, \phi^{-1}\boldsymbol{C}_0)$ and $\phi|D_0 \sim G(n_0/2, n_0s_0/2)$.

Proof (1) Item (a) follows immediately using the parameter evolution equation and standard facts from the Normal theory. With respect to (b), using the prior distribution in (a), it follows that

$$
\begin{aligned}
f_t &= E[E(Y_t|\boldsymbol{\beta}_t)|\sigma^2, D_{t-1}] = E(\boldsymbol{x}_t'\boldsymbol{\beta}_t|\sigma^2, D_{t-1}) = \mathbf{x}_t'\mathbf{a}_t \\
Q_t &= V[E(Y_t|\boldsymbol{\beta}_t)|\sigma^2, D_{t-1}] + E[V(Y_t|\boldsymbol{\beta}_t)|\sigma^2, D_{t-1}] \\
&= V[\boldsymbol{x}_t'\boldsymbol{\beta}_t|\sigma^2, D_{t-1}] + \sigma^2 = \sigma^2(\mathbf{x}_t'R_t\mathbf{x}_t + 1)
\end{aligned}
$$

and the normality is a consequence of the fact that all the distributions involved are Normal.

To prove part (c), suppose that the posterior distribution at $t - 1$ is as given in the theorem. We wish to show that (c) follows from the application of Bayes theorem, that is,

$$
p(\boldsymbol{\beta}_t|\sigma^2, D_t) \propto p(\boldsymbol{\beta}_t|\sigma^2, D_{t-1})p(y_t|\boldsymbol{\beta}_t, \sigma^2).
$$

The proof of part (c) is completed by using Theorem 2.1 and the identity

$$
C_t^{-1} = R_t^{-1} + \mathbf{x}_t'\mathbf{x}_t\sigma^{-2}.
$$

If σ^2 is unknown and defining $\phi = \sigma^{-2}$ it will follow that

- by hypothesis, $\phi|D_{t-1} \sim G(n_{t-1}/2, n_{t-1}s_{t-1}/2)$ and $y_t|\phi, D_{t-1} \sim N(f_t, Q_t/\phi)$. Then, by Bayes theorem,

$$
p(\phi|D_t) \propto \phi^{(n_{t-1}+1)/2-1} \exp\left\{-\frac{\phi}{2}\left(n_{t-1}s_{t-1} + \frac{e_t^2}{Q_t}\right)\right\}
$$

  and, therefore, $\phi|D_t \sim G(n_t/2, n_ts_t/2)$.

- for Part 3 of the theorem, the proofs of items (a)–(c) follow from the results about conjugacy of the Normal-χ^2 to the Normal model and from the marginal distributions obtained in Sections 3.3 and 8.3.

$\square$

The on-line estimation procedure of the state parameters in Part 1 is sometimes referred to as the Kalman filter.

8.5.3 Special aspects

Among the special aspects involved in the dynamic modeling, the fact that it is possible to model the observational variance deserves special attention. For example, it can be modeled as a power law, $\sigma_t^2 = \sigma^2 \mu_t^b$, where $\mu_t = \mathbf{x}_t' \boldsymbol{\beta}_t$ is the process mean level. The constant b can be chosen in parallel to the well-known Box-Cox family of transformation. The scale factor σ^2 can be sequentially estimated as stated in the theorem. The main advantage in this form of modeling is to avoid the transformation of the original data keeping and, in this way, the interpretation of the parameters, which is very useful when we wish to perform some subjective intervention.

The discount factors may be used to avoid direct setting or estimation of the evolution matrices $\mathbf{W}_t$. These factors are fixed numbers in interval $(0, 1]$, describing subjectively the loss of information through time. Remember that $\mathbf{R}_t = \mathbf{P}_t + \mathbf{W}_t$ where $\mathbf{P}_t = \mathbf{G}_t \mathbf{C}_{t-1} \mathbf{G}_t'$. Denoting the discount factor by δ, we can rewrite $\mathbf{R}_t = \mathbf{P}_t/\delta$, showing clearly that there is a relationship between $\mathbf{W}_t$ and δ. This is given by $\mathbf{W}_t = (\delta^{-1} - 1)\mathbf{P}_{t-1}$, showing that the loss of information is proportional to the posterior variance of the state parameters. For example, if $\delta = 0.9$, only 90% of the information passes through time.

Other relevant aspects of dynamic linear models are the abilities to easily take care of missing observations and to automatically implement subjective interventions. In the first case, it suffices not to use the updating equations at the time the observations are missing. In this way, the uncertainties increase with the evaluation of the new prior distribution, and the recurrence equation continues to be valid without any additional problem. From the intervention point of view, the simplest proposal is to use a small discount factor, close to 0, at the time of announced structural changes in the data generation process. In this way the more recent observations will be strongly considered in the updating of the prior distribution and the system can be more adaptive to possible changes.

Dynamic models can be combined with the hierarchical models of the previous section. Hierarchical models with K levels are built at each observation time and the time evolution is placed at the higher K-th level of the hierarchy. Details are provided in Gamerman and Migon (1993).

Finally, it is worth mentioning that parameter distributions at time t can be revised with the arrival of the new observations. We can generically obtain the parameter distributions $p(\boldsymbol{\beta}_t|D_{t+k})$, $\forall t, k$ integer. If $k > 0$, this is named the smoothed distribution; if $k = 0$, it is just the posterior; and if $k < 0$ it is the prior distribution. In dynamic model it is common to use the distributions $p(\boldsymbol{\beta}_t|D_n)$, $\forall t = 1, \ldots, n$, where n is the size of the series, to retrospectively an-

alyze the parameter behavior. For example, one may want to quantify the
effect of a behavior change induced by some measure of economic policy. The
future data would inform about the change which occurred in any particular parameter of the model describing the behavior of the involved economic
agents.

8.5.4   Hyperparameter estimation

We can now return to the situation where σ_t^2, $\mathbf{W}_t$, and possibly some components of $\mathbf{G}_t$ are also unknown and must be estimated. Let us denote all these
unknown parameters by $\boldsymbol{\theta}$.

Their marginal likelihood $l(\boldsymbol{\theta}; \boldsymbol{y})$ must be obtained. This task is simplified
by noting that

$$
\begin{aligned}
l(\boldsymbol{\theta}; \boldsymbol{y}) &= p(\boldsymbol{y}|\boldsymbol{\theta}) \\
&= \prod_{i=1}^{n} p(y_t|D_{t-1}, \boldsymbol{\theta}).
\end{aligned}
$$

The last densities are the one-step-ahead predictive densities, obtained in
Part 3(b) of Theorem 8.1. The parameters of these distributions are functions
of $\boldsymbol{\theta}$. Their dependence on $\boldsymbol{\theta}$ is quite involved given the iterative nature of
Theorem 8.1, but it can easily be computed. This is the key for the calculation
of relevant quantities.

From the Bayesian perspective, the above likelihood must be combined
with a prior distribution for $\boldsymbol{\theta}$. Then, application of Bayes theorem provides
their posterior distribution. This will typically be intractable analytically and
methods such as MCMC may be used to obtain approximate versions of the
relevant posterior distribution. Samples of the state parameters $\boldsymbol{\beta}_t$ can be
obtained once posterior samples from the parameters are available. Alternatively, samples from the full posterior distribution of all unknown parameters
may also be jointly obtained. MCMC methods may also in this setup. These
methods are described in detail by West and Harrison (1997).

From the classical perspective, estimation may be based on the above likelihood. A number of methods may be used. The most direct is the calculation
of the maximum likelihood estimator. Note that the iterative nature of the
likelihood prevents direct calculation and numerical routines must be used.
Approximate confidence intervals for $\boldsymbol{\theta}$ may be built based on the asymptotic normality of its maximum likelihood estimator. Alternatively, the EM
algorithm may be applied to the *complete* likelihood $l(\boldsymbol{\theta}; \boldsymbol{y}, \boldsymbol{\beta}_0, \ldots, \boldsymbol{\beta}_n)$. These
methods are described in detail by Harvey (1989).

8.6   Linear models with constraints

It is common in practice to face situations where a low ratio between the number of explanatory variables and the number of observations or high collinearity among the explanatory variables is observed. Variable selection methods are usual approaches but are not feasible when the number of explanatory variables exceeds the number of observations, a phenomenon that is frequently observed nowadays. Some practical examples where the number of explanatory variables greatly exceed the number of observations are the analysis of microarray data in Genetics and analysis of credit scoring in Finance. Whenever one of the above situations is considered, an ill conditioning problem is observed in classical inference, since now a singular or nearly singular matrix $\mathbf{X}'\mathbf{X}$ must be inverted in order to obtain the estimates of the regression coefficients, and small changes in $\mathbf{X}$ may lead to large changes in the $(\mathbf{X}'\mathbf{X})^{-1}$.

In this context, a shrinkage can be considered in order to overcome the ill conditioning problem. A first possibility is to consider the ridge regression approach (Hoerl and Kennard, 1970), where the idea is to enforce a squared constraint on the coefficients leading to minimization of the quadratic form $S(\boldsymbol{\beta}) = (\hat{\mathbf{y}} - \mathbf{X}\boldsymbol{\beta})'(\hat{\mathbf{y}} - \mathbf{X}\boldsymbol{\beta})$ subject to $\sum_{j=1}^{p} \beta_j^2 < c$, for $c > 0$. As a result, following James, Witten, Hastie and Tibshirani (2013), the ridge regression estimate is given by

$$\hat{\boldsymbol{\beta}}_r = (\mathbf{X}'\mathbf{X} + \lambda \mathbf{I}_p)^{-1}\mathbf{X}'\mathbf{y},$$

where λ is a pre-chosen penalty term, also referred to as a tuning parameter.

This is the solution of the minimization of the modified log likelihood $S(\boldsymbol{\beta}) + \lambda \sum_{j=1}^{p} \beta_j^2$. The second term is a penalization as described in Section 4.4 and is refereed to as the ridge regression shrinkage penalization. In this case, it is common practice to assume that the explanatory variables are standardized in order to have them in the same scale and also that $\mathbf{y}$ is a centered variable. The shrinkage penalty is applied to the coefficients related to the explanatory variables but not to the intercept which only represents a general mean. A critical point here is the choice of λ. At least in principle, resampling methods, such as the cross-validation (see Section 5.6.3), can be used for choosing the value of λ that optimizes a chosen criterion, e.g., minimization of prediction errors.

A direct consequence of the minimization subject to a constraint is to penalize large values of β_js. Larger λ implies β_j's closer to zero. The usual linear regression approach (see Section 8.2) is obtained for $\lambda = 0$, a situation where no penalization is imposed to the minimization process. The matrix

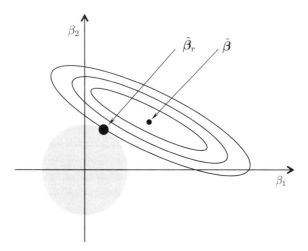

Figure 8.1 *Geometrical view of a ridge regression for the case $p = 2$ (based on Tibshirani, 1996).*

$\mathbf{X}'\mathbf{X}$ is equal to $n\mathbf{I}_p$ for orthogonal explanatory variables. In this case, $\hat{\boldsymbol{\beta}}_r = [n/(n+\lambda)]\,\hat{\boldsymbol{\beta}}$, where $\hat{\boldsymbol{\beta}}$ was obtained in Section 8.2. The covariance matrix of the estimates is easily obtained as $\sigma^2(\mathbf{X}'\mathbf{X}+\lambda\mathbf{I}_p)^{-1}(\mathbf{X}'\mathbf{X})(\mathbf{X}'\mathbf{X}+\lambda\mathbf{I}_p)^{-1}$. (The details are left as an exercise.)

Example 8.17 *Consider a simple situation where $p = 2$, where we have only 2 explanatory variables. In this case, the ridge constraint region is $\beta_1{}^2 + \beta_2{}^2 < c$. Figure 8.1 provides a simple geometrical view of a ridge regression for this case. The ellipses correspond to the contour of the $S(\boldsymbol{\beta})$, which is minimized at $\hat{\boldsymbol{\beta}} = (\hat{\beta}_1, \hat{\beta}_2)$, and the circle $\beta_1{}^2 + \beta_2{}^2 < c$ corresponds to the constraint in the ridge regression. The ellipse and circle sizes are minimized simultaneously in ridge regression. The ridge coefficient estimate $\hat{\boldsymbol{\beta}}_r = (\hat{\beta}_{r,1}, \hat{\beta}_{r,2}$ is equal to the point at which the ellipse and the circle intersect.*

As we already discussed, penalizations used in frequentist inference can be incorporated into Bayesian analysis as prior densities. Considering standardized explanatory variables, assuming that $\boldsymbol{\beta}$ has prior distribution $N(\mathbf{0}, \sigma^2\lambda^{-1}\mathbf{I}_p)$ implies that the posterior of $\boldsymbol{\beta}|\mathbf{Y}, \mathbf{X}$ is an $N[\hat{\boldsymbol{\beta}}_r, \sigma^2(\mathbf{X}'\mathbf{X}+\lambda\mathbf{I}_p)^{-1}(\mathbf{X}'\mathbf{X})(\mathbf{X}'\mathbf{X}+\lambda\mathbf{I}_p)^{-1}]$. That is, the posterior mean is equal to the ridge regression estimate and the posterior covariance matrix is equal to the covariance matrix of the ridge regression estimator. Unless $\lambda = \infty$, the ridge regression shrinkage penalization $\lambda\sum_{j=1}^{p}\beta_j{}^2$ does not set the regression coefficients at 0, though it shrinks them towards this value. Model interpretation is directly impacted since all explanatory variables will be retained in the model, even in the presence of reduced magnitude of the coefficients.

It may be argued that the penalization based on the squared value of the coefficients shrinks them too strongly towards 0. This apparent disadvantage is overcome by the least absolute shrinkage and selection operator, also referred to as the Lasso estimation (Tibshirani, 1996), which penalizes the absolute size of the coefficients (L_1 norm) instead of their squared sizes (L_2 norm). The Lasso enforces an absolute constraint on the coefficients leading to minimization of the quadratic form $S(\boldsymbol{\beta}) = (\hat{\mathbf{y}} - \mathbf{X}\boldsymbol{\beta})'(\hat{\mathbf{y}} - \mathbf{X}\boldsymbol{\beta})$ subject to $\sum_{j=1}^{p} |\beta_j| < c$, for $c > 0$. That is, the constraint $\lambda \sum_{j=1}^{p} |\beta_j|$ is imposed by the Lasso to the minimization procedure.

Although a Lasso solution $\hat{\boldsymbol{\beta}}_l$ is usually feasible only via computational methods, an approximate closed form estimate is derived by rewriting $\sum |\beta_j| \leq c$ as $\sum \beta_j^2/|\beta_j| \leq c$. In this case, the Lasso will solve the L_1-penalized regression problem, finding the estimator $\hat{\boldsymbol{\beta}}_l$ which minimizes

$$(\hat{\mathbf{y}} - \mathbf{X}\boldsymbol{\beta})'(\hat{\mathbf{y}} - \mathbf{X}\boldsymbol{\beta}) + \lambda \sum_{j=1}^{p} \frac{\beta_j^2}{|\beta_j|}.$$

As a result, the Lasso regression estimator $\hat{\boldsymbol{\beta}}_l$ can be approximated by a ridge regression estimator given by

$$\hat{\boldsymbol{\beta}}_l = (\mathbf{X}'\mathbf{X} + \lambda \mathbf{W}_p)^{-1}\mathbf{X}'\mathbf{y},$$

where $\mathbf{W}_p = diag(1/|\hat{\beta}_j|)$ is a diagonal matrix with elements equal to $1/|\hat{\beta}_j|$ and λ is the tuning parameter defined as before but now chosen so that $\sum_{j=1}^{p} |\beta_j| = c$. The covariance matrix is approximated by $\sigma^2(\mathbf{X}'\mathbf{X} + \lambda \mathbf{W}_p)^{-1}(\mathbf{X}'\mathbf{X})(\mathbf{X}'\mathbf{X} + \lambda \mathbf{X})^{-1}$.

Example 8.18 *(Example 8.17 continued) Consider the situation where $p = 2$ and the constraint Lasso region is $|\beta_1| + |\beta_2| < c$. Figure 8.2 provides a simple geometrical view of a Lasso approach for this case. The ellipses correspond to the contour of the $S(\boldsymbol{\beta})$ which is minimized at $\hat{\boldsymbol{\beta}} = (\hat{\beta}_1, \hat{\beta}_2)$ and the rhombus $|\beta_1| + |\beta_2| < c$ corresponds to the constraint in the Lasso approach. Since the ellipse and rhombus sizes are minimized simultaneously in the Lasso, the resulting estimator $\hat{\boldsymbol{\beta}}_l = (\hat{\beta}_{l1}, \hat{\beta}_{l2})$ is equal to the point at which the ellipse and the rhombus intersect. If this happens in any of its vertices, the corresponding coefficient is estimated as 0. The figure indicates that the Lasso yields more 0-valued estimates than ridge estimation. This feature is specially attractive in situations where p is large.*

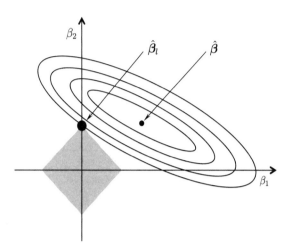

Figure 8.2 *Geometrical view of a Lasso for the case $p = 2$ (based on Tibshirani, 1996).*

A disadvantage of the above approach is that it leads to an estimator variance equal to zero for the corresponding explanatory variables with $\hat{\beta}_j = 0$. The same problem is observed for other proposed alternatives, such as the estimator for the covariance of the estimates derived by Fan and Li (2001) from a sandwich formula. The problem is, however, overcome from the Bayesian point of view via the so-called Bayesian Lasso, proposed by Park and Casella (2008).

We first note that the form of the Lasso regression penalty $\lambda \sum_{j=1}^{p} |\beta_j|$ suggests that the resulting estimates are equivalent to the posterior mode estimates under a Normal likelihood with the regression coefficients having independent and identical Laplace prior distributions given by $p(\beta_j) = (\lambda/2) \exp(-\lambda |\beta_j|)$. The Laplace prior distribution has its mode at zero. Consequently, the posterior mode may also be 0.

In order to guarantee a unimodal posterior density, Park and Casella (2008) consider a conditional Laplace prior distribution for the regression coefficients given by

$$p(\boldsymbol{\beta}|\sigma^2) = \prod_{j=1}^{p} \frac{\lambda}{2\sqrt{\sigma^2}} \exp\left(-\frac{\lambda|\beta_j|}{\sqrt{\sigma^2}}\right)$$

and a noninformative marginal prior $p(\sigma^2) = 1/\sigma^2$ for σ^2.

Theorem 8.2 *The joint posterior distribution for $\boldsymbol{\beta}$ and σ^2 is unimodal un-*

der the prior distribution

$$p(\boldsymbol{\beta}, \sigma^2) = \frac{1}{\sigma^2} \times \prod_{j=1}^{p} \frac{\lambda}{2\sqrt{\sigma^2}} \exp\left(-\frac{\lambda|\beta_j|}{\sqrt{\sigma^2}}\right).$$

Proof The approximate log posterior distribution for $\boldsymbol{\beta}$ and σ^2 is given by

$$\log\left(\frac{1}{\sigma^2}\right) - \frac{n+p-1}{2}\log(\sigma^2) - \frac{1}{2\sigma^2}(\mathbf{y} - \mathbf{X}\boldsymbol{\beta})'(\mathbf{y} - \mathbf{X}\boldsymbol{\beta}) - \frac{\lambda}{\sqrt{\sigma^2}}\sum_{j=1}^{p}|\beta_j|.$$

Considering the coordinate transformation $\boldsymbol{\gamma} = \boldsymbol{\beta}/\sqrt{\sigma^2}$ and $\delta = 1/\sqrt{\sigma^2}$, the approximate log posterior distribution for $\boldsymbol{\gamma}$ and δ is given by

$$\log(\delta^2) + (n+p-1)\log(\delta) - \frac{1}{2}(\delta\mathbf{y} - \mathbf{X}\boldsymbol{\beta})'(\delta\mathbf{y} - \mathbf{X}\boldsymbol{\beta}) - \lambda\sum_{j=1}^{p}|\gamma_j|.$$

All terms are concave in $\boldsymbol{\gamma}$ and δ. Then, the posterior distribution is unimodal since the coordinate transformation considered is continuous, leading to equivalent results in both coordinates. □

A hierarchical representation of the alternative Bayesian Lasso model is given by

$$
\begin{aligned}
\mathbf{Y} \mid \mathbf{X}, \boldsymbol{\beta}, \sigma^2 &\sim N(\mathbf{X}\boldsymbol{\beta}, \sigma^2 \mathbf{I}_n), \\
\boldsymbol{\beta} \mid \sigma^2, \phi_1^2, \ldots, \phi_p^2 &\sim N(\mathbf{0}, \sigma^2 \mathbf{D}_\phi), \\
\phi_1^2, \ldots, \phi_p^2 &\sim \prod_{j=1}^{p} f(\phi_j^2) \text{ where } f(\phi^2) \propto \frac{\lambda^2}{2} \exp(-\frac{\lambda^2 \phi^2}{2}), \\
\sigma^2 &\sim p(\sigma^2),
\end{aligned}
$$

where $\mathbf{D}_\phi = diag(\phi_1^2, \ldots, \phi_p^2)$ is the covariance matrix of the regression coefficients and $\sigma^2, \phi_1^2, \ldots, \phi_p^2 > 0$. The conditional prior distribution for $\boldsymbol{\beta}$, $p(\boldsymbol{\beta}|\sigma^2)$, is recovered after integrating out $\phi_1^2, \ldots, \phi_p^2$.

Other Lasso-type regressions are presented in Kyung, Gill, Ghosh and Casella (2010), including a general version of the Lasso model in the presence of two tuning parameters. Finally, Zou and Hastie (2005) proposed the elastic net, a mixture designed to compromise the ridge and lasso penalizations. A Bayesian version is readily obtained by using a mixture of Normal and Laplace prior distributions for the regression coefficients.

Exercises

§8.2

1. Show that the derivative of $S(\boldsymbol{\beta})$ with respect to $\boldsymbol{\beta}$ is given by $2(\mathbf{X}'\mathbf{X}\boldsymbol{\beta} - \mathbf{X}'\mathbf{y})$. Verify that $\hat{\boldsymbol{\beta}}$ does in fact minimize $S(\boldsymbol{\beta})$ by calculating the second derivative of $S(\boldsymbol{\beta})$ with respect to $\boldsymbol{\beta}$.

2. Prove that $S(\boldsymbol{\beta}) = (\boldsymbol{\beta} - \hat{\boldsymbol{\beta}})'\mathbf{X}'\mathbf{X}(\boldsymbol{\beta} - \hat{\boldsymbol{\beta}}) + S_e$ where

$$
\begin{aligned}
S_e &= \mathbf{y}'\mathbf{y} - \hat{\boldsymbol{\beta}}'\mathbf{X}'\mathbf{X}\hat{\boldsymbol{\beta}} \\
&= (\mathbf{y} - \mathbf{X}\hat{\boldsymbol{\beta}})'(\mathbf{y} - \mathbf{X}\hat{\boldsymbol{\beta}}).
\end{aligned}
$$

3. Obtain a $100(1 - \alpha)\%$ confidence region for $\boldsymbol{\beta}$ based on $[S(\boldsymbol{\beta}) - S_e]/ps^2$. What is the form of the region?

4. Consider the model $Y_i = \beta_1 + \beta_2 x_{2i} + \ldots + \beta_p x_{pi} + e_i$, $i = 1, \ldots, n$ with the e_i's iid $N(0, \sigma^2)$. Construct a $100(1-\alpha)\%$ confidence region for $(\beta_2, \ldots, \beta_p)$ from the maximum likelihood ratio test of validity of the model with level α.

5. Consider the model $Y_i = \beta_1 + \beta_2(x_{2i} - \bar{x}_2) + \ldots + \beta_p(x_{pi} - \bar{x}_p) + e_i$, $i = 1, \ldots, n$ with the e_i's iid $N(0, \sigma^2)$ and $\bar{x}_j = \sum_{i=1}^{n} x_{ji}$, for $j = 2, \ldots, p$. Show that the maximum likelihood ratio tests of level α for the hypotheses $H_j \colon \beta_j = 0$, for $j = 2, \ldots, p$, reject H_j if

$$
| \hat{\beta}_j | > t_{\alpha/2, n-p}\, s \Big/ \sqrt{\sum_{j=1}^{n}(x_{ji} - \bar{x}_j)^2}, \text{ for } j = 2, \ldots, p.
$$

6. Obtain the expression of the $100(1 - \alpha)\%$ confidence interval for the prediction of a future observation Y^*

   (a) in the simple linear regression with explanatory variable x^*;

   (b) in the analysis of variance model with a single classification factor, for an observation for group j, for $j = 1, \ldots, k$.

7. (Dose-response problem) Assume $d_1, \ldots, d_k$ doses are applied to $m_i, \ldots, m_k$ individuals and observe the numbers $Y_1, \ldots, Y_k$ of individuals who responded to the dosage in each group. Let $\pi_i, \ldots, \pi_k$ be the probability

of success, so $Y_i \sim Binomial(m_i, \pi_i)$, for $i = 1, \ldots, k$. For each individual there is a minimum dose U, below which the individual will not respond to the treatment. Let U follow the Normal distribution with mean μ and variance σ^2. So, the proportion of individuals that respond to a dose d_i is $\pi_i = P(U \le d_i) = \Phi(\beta_1 + \beta_2 d_i)$, where $\Phi(.)$ is the distribution function of the standard Normal distribution, $\beta_1 = -\mu/\sigma$ and $\beta_2 = 1/\sigma$. Let us consider the probit function $\Phi^{-1}(\pi_i) = \beta_1 + \beta_2 d_i$, for $i = 1, \ldots, k$.

(a) Obtain the lethal dose $d(\pi_i)$ at $\pi_i = 0.50$ (lethal dose at 50%).

(b) Obtain the expression of the $100(1 - \alpha)\%$ confidence interval for the prediction of a future lethal dose $d(\pi)$.

8. Let $Y_1, \ldots, Y_n$ be a random sample from Poisson distributions with parameter $\theta_i > 0$, for $i = 1, \ldots, n$. Consider the link function $\log(\theta_i) = \boldsymbol{x}_i \boldsymbol{\beta}$, where $\boldsymbol{x}$ are explanatory variables, for $i = 1, \ldots, n$. Construct a $100(1-\alpha)\%$ confidence region for $\boldsymbol{\beta}$.

§8.3

9. Under the conditions of the Bayesian regression model of Section 8.3, show that

$$(\boldsymbol{\beta} - \boldsymbol{\mu}_0)' \mathbf{C}_0 (\boldsymbol{\beta} - \boldsymbol{\mu}_0) + (\boldsymbol{\beta} - \hat{\boldsymbol{\beta}})' \mathbf{X}' \mathbf{X} (\boldsymbol{\beta} - \hat{\boldsymbol{\beta}})$$
$$= (\boldsymbol{\beta} - \boldsymbol{\mu}_1) \mathbf{C}_1 (\boldsymbol{\beta} - \boldsymbol{\mu}_1) + \boldsymbol{\mu}_0' \mathbf{C}_0 \boldsymbol{\mu}_0 + \hat{\boldsymbol{\beta}}' \mathbf{X}' \mathbf{X} \hat{\boldsymbol{\beta}} + \boldsymbol{\mu}_1' \mathbf{C}_1 \boldsymbol{\mu}_1,$$

where $\boldsymbol{\mu}_1 = \mathbf{C}_1^{-1}(\mathbf{C}_0 \boldsymbol{\mu}_0 + \mathbf{X}' \mathbf{y})$ and $\mathbf{C}_1 = \mathbf{C}_0 + \mathbf{X}' \mathbf{X}$.

10. Obtain the expressions of $\boldsymbol{\mu}_1$, $\mathbf{C}_1$, n_1 and σ_1^2 for the particular case of a simple linear regression, showing that they coincide with the expressions obtained in Example 8.9.

11. Obtain the joint posterior distribution of β_0 and β_1 in the simple linear regression model. Are these parameters conditionally independent given ϕ a posteriori? Are these parameters marginally independent a posteriori? Repeat the exercise for the one-way analysis of variance model.

12. Perform Bayesian inference for the simple linear regression model with the Normal prior with same marginal distribution as earlier but with prior correlation ρ conditional on ϕ, instead of 0.

13. Prove that if

(a) $\mathbf{Y} = \mathbf{X}\boldsymbol{\beta} + \mathbf{e}$, and $\mathbf{e} \sim N(\mathbf{0}, \phi^{-1}\mathbf{I}_n)$ where $\phi = 1/\sigma^2$,

(b) $\boldsymbol{\beta} \mid \phi \sim N(\boldsymbol{\mu}, \phi^{-1}\mathbf{C})$ and

(c) $\nu \sigma_0 \phi \sim \chi_\nu^2$,

then $\mathbf{Y} \sim t_\nu(\mathbf{X}\boldsymbol{\mu}, \sigma_0^2(\mathbf{I}_n + \mathbf{XCX}'))$, using only the formula $p(\mathbf{z}) = \int p(\mathbf{z} \mid \mathbf{w})p(\mathbf{w})d\mathbf{w}$.

14. Obtain the expression of the Bayes factor to test the hypothesis of validity of the model for

  (a) the simple linear regression model;

  (b) the analysis of variance model with a single classification factor with k levels.

15. (One way analysis of variance) Let $y_{ji} = \beta_j + e_{ji}$, $i = 1, \ldots, n_j$, for $j = 1, \ldots, k$ be the model, that is, the n_j observations in group j have the same mean, for $j = 1, \ldots, k$. The total number of observations n is given by $\sum_{j=1}^{k} n_j$. Obtain the predictive distribution of the vector $\boldsymbol{Y}^*$, consisting of collection of future observations form the k groups.

§8.4

16. Consider the hierarchical model with K stages given by

$$
\begin{aligned}
\mathbf{Y} \mid \boldsymbol{\beta}_1, \phi &\sim N(\mathbf{X}_1\boldsymbol{\beta}_1, \phi^{-1}\mathbf{I}_n) \\
\boldsymbol{\beta}_k \mid \boldsymbol{\beta}_{k+1}, \phi &\sim N(\mathbf{X}_{k+1}\boldsymbol{\beta}_{k+1}, \phi^{-1}\mathbf{C}_k^{-1})\, k = 1, \ldots, K-1 \\
\boldsymbol{\beta}_K \mid \phi &\sim N(\boldsymbol{\mu}, \phi^{-1}\mathbf{C}_K^{-1}) \\
n_0\sigma_0^2\phi &\sim \chi_{n_0}^2.
\end{aligned}
$$

  (a) Obtain the prior distribution of $\boldsymbol{\beta}_k \mid \phi$, for $k = 1, \ldots, K-1$.

  (b) Obtain the marginal prior distribution of $\boldsymbol{\beta}_k \mid \phi$, for $k = 1, \ldots, K-1$.

  (c) Obtain the marginal distribution of $\mathbf{Y}$.

17. Show that in the hierarchical analysis of variance model with a single classification factor with k levels, $Cov(\beta_j, \beta_{j'} \mid \phi) = Cov(\beta_j, \beta_{j'}) = c_2^{-1}$ and $Cor(\beta_j, \beta_{j'} \mid \phi) = Cor(\beta_j, \beta_{j'}) = (1 + c_2/c_1)^{-1}$, $\forall(j, j')$, where Cor denotes the correlation.

18. Consider the hierarchical model with two stages. Show that the posterior distribution of $\boldsymbol{\beta}_2$ and ϕ is multivariate Normal-χ^2 with parameters $\boldsymbol{\mu}_2$, $\mathbf{C}_2^*$, n_2 and σ_2^2 given by

$$
\begin{aligned}
\boldsymbol{\mu}_2 &= \mathbf{C}_2^{*-1}(\mathbf{C}_2^{-1}\boldsymbol{\mu}_0 + \mathbf{X}_1'\mathbf{C}_0\mathbf{y}), \\
\mathbf{C}_2^* &= \mathbf{C}_2 + \mathbf{X}_1\mathbf{X}_2\mathbf{C}_0\mathbf{X}_1'\mathbf{X}_2', \\
n_1 &= n + n_0, \text{ and} \\
n_1\sigma_1^2 &= n_0\sigma_0^2 + (\mathbf{y} - \mathbf{X}_1\mathbf{X}_2\boldsymbol{\mu}_2)'\mathbf{C}_0\mathbf{y} + (\boldsymbol{\mu}_0 - \boldsymbol{\mu}_2)'\mathbf{C}_2^{*-1}\boldsymbol{\mu}_0.
\end{aligned}
$$

19. Show that in the hierarchical analysis of variance model with a single classification factor, m observations in each group and non-informative prior at the second level,

$$E(\beta_j \mid \phi, \mathbf{y}) = w_j \bar{y}_j + (1 - w_j)\bar{\bar{y}},$$

with $0 \leq w_j \leq 1$, for $j = 1, \ldots, k$ and $\bar{\bar{y}}$ is the average of group averages. Obtain the expressions for w_j, for $j = 1, \ldots, k$.

20. Show that the mixed linear model of Example 8.14 can be written in the standard hierarchical model framework, identifying all model components.

§8.5

21. Consider the 1st order polynomial dynamic model with $\sigma_t^2 = \sigma^2$ and $W_t = W$, $\forall t$. Derive the predictive distribution for an horizon $k > 0$ and show that

$$y_{t+k}|D_t \sim N(m_t, C_t + kW + \sigma^2).$$

Obtain the distribution of $(y_{t+1} + \ldots + y_{t+k} \mid D_t)$ from the joint predictive distribution of $(y_{t+1}, \ldots, y_{t+k} \mid D_t)$. For example, take $k = 2$, and prove that

$$y_{t+1} + y_{t+2}|D_t \sim N(2m_t, 4C_t + 2\sigma^2 + 5W).$$

22. Consider again the 1st order polynomial dynamic model. Show that

$$\beta_{t-1}|D_t \sim N(m_{t-1}^*, C_{t-1}^*),$$

with

$$
\begin{aligned}
E(\beta_{t-1}|D_t) &= m_{t-1} + (C_{t-1}/R_t)(m_t - m_{t-1}) \text{ and} \\
V(\beta_{t-1}|D_t) &= C_{t-1} - (C_{t-1}/R_t)^2(R_t - C_t).
\end{aligned}
$$

23. Suppose the series was not observed at time t so that Y_t was missing and, therefore, $D_t = D_{t-1}$. Obtain the distributions of $\beta_t|D_t$ and $y_{t+1}|D_t$ assuming knowledge of m_{t-1} and C_{t-1} for 1st order polynomial dynamic model.

§8.6

24. Assume we have orthogonal explanatory variables.

(a) Show that the ridge regression estimator is given by $\hat{\beta}_r = [n/(n+\lambda)]\,\hat{\beta}$, where $\hat{\beta}$ is the vector of least square estimates.

(b) Show that the covariance matrix of the estimator $\hat{\beta}_r$ is $\sigma^2(\mathbf{X}'\mathbf{X} + \lambda\mathbf{I}_p)^{-1}(\mathbf{X}'\mathbf{X})(\mathbf{X}'\mathbf{X} + \lambda\mathbf{I}_p)^{-1}$.

Sketched solutions to selected exercises

Chapter 2

5. The test X is such that $P(X = 1|\theta = 1) = 0.95, P(X = 1|\theta = 0) = 0.40$ and the disease prevalence is $P(\theta = 1) = 0.70$.

   (a) From Example 2.4 on page 24, $P(\theta = 1|X = 1) = 0.847$. By Bayes theorem, $P(\theta = 1|X = 0) \propto l(\theta = 1; X = 0)P(\theta = 1) = 0.035$, and $P(\theta = 0|X = 0) \propto l(\theta = 0; X = 0)P(\theta = 0) = 0.180$. Therefore, $P(\theta = 1|X = 0) = 0.035/(0.035 + 0.180) = 0.163$. The result $X = 1$ makes the doctor more certain about John's illness because $P(\theta = 1|X = 1) = 0.847 > 0.163 = P(\theta = 1|X = 0)$.

   (b) Using $P(\theta|X = 1)$ as the prior for the second experiment, it follows from the Bayes theorem that $P(\theta = 1|X_1 = 1, X_2 = 1) \propto 0.95 \times 0.847 = 0.805$ and $P(\theta = 0|X_1 = 1, X_2 = 1) \propto 0.40 \times 0.153 = 0.063$. So, the probability that John is ill is $P(\theta = 1|X_1 = 1, X_2 = 1) = 0.805/(0.063 + 0.805) = 0.927$.

   (c) The likelihood for n *positive* results is $(0.95)^n$. Therefore, the solution of $P(\theta = 1|X_1 = 1, \ldots, X_n = 1) = (0.95)^n 0.70/((0.4)^n 0.3 + (0.95)^n 0.7) \geq 0.99$ is obtained by trial and error as $n = 8$.

8. Let θ represent the event *the driver is drunk* with $P(\theta) = 0.75$. The test X_1 will be positive ($=1$) if the level of alcohol in his/her blood is high and zero otherwise. It is known that $P(X_1 = 1|\theta = 1) = 0.8$, and for a second test $P(X_2 = 0|\theta = 0) = 1$ and $P(X_2 = 0|\theta = 1) = 0.10$.

   (a) $P(X_1 = 1) = (0.8)(0.25) + (0.2)(0.75) = 0.35$ can be interpreted as the proportion of drivers stopped who have to be submitted to a second test.

   (b) By Bayes theorem, $P(\theta = 1|X_1 = 1) \propto (0.8)(0.25)/0.35 = 0.571$. Also, $P(X_2 = 1|X_1 = 1) = 0(1 - 0.571) + (0.9)(0.571) = 0.514$. Therefore, $P(\theta = 1|X_1 = 1, X_2 = 1) = (0.9)(0.571)/0.514 = 1$.

   (c) Obviously, it is $P(X_1 = 0) = 1 - P(X_1 = 1) = 0.65$.

11. (a) $p(\theta|x,\mu) \propto p(\theta,\mu,x) = p(x|\theta,\mu)p(\theta|\mu)p(\mu)$. But, $p(\theta|x,\mu) \propto \exp\{-(0.5)[(x-\theta)^2/\sigma^2 + (\theta-\mu)^2/\tau^2 + \mu^2]\}$. After some algebra we get $(\theta|x,\mu) \sim N(\mu_1,\tau_1^2)$ with $\mu_1 = \tau_1^2(x/\sigma^2 + \mu/\tau^2)$ and $\tau_1^{-2} = (\sigma^{-2} + \tau^{-2})$. This is in fact an application of Theorem 2.1.

(b) In order to apply Bayes theorem we need the likelihood $l(\mu;x) = p(x|\mu) = \int p(x|\theta,\mu)p(\theta|\mu)d\theta \propto \int \exp\{-(0.5)[(x-\theta)^2/\sigma^2 - (\theta-\mu)^2/\tau^2)]\}d\theta$. After some calculations we get $p(x|\mu) \propto \exp[-(0.5)(x-\mu)^2/(\tau^2+\sigma^2)]$. Using Theorem 2.1 once again, it follows that $\mu|x \sim N(\mu_1,\tau_1^2)$ with $\mu_1 = x/(\sigma^2+\tau^2+1)$ and $\tau_1^2 = (\sigma^2+\tau^2)/(\sigma^2+\tau^2+1)$.

(c) $p(\theta) \propto \int p(\theta|\mu)p(\mu)d\mu$ or $\theta \sim N(0,1+\tau^2)$. It follows immediately by Bayes theorem that $\theta|x \sim N(\theta_1,\tau_2^2)$ where $\theta_1 = \tau_2^2 x\sigma^2$ and $\tau_2^2 = \sigma_2(\tau^2+1)/(\sigma^2+\tau^2+1)$.

In fact the above results can be easily obtained after rewriting the exercise as $X = \theta + \epsilon$, $\epsilon \sim N(0,\sigma^2)$, $\theta = \mu + \omega$, $\omega \sim N(0,\tau^2)$ and $\mu \sim N(0,1)$, where ϵ, ω and μ are independent. Using results about linear combination of Normal variates, we get $X = \mu + \omega + \epsilon$ or $X|\mu \sim N(\mu,\tau^2+\sigma^2)$. Analogously, $\theta = \mu + \omega$ or $\theta \sim N(0,\tau^2+1)$.

16. From exchangeability of the X_i's and Theorem 2.2, any sequence of n X_i's having k values of 1 and $n-k$ values 0 has probability given by $\int_0^1 \theta^k(1-\theta)^{n-k}dF(\theta)$, $\forall k \leq n > 0$.

(a) Since $T = \sum X_i$, $P(T=t) = \sum_{\mathbf{x}\in A}\int_0^1 \theta^k(1-\theta)^{n-k}dF(\theta)$ where $A = \{\mathbf{x}|\sum X_i = t\}$. The number of n-tuples in A is $\binom{n}{t}$ and they all have the same probability. Then, $P(T=t) = \int \binom{n}{t}\theta^t(1-\theta)^{n-t}dF(\theta)$.

(b) $E(T) = \sum_{i=1}^n E(X_i) = \sum_{i=1}^n E(X_1) = nE(X_1) = nE[E(X_1|\theta)] = nE(\theta)$ since $X_1|\theta \sim Ber(\theta)$.

22. (a) The sample space depends on the unknown parameter; hence, the distribution does not belong to the exponential family.

(c) $p(x|\theta) = (2\pi)^{-1/2}\exp\{-0.5[x^2 - 2\theta x + \theta^2]/\theta\} = (2\pi)^{-1/2}\exp(x)\exp\{-0.5\ [x^2/\theta - \theta - log(\theta)]\}$. Identifying $a(x) = (2\pi)^{-1/2}exp(x)$, $u(x) = x^2/2$, $\phi(\theta) = \theta^{-1}$ and $b(\theta) = -(\theta + log(\theta))/2$, we conclude that it is a member of the one parameter exponential family and, by the Neyman factorization criterion, X^2 is a sufficient statistic.

(e) $p(x|x \neq 0,\theta) = p(x|\theta)/P[X \neq 0|\theta] = \binom{n}{x}\theta^x(1-\theta)^{n-x}/[1-(1-\theta)^n]$, $x = 1,\ldots,n$ or $p(x|x \neq 0,\theta) = \binom{n}{x}\exp\{x\log[\theta/(1-\theta)] + n\log(1-\theta) - log[1-(1-\theta)^n]\}$. So, $a(x) = \binom{n}{x}$, $u(x) = x$, $\phi(\theta) = \log[\theta/(1-\theta)]$ and $b(\theta) = n\log(1-\theta) - \log[1-(1-\theta)^n]$. It is clear that X is a sufficient statistic for θ.

(g) $p(x|\theta) = \theta^x \log(\theta)/(\theta-1) = \exp[x \log \theta + \log[\log \theta/(\theta-1)]$. So, $a(x) = 1$, $u(x) = x$, $\phi(\theta) = \log(\theta)$ and $b(\theta) = \log[\log(\theta)/(\theta-1)]$. So, it is a member of the exponential family and $T(X) = X$ is a minimal sufficient statistic.

25. Remember that $p(x|\theta) = a(x)\exp[U(x)\phi(\theta) + b(\theta)]$ and $\int p(x|\theta)dx = 1$.

(a) Differentiating both sides, with respect to θ follows that $\int[U(x)\phi'(\theta) + b'(\theta)]p(x|\theta)dx = 0$. It is then easy to get $E[U(X)] = -b'(\theta)/\phi'(\theta)$.

(b) Differetiating again $\int[U(x)\phi'(\theta) + b'(\theta)]p(x|\theta)dx = 0$, we get $\int[U(x)\phi''(\theta)p(x|\theta) + U(x)\phi'(\theta)p'(x|\theta) + b''(\theta)p(x|\theta)dx + b'(\theta)p'(x|\theta)]dx$ where $p'(x|\theta) = [U(x)\phi'(\theta) + b'(\theta)]p(x|\theta)$. After some calculation it follows $E[U^2(X)] = [\phi''(\theta)b'(\theta) - \phi'(\theta)b''(\theta)]/[\phi'(\theta)]^3$. The variance follows from the above calculations.

31. (a) It is easy to get $p(\psi, \xi) = p_{\theta,\phi}(\psi, \xi)|J| \propto \xi$, because the Jacobian of the transformation is just ξ.

(b) $p(\psi, \xi|x, y) \propto \theta^x \exp(-\theta) \phi^y \exp(-\phi)$ where $\theta = \psi\xi$ and $\phi = \xi(1-\psi)$. So $p(\psi, \xi|x, y) \propto \xi^{x+y+1} \exp(-\xi) \psi^x (1-\psi)^y$ or proportional to the product of a $Ga(x+y+2, 1)$ and a $Be(x+1, y+1)$ distribution.

(c) First of all it is easy to see that $X+Y \sim Pois(\xi)$, where $\xi = \theta+\phi$, since x and y are independent Poisson distributed with parameters θ and ϕ, respectively. Since $p(x|x + y, \psi, \xi) = p(x|\psi, \xi)p(x + y|x, \psi, \xi)/p(x + y|\psi, \xi)$ it follows that $p(x|x + y, \psi, \xi) = \binom{x+y}{x}\psi^x(1-\psi)^y$ which is only a function ψ.

(d) Using the distributions obtained in (c) and the factorization criterion, it is simple to obtain the results.

(e) The marginal likelihoods of ψ and ϕ are given by $l(\psi; x, y) = p(x, y|\psi) = \int p(x, y|\psi, \xi)p(\xi|\psi)d\xi$ and $l(\xi; x, y) = p(x, y|\xi) = \int p(x, y|\psi, \xi)p(\psi|\xi)d\psi$. Since $p(\xi|\psi) \propto \xi$ and $p(\psi|\xi) \propto k$, simple integrations lead to $l(\psi; x, y) \propto \psi^x(1-\psi)^y$ and $l(\xi; x, y) \propto \xi^{x+y}\exp(-\xi)$.

34. (a) The inverse transformations are $\theta_1 = \lambda\psi$ and $\theta_2 = \psi(1-\lambda)$ and the Jacobian is $J = \psi$. The distribution of λ, ψ follows as $p(\lambda, \psi) \propto \psi$, since $p(\theta) \propto k$.

(b) $p(\mathbf{x}|\lambda, \psi) = \lambda^{x_1}(1-\lambda)^{x_2}\psi^{x_1+x_2}(1-\psi)^{1-x_1-x_2}$. Then the marginal likelihood for ψ is $p(\mathbf{x}|\psi) = \int p(\mathbf{x}|\lambda, \psi)p(\lambda|\psi)d\lambda$. Since $p(\lambda|\psi) \propto k$, $p(\mathbf{x}|\psi) \propto \psi^{x_1+x_2}(1-\psi)^{1-x_1+x_2}$.

(c) By the Neyman factorization criterion with $f(x_1+x_2, \psi) = \psi^{x_1+x_2}(1-\psi)^{1-x_1+x_2}$ and $g(\mathbf{x}) = k$ follows that $X_1 + X_2$ is sufficient for ψ.

Chapter 3

1. (a) My experience of living at Rio gives me confidence to assess the first quartile as 20^oC and the median as 28^oC.

   (b) A good approximation for the standard deviation is $\sigma \simeq 1.25|Q_1 - Q_2|$, where Q_i is the $i-th$ quartile. So my assessment is that the temperature is normally distributed with mean 28^oC and standard deviation equal to 10^oC.

   (c) The first quantile determined from the Normal distribution corresponds to 2.7^oC conflicting with the subjective assessment of 8^oC. The assessment of normality must be revised.

5. (a) $l(\theta; \boldsymbol{y}) = \prod_{i=1}^{4} \binom{n_i}{y_i} \theta^{y_i}(1-\theta)^{n_i-y_i} \propto \theta^9(1-\theta)^3$.

   (b) The class of the Beta distributions can be used as prior, say, with $a = 1$ and $b = 1$ representing a vague initial information. So, $p(\theta|\boldsymbol{y}) \propto \theta^9(1-\theta)^3$ or a $\theta|\boldsymbol{y} \sim Beta(10, 4)$.

   (c) Since conditions are similar, it is natural to assume that $y_5|\theta \sim Bin(n_5, \theta)$ with $n_5 = 3$. The predictive distribution will be a Beta-Binomial distribution with parameter $(3, 10, 4)$ providing the following probabilities

Y_5	0	1	2	3
$p(y_5|\theta)$	0.036	0.178	0.393	0.393

10. (a) Assuming that $\theta \sim G(a, b)$, $\mu = E[\theta] = a/b = 4$ and the coefficient of variation $CV(\theta) = \sigma/|\mu| = 1/a^{1/2} = 1/2$. It follows that $a = 4$ and $b = 1$. The posterior variance will be less or equal to 0.01 if and only if $(t+a)/(n+b)^2 \leq 0.01$, where $t = \sum x_i$. Solving the quadratic inequality in n, it follows that $n \geq 10(4 + t)^{1/2} - 1$.

   (b) The posterior mean can be written as $\mu_1 = (a+t)/(b+n) = \bar{x}_n\gamma_n\bar{x}_n + (1 - \gamma_n)\mu_0$ where $\mu_0 = a/b$ and $\gamma_n = n/(n + b)$. The limit of γ_n, when $n \to \infty$, will be 1.

   (c) The posterior for θ is $Beta(a + t, b + n - t)$ where $t = \sum x_i$. This distribution has mean $(a + t)/(a + b + n) = \gamma_n\bar{x}_n + (1 - \gamma_n)\mu_0$ where $\mu_0 = a/(a + b)$ and $\gamma_n = n/(a + b + n)$. The limit of γ_n, when $n \to \infty$, will be 1.

16. (a) The Pareto distribution is a member of the one parameter exponential family with $a(\mathbf{x}) = 1$, $\phi(\theta) = -(\theta + 1)$, $U(\mathbf{x}) = \sum \log(x_i)$ and $b(\theta) = \log(\theta) + \theta\log(b)$. So the sufficient statistic for θ is $U(\mathbf{x})$.

   (b) The observed information is $-d^2\log(p(x|\theta))/d\theta^2 = n/\theta^2$ which coincides with the Fisher information for a sample of size n. The Jeffreys prior will be $p(\theta) \propto I(\theta)^{1/2} \propto \theta^{-1}$. It is obviously improper since $\int p(\theta)d\theta = \int k\theta^{-1}d\theta$ diverges.

(c) The posterior distribution is $p(\theta|\mathbf{x}) \propto \theta^{n-1}b^{n\theta}/(\{\prod_{i=1}^{n} x_i\}^{1/n})^{n\theta}$. Let $z = \{\prod_{i=1}^{n} x_i\}^{1/n}$ be the sample geometric mean. It is clear that $p(\theta|\mathbf{x}) \propto \theta^{n-1}exp(-n\ \theta \log(z/b))$ which corresponds to a $G(n, n\log(z/b))$.

20. The non-informative prior for θ is $p(\theta) \propto k$. The density for $\phi = a\theta + b$, $a \neq 0$, is $p(\phi) = p_\theta[(\phi - b)/a]\ |d\theta/d\phi| \propto k\ d[(\phi - b)/a]/d\phi \propto k$. If $p(\theta) \propto \theta^{-1}$, $\theta > 0$, then $p(\phi) = p_\theta(\phi^{1/a})\ |d\theta/\phi|$ with $\phi = \theta^a$, $a \neq 0$ or $p(\phi) \propto \phi^{-1}$. If $\psi = \log\theta$, then $p(\psi) \propto \exp(-\psi)exp(\psi) = 1$.

29. (a) Let $\theta_i|\mu \sim N(\mu, b)$, b known and suppose that the θ_i's are independent given μ. Assuming that $p(\mu) \propto k$, $p(\boldsymbol{\theta}, \mu) \propto \prod_{i=1}^{n} p(\theta_i|\mu, b)$, or $\boldsymbol{\theta}|\mu \sim N(\mu\mathbf{1}_n, b\mathbf{I}_n)$.

(b) Since $p(\mu|\boldsymbol{y}) \propto p(\boldsymbol{y}|\mu)p(\mu)$ where $Y_i|\mu \sim N[\mu, (a + b)]$, $i = 1, \ldots, n$ independent, it follows from Theorem 2.1 that $\mu|\boldsymbol{y} \propto N[\bar{y}, (a + b)/n]$.

(c) Note that $p(\theta_i|\mu, \boldsymbol{y}) \propto p(\theta_i|\mu)p(\boldsymbol{y}|\theta_i)$. Once again, it follows from Theorem 2.1 that $\theta_i|\mu, \boldsymbol{y} \sim N[\mu_i^*, b^*]$, where $\mu_i^* = (a\mu + by_i)/(a + b)$ and $b^* = ab/(a + b)$. Hence, $E(\theta_i|\mu, \boldsymbol{y}) = \mu_i^*,\ ,\ i = 1, \ldots, n$.

(d) $E(\theta_i|\boldsymbol{y}) = E[E(\theta_i|\boldsymbol{y}, \mu)] = E(\mu_i^*) = by_i/(a + b)$.

Chapter 4

4. The posterior $p(\theta|\mathbf{x}) \propto \theta^t(1 - \theta)^{n-t}I_\theta(0, 1)$ where $t = \sum_{i=1}^{n} x_i$, that is, a $Beta(t + 1, n - t + 1)$ distribution.

(a) $E[L(\theta, d)|\mathbf{x}] \propto \int_0^1 \{(\theta-d)^2/[\theta(1-\theta)]\}\theta^t(1-\theta)^{n-t}d\theta = \int_0^1 (\theta-d)^2\theta^{t-1}(1-\theta)^{n-t-1}d\theta$. This integral is proportional to the expected value of the square loss with respect to the $Beta(t, n - t)$ distribution. So, it is minimized when $d = (t/n) = \bar{x}$. The risk of the Bayes estimator is $R(\bar{x}) = B(t + 1, n - t + 1)\int_0^1 (\theta - \bar{x})^2\theta^t(1 - \theta)^{n-t}d\theta$. Multiplying and dividing by $B(t, n-t)$, we get $R(\bar{x}) = [B(t+1, n-t+1)/B(t, n)]V(\theta|\mathbf{x})$. After some simplifications, it follows that $R(\bar{x}) = 1/n^2$.

(b) The predictive distribution is obtained via $p(x_{n+1}|\mathbf{x}) = \int_0^1 p(x_{n+1}|\theta)\ p(\theta|\mathbf{x})\ d\theta$. Assuming that $X_{n+1} \sim Bern(\theta)$ independent of $\mathbf{x}$, it follows that $p(x_{n+1}|\mathbf{x}) = B(t + 1, n - t + 1)\int_0^1 \theta^{x_{n+1}+t}(1 - \theta)^{n-x_{n+1}-t+1}d\theta$. Solving the integral gives $p(x_{n+1}|\mathbf{x}) = B(t + 1, n - t + 1)/B(t + x_{n+1} + 1, n - x_{n+1} - t + 2)$. Finally, $p(x_{n+1}|\mathbf{x}) = (t+1)^{x_{n+1}}(n-t+1)^{1-x_{n+1}}/(n+2)$. The mean and variance of the predictive distribution are $(t+1)/(n+2)$ and $(t + 1)(n - t + 1)/(n + 2)^2$, respectively.

(c) Let $t_1, \ldots, t_k$ be the counts associated with each of the k possible values of X. Then, straightforward generalizations give for each θ_i the Bayes estimator t_i/n, $i = 1, \ldots, k$, with associated risks $1/n^2$.

The predictive probability function generalizes to $p(x_{n+1} = i|\mathbf{x}) = (t_i + 1)/(n + k)$, $i = 1, \ldots, k$.

7. (a) $p(\theta|x) \propto \theta^{-1} I_\theta(0, 1)$. The proportionality constant is obtained making $1 = \int p(\theta|x)d\theta = \int_{x-1}^{x+1} k\theta^{-1}d\theta = k\log\theta|_{x-1}^{x+1}$. So, $k^{-1} = \log(x+1)/(x-1)$, $x > 1$.

(b) Posterior mean, mode and median are easily obtained. $E(\theta|x) = \int_{x-1}^{x+1} k\,\theta\,\theta^{-1}d\theta = k[(x+1) - (x-1)] = 2k = 2\log(x-1)/(x+1)$. To evaluate the mode, it is enough to observe that $p(\theta|x)$ is a strictly decreasing function of θ. So its maximum occurs at $x-1$. The median, by its turn, is the solution of $\int_{x-1}^{m} k\theta^{-1}d\theta = 1/2$ or $k\log[m/(x-1)] = 1/2$. After some simplifications it follows that $m = [(x-1)(x+1)]^{1/2}$.

12. The density functions for each type of bulb are $p(x|\psi) = \psi\exp(-\psi x)$, $\psi^{-1} = \theta$, 2θ and 3θ.

(a) Observing one bulb of each type, it follows that $l(\theta; \mathbf{x}) \propto \exp[-(x_1 + x_2/2 + x_3/3)/\theta]/6\theta^3$. The MLE is the solution of the equation $d\log l(\theta; \mathbf{X})/d\theta = -3/\theta + (X_1 + X_2/2 + X_3/3)/\theta^2 = 0$. That is, $\hat{\theta} = (X_1 + X_2/2 + X_3/3)/3$.

(c) Assuming the prior $\psi \sim G(\alpha, \beta)$, it follows that the posterior distribution is $p(\psi|\mathbf{x}) \propto \psi^{3+\alpha-1} \exp[-\psi(x_1 + x_2/2 + x_3/3 + \beta)]$.

(d) The Bayes estimator is the posterior mean, $E[\psi|\mathbf{x}] = (\alpha+3)/(\beta+x_1+x_2/2+x_3)$, if a square loss function is assumed. For the $0-1$ loss function the Bayes estimator is the posterior mode, $(\alpha+2)/(\beta+x_1+x_2/2+x_3/3)$.

21. It is worth remembering that the moment generating function for the $Pois(\theta)$ is $\varphi_X(t) = E[\exp(-tX)] = \exp[-\theta(1 - \exp(-t))]$.

(a) Since $\varphi_Y(t) = \prod_{i=1}^{n} \varphi_{X_i}(t) = \exp[-n\theta(1 - \exp(-t))]$, then $E[\exp(-cY)] = \varphi(c) = \exp[-n\theta(1 - \exp(-c))]$. The estimator $\exp(-cY)$ will be unbiased for $\exp(-\theta)$ iff $n\theta(1 - \exp(-c)) = \theta$ or $c = \log[n/(n-1)]$. Therefore, $(1 - 1/n)^Y$ is unbiased estimator of $e^{-\theta}$.

(b) By Cramer-Rao inequality, the variance lower bound of an unbiased estimator of $h(\theta)$ is given by $h'(\theta)^2/I(\theta)$. In the present case $h(\theta) = \exp(-\theta)$, so $h'(\theta) = -\exp(-\theta)$ and the expected information is obtained as $E[-d^2 \log l(\theta; x)/d\theta^2] = E[\sum x_i/\theta^2] = n/\theta$. Therefore, $V[\exp(-cY)] \geq \theta\exp(-2\theta)/n$, for $c = \log[n/(n-1)]$.

(c) The variance of $\exp(-cY)$ can be calculated as $V[\exp(-cY)] = \varphi_Y(2c) - \varphi_Y^2(c) = \exp[-n\theta(1 - exp(-2c))] - \exp[-2n\theta(1 - \exp(-c))]$, for $c = \log[n/(n-1)]$. After some algebra, it follows that $V[\exp(-cY)] = \exp[-\theta(2 - n)] - \exp(-2\theta)$. The ratio between the Cramer-Rao lower bound and the variance of the estimator is less than 1, so the estimator is not efficient for $\exp(-\theta)$.

29. (a) $p(\theta|x) \propto p(x|\theta) \propto \theta^x(1 - \theta)^{n-x}$. Assuming that $X = n$, it follows that $p(\theta|x) \propto \theta^n$, which is a monotonic increasing function of θ. So the maximum posterior density interval will be in the form $[a, 1]$, for some $a < 1$ such that $P[\theta \geq a|x] = 1 - \alpha$.

(b) Let $\psi = \theta/(1 - \theta)$. Then, $P[a/(1 - a) \leq \psi|x] = P[a/(1 - a) \leq \theta/(1 - \theta)|x] = P(a \leq \theta \leq 1|x) = 1 - \alpha$.

(c) It is easy to evaluate $p(\psi|x) = p_\theta(\psi/(1 + \psi)|x) \, |d\theta/d\psi|$ or $p(\psi|x) \propto \psi^n/(1 + \psi)^{n+2}$, which is in the form of an $F[2(n + 1), 2]$ distribution.

(d) The interval obtained in (b) is not an HDP because the F density is not monotonically decreasing in θ.

32. (a) a exponentially distributed lifetimes with observed mean lifetime equal to b lead to a likelihood $l(\theta) \propto \theta^a \exp(-\theta ab)$. The prior distribution for θ can then be obtained as $p(\theta) \propto \theta^a exp(-\theta ab)$. This distribution is in the form of a $Gamma(a + 1, ab)$.

(b) Using results of the conjugate analysis, it follows that $p(\theta|\mathbf{x})$ is a $Gamma(a + t + 1, n + ab)$ where $t = \sum x_i$.

(c) Since $T \sim Pois(n\theta)$, $p(\theta|t) \propto [(n\theta)^t \exp(-n\theta)][\theta^a exp(-ab\theta)]$ or $\theta|t \sim Gamma(a + t + 1, n + ab)$.

(d) The posterior distributions obtained in (b) and (c) are the same, which is not surprising since T is a sufficient statistic for θ.

36. $P((\theta - \mu_1)^2 \leq 4V_1|\mathbf{x}) = P(|\theta - \mu_1|/\sqrt{V_1} \leq 2|\mathbf{x}) = P(t_{n_1}(0, 1) \leq 2\sqrt{n_1/(n_1 - 2)})$. By trial and error, $n_1 = 11$ is the largest integer ensuring that the above probability is larger than 0.95. Therefore, $n = n_1 - n_0 = 6$.

Chapter 5

2. The log-likelihood function is given by $L(\alpha, \beta; \mathbf{x}) = n[\alpha \log \beta - \log \Gamma(\alpha)] + (\alpha - 1)T_1 - \beta T_2$, where $T_1 = \sum_i \log X_i$ and $T_2 = \sum_i X_i$. The maximum likelihood equations will be $\partial L(\alpha, \beta; \mathbf{X})/\partial \alpha = n[\log \beta - \Gamma'(\alpha)/\Gamma(\alpha)] + T_1$ and $\partial L(\alpha, \beta; \mathbf{X})/\partial \beta = n\alpha/\beta - T_2$. The MLE of β as a function of α will be $\hat\beta(\alpha) = \alpha/\bar{X}$, where $\bar{x} = T_2/n$. The profile log-likelihood of α is

$L(\alpha, \hat{\beta}(\alpha)) = n[\alpha \log(\alpha/\bar{X}) - \log(\Gamma(\alpha))] + (\alpha - 1)T_1 - n\alpha$. Differentiating the profile log-likelihood with respect to α we get $\partial L(\alpha, \hat{\beta}(\alpha); \mathbf{X})/\partial \alpha = n[\log \alpha - \Gamma'(\alpha)/\Gamma(\alpha)] + T_1 - n \log \bar{X}$. A numerical optimization method, such as Newton-Raphson, can then be used after a numerical approximation for the digamma function, to obtain the MLE for α. The MLE estimator for β follows from the equation $\hat{\beta} = \hat{\alpha}/\bar{X}$ and the invariance of MLE's.

10. We have $\theta = \lambda^a$ and $p(\mathbf{X}|\theta) \propto e^{-n\theta}\theta^T$ where $T = \sum_i X_i$.

   (a) The likelihood function is $l(\lambda; \mathbf{X}) = k \exp(-n\lambda^a)\lambda^{aT}$.

   (b) Jeffreys prior is defined as $p(\lambda) \propto I(\lambda)^{1/2}$ where $I(\lambda) = E[-L''(\lambda; \mathbf{X})]$, L is the log-likelihood and derivatives are taken with respect to λ. Then, $L(\lambda; \mathbf{X}) = -n\lambda^a + aT \log \lambda$. Its first and second derivative are respectively given by $L'(\lambda; \mathbf{X}) = -na\lambda^{a-1} + aT/\lambda$ and $L''(\lambda; \mathbf{X}) = -na(a-1)\lambda^{a-2} - aT/\lambda^2$, and the MLE is $\hat{\lambda} = (T/n)^{1/a}$. Also, $I(\lambda) = na^2\lambda^{a-2}$ and Jeffreys prior is $p(\lambda) \propto \lambda^{a-1}$.

   (c) The third derivative is $L'''(\lambda; \mathbf{X}) = -na(a-1)(a-2)\lambda^{a-3} + 2aT/\lambda^3$. Evaluating it at $\hat{\lambda} = \bar{x}^{1/a}$ gives $L'''(\hat{\lambda}; \mathbf{X}) = [-a(a-1)(a-2)T + 2aT]/(T/n)^{3/a}$. It is easy to verify that $L'''(\hat{\lambda}; \mathbf{X}) = 0$ iff $a = 3$ or $a = 0$ (degenerate solution).

   (d) The transformation that makes the third derivative null improves the asymptotic approximation. So, with respect to skewness, the parametrization $\lambda = \theta^{1/3}$ should be used to improve approximations.

18. We have already obtained that $V(\hat{\theta}) = 0.126/n$. Now, $V(\tilde{\theta}) = V[w(X)]/n$ where $w(X) = (2\pi)^{-1}X^2/(1 + X^2)^2$ and $V[w(X)] = E[w^2(X)] - \theta^2$, since the importance sampling estimator is unbiased. $E[w^2(X)] = \int_2^\infty (2\pi)^{-2}[x^2/(1 + x^2)]^2(2/x^2)dx = 0.021875$. Therefore, $V(\tilde{\theta}) = [0.021875 - (0.1476)^2]/n = 9 \times 10^{-5}/n$, which is substantially smaller than $V(\hat{\theta})$.

22. First, assume that $\mathbf{y} \neq \mathbf{x}$. The transition kernel of the chain is given by $q^*(\mathbf{x}, \mathbf{y}) = q(\mathbf{x}, \mathbf{y})\alpha(\mathbf{x}, \mathbf{y})$. Then, suppose that $\alpha(x, y) < 1$ so that $\alpha(\mathbf{x}, \mathbf{y}) = [p(\mathbf{y})q(\mathbf{y}, \mathbf{x})]/[p(\mathbf{x})q(\mathbf{x}, \mathbf{y})]$. Then, $\alpha(\mathbf{y}, \mathbf{x}) = 1$ and $p(\mathbf{x})q^*(\mathbf{x}, \mathbf{y}) = p(\mathbf{x})q(\mathbf{x}, \mathbf{y})\alpha(\mathbf{x}, \mathbf{y}) = p(\mathbf{y})q(\mathbf{y}, \mathbf{x})\alpha(\mathbf{y}, \mathbf{x}) = p(\mathbf{y})q^*(\mathbf{y}, \mathbf{x})$. This equality is also trivially satisfied when $\mathbf{y} = \mathbf{x}$. Integrating both sides with respect to $\mathbf{x}$ gives $p(\mathbf{y}) = \int p(\mathbf{x})q^*(\mathbf{x}, \mathbf{y})d\mathbf{y}$, which means that $p(\mathbf{y})$ is the stationary distribution of the chain.

Chapter 6

3. (a) The p-value is the statistic $\alpha(T) = F_T(T)$ where F_T is the distribution function of the test statistic T. By the probability integral transform (see, for example, DeGroot (1985, pg. 154)), it follows that $\alpha(T) = F_T(T) \sim U(0, 1)$. Since $\alpha(T_i)$, $i = 1, \ldots, k$ are function of independent statistics and all with the same distribution, it constitutes a random sample of the $U(0, 1)$ distribution.

   (b) The hypothesis is rejected if $\alpha(T) > \alpha$, the significance level. So, small values of $\alpha(T)$ should lead to rejection, and hence, large values of $-2 \log \alpha(T)$ should lead to rejection. Therefore, large values of F must lead to rejection of the hypothesis.

   (c) It is well known that if $X \sim U(0, 1)$, then $Y = -\log X \sim Exp(1) = G(1, 1)$ or $2Y \sim G(1, 1/2)$. Since F is the sum of k iid Gamma distributed random quantities, then $F \sim G(k, 1/2)$ or χ^2_{2k}.

10. (a) The maximum likelihood ratio is $\lambda(X) = \theta_0 \exp[-X(\theta_0 - \theta_1)]/\theta_1$, and the null hypothesis is accepted iff $\lambda(X) > c$ or $X < (\theta_0 - \theta_1)^{-1} \log[\theta_0/(c\theta_1)] = t$, where t is such that $P(X > t|\theta_0) = \alpha$. Then, $\exp(-\theta_0 t) = \alpha$ or $t = -\theta_0^{-1} \log \alpha$.

   (b) The p-value is $P[X > 3|\theta_0] = \exp(-3\theta_0) = 0.0498$.

   (c) The Bayes factor is $BF(H_0, H_1) = 2e^{-3/2} = 0.45$.

   (d) The odds ratio is $p(H_0|x)/p(H_1|x) = p(H_0)/p(H_1)BF(H_0, H_1) = BF(H_0, H_1)$, since $p(H_0) = p(H_1)$. So, $p(H_0|x) = \{[BF(H_0, H_1)]^{-1} + 1\}^{-1} = [(0.45)^{-1} + 1]^{-1} = 0.31$.

   (e) Using the result in (b), the null hypothesis is rejected at the $\alpha = 5\%$ level because the p-value is less than α, which is somewhat conflicting with the posterior probability of 0.31 for the null hypothesis, because the latter is smaller than 0.5 but is far from recommending rejection.

13. (a) The likelihood function under H_0 is $p(x_1, x_2|\theta) = \theta^t(1 - \theta)^{2-t}$, $t = x_1 + x_2$ and $\hat{\theta} = t/2$ is the MLE. Under H_1, $p(x_1, x_2|\theta) = \theta^{x_1}(1 - \theta)^{1-x_1}\theta^{x_2}(1 - \theta)^{1-x_2}$ and MLE of (θ_1, θ_2) is (x_1, x_2).

   (b) Using a uniform prior, the posterior distribution follows, from Bayes theorem, as $p(\theta|\mathbf{x}, H_0) \propto \theta^t(1 - \theta)^{2-t}$ and $p(\theta_1, \theta_2|\mathbf{x}, H_1) \propto \theta^{x_1}(1 - \theta)^{1-x_1}\theta^{x_2}(1 - \theta)^{1-x_2}$.

   (c) The GMLE is the mode of the posterior distribution. In this exercise, the GMLE's coincide with the MLE's obtained in (b).

   (d) The predictive distribution is given by $p(\mathbf{x}) = E_\theta[p(\mathbf{x}|\theta)]$. So $p(x_1, x_2|H_0) = \int_0^1 \theta^t(1 - \theta)^{2-t}d\theta = \Gamma(t + 1)\Gamma(3 - t)/\Gamma(4)$ and

$p(x_1, x_2|H_1) = \Gamma(x_1+1)\Gamma(2-x_1)\Gamma(1+x_2)\Gamma(2-x_1)/\Gamma(3)^2$. Numerically, we get the table of probabilities below

(x_1, x_2)	$(0,0)$	$(0,1)$	$(1,0)$	$(1,1)$
H_0	1/3	1/6	1/6	1/3
H_1	1/4	1/4	1/4	1/4

(e) The result follows immediately from the table above. The result states that the data favors H_0 (equality of distributions) twice more when $x_1 = x_2$.

15. The MLE of θ is $\hat{\theta} = \sum_i x_i / \sum_i t_i$ and the Fisher information is $I(\theta) = \sum t_i/\theta$. From Section 5.3.1, the asymptotic distribution of the MLE is $\hat{\theta} \sim N[\theta, I^{-1}(\theta)]$. A $100(1 - \alpha)\%$ asymptotic confidence interval is $\hat{\theta} - z_{\alpha/2}\hat{I}^{-1/2}(\theta) \leq \theta \leq \hat{\theta} + z_{\alpha/2}\hat{I}^{-1/2}(\theta)$ where $\hat{I}$ is an estimator of I given by $\sum t_i/\hat{\theta}$. If $\sum_i x_i = 10$ and $\sum_i t_i = 5$, then $\hat{\theta} = 2$ and $\hat{I}(\theta) = 5/2$. A 95% asymptotic confidence interval for θ is $(0.76, 3.24)$. Since $\theta_0 = 1$ belongs to the confidence interval described above, we accept the null hypothesis at the 5% level. With a 1% level, the corresponding interval is $(0.37, 3.63)$ and $\theta_0 = 1$ will again be accepted.

21. Let the cell frequencies be denoted by $N_1, \ldots, N_p$, with $N_i > 0$ and $\sum_{i=1}^p N_i = n$ and parameter $\theta_1, \ldots, \theta_p$, $\theta_i > 0$ with $\sum_{i=1}^p \theta_i = 1$. The log-likelihood function is $L(\boldsymbol{\theta}; \mathbf{N}) = k + \sum_{i=1}^p N_j \log \theta_j$, the MLE of $\boldsymbol{\theta}$ is $\hat{\boldsymbol{\theta}} = \mathbf{N}/n$ and the score function is $U(\mathbf{N}; \boldsymbol{\theta}) = (N_1/\theta_1, \ldots, N_{p-1}/\theta_{p-1})' - 1N_p/\theta_p$. The Fisher information is $I(\boldsymbol{\theta}) = n \, diag(1/\theta_1, \ldots, 1/\theta_{p-1}) - n11'/\theta_p$, and its inverse is $n^{-1}[diag(\theta_1, \ldots, \theta_{p-1}) - (\theta_1, \ldots, \theta_{p-1})(\theta_1, \ldots, \theta_{p-1})']$. Then, it follows that $W_E = \sum_{i=1}^p (N_i - n\theta_{i,0})^2/N_i$ and $W_U = \sum_{i=1}^p (N_i - n\theta_{i0})^2/n\theta_{i,0}$.

Chapter 7

1. The prior has density given by $p(\theta) = \alpha\beta^\alpha/\theta^{\alpha+1} I_\theta(\beta, \infty)$, and the likelihood is given by $l(\theta; \mathbf{X}) = I_\theta(T, \infty)/\theta^n$, for $T = \max_i X_i$. Then, $\theta|\mathbf{x} \sim Pa(\beta_1, \alpha_1)$, where $\alpha_1 = \alpha + n$ and $\beta_1 = \max\{t, \beta\}$.

(a) From Bayes theorem it follows that $p(\mathbf{x}) = p(\mathbf{x}|\theta)p(\theta)/p(\theta|\mathbf{x}) = \alpha\beta^\alpha/\alpha_1\beta_1^{\alpha_1}$. Assuming that Y and $\mathbf{X}$ are conditionally independent given θ, it follows that $p(y|\mathbf{x}) = \alpha_1\beta_1^{\alpha_1}/\alpha_2\beta_2^{\alpha_2}$, with $\alpha_2 = \alpha_1 + 1$ and $\beta_2 = \max\{y, \beta_1\}$.

(b) If $T > \beta$, then $\beta_1 = t$, $\beta_2 = \max\{t, y\}$ and $p(y|\mathbf{x}) = \alpha_1 t^{\alpha_1}/[\alpha_2(\max\{t, y\})^{\alpha_2}]$. So, $P(Y > t|\mathbf{x}) = (\alpha_1/\alpha_2)t^{\alpha_1} \int_t^\infty y^{-\alpha_2} dy = \alpha_1/[\alpha_2(\alpha_2 - 1)]$.

(c) If $\alpha \to 0$, then $\alpha_1 \to n$, $\alpha_2 \to n+1$, $p(y|\mathbf{x}) \to n\beta_1^n/[(n+1)\beta_2^{n+1}]$ and $P(Y > t|\mathbf{x}) \to n/[(n+1)n] = 1/(n+1)$. If $\alpha, \beta \to 0$, then $p(y|\mathbf{x}) \to nt^n/[(n+1)(\max\{t,y\})^{n+1}]$ or $p(y|\mathbf{x}) \to n/[(n+1)t]$, if $y < t$ and $nt^n/[(n+1)y^{n+1}]$, if $y > T$. Also, $P(Y > t|\mathbf{x}) \to 1/(n+1)$.

4. The easiest way to predicte Y in a classical way is to substitute θ by its estimator $\hat{\theta}$. In this example, $p(y|\hat{\theta}) = \binom{m}{y}(x/n)^y(1 - x/n)^{m-y}$, where $\hat{\theta} = x/n$. This distribution has mean mx/n and variance $mx(n - x)/n^2$. (This can be contrasted with the Bayesian prediction under the improper prior $p(\theta) \propto \theta^{-1}(1 - \theta)^{-1}$ that leads to $E(Y|x) = mx/n$ and $V(Y|x) = mx(n - x)(1 + c)/n^2$ where $c = (m - 1)/(n + 1) \geq 0$. This again is similar but overdispersed with respect to the classical result.)

Chapter 8

3. Define $F(\boldsymbol{\beta}) = [S(\boldsymbol{\beta}) - S_e]/ps^2 \sim F(p, n - p)$. Therefore, $Pr[F(\boldsymbol{\beta}) \leq \bar{F}_\alpha(p, n-p)] = 1-\alpha$. Solving for $\boldsymbol{\beta}$, it means that $\{\boldsymbol{\beta} : (\boldsymbol{\beta}-\hat{\boldsymbol{\beta}})'\mathbf{X}'\mathbf{X}(\boldsymbol{\beta}-\hat{\boldsymbol{\beta}}) \leq ps^2\bar{F}_\alpha(p, n-p)\}$ defines a $100(1-\alpha)$ % confidence region for $\boldsymbol{\beta}$. This region has the form of a p-dimensional ellipsoid centered around $\hat{\boldsymbol{\beta}}$.

9. Developing the product and conveniently collecting the terms, $Q = (\boldsymbol{\beta} - \boldsymbol{\mu}_0)'\mathbf{C}_0(\boldsymbol{\beta} - \boldsymbol{\mu}_0) + (\boldsymbol{\beta} - \hat{\boldsymbol{\beta}})'\mathbf{X}'\mathbf{X}(\boldsymbol{\beta} - \hat{\boldsymbol{\beta}}) = \boldsymbol{\beta}'[\mathbf{C}_0 + \mathbf{X}'\mathbf{X}]\boldsymbol{\beta} - 2\boldsymbol{\beta}'[\mathbf{C}_0\boldsymbol{\mu}_0 + \mathbf{X}'\mathbf{X}\hat{\boldsymbol{\beta}}]+\boldsymbol{\mu}_0'\mathbf{C}_0\boldsymbol{\mu}_0+\hat{\boldsymbol{\beta}}'\mathbf{X}'\mathbf{X}\hat{\boldsymbol{\beta}}$. Using $\boldsymbol{\mu}_1$ and $\mathbf{C}_1$ as defined and completing the squares, it follows that $Q = Q_1 + Q_2$ where $Q_1 = \boldsymbol{\beta}'\mathbf{C}_1^{-1}\boldsymbol{\beta} - 2\boldsymbol{\beta}'\mathbf{C}_1^{-1}\boldsymbol{\mu}_1 + \boldsymbol{\mu}_1'\mathbf{C}_1^{-1}\boldsymbol{\mu}_1$ and $Q_2 = \boldsymbol{\mu}_0'\mathbf{C}_0\boldsymbol{\mu}_0 + \hat{\boldsymbol{\beta}}'\mathbf{X}'\mathbf{X}\hat{\boldsymbol{\beta}} - \boldsymbol{\mu}_1'\mathbf{C}_1^{-1}\boldsymbol{\mu}_1$. Now, $Q_1 = (\boldsymbol{\beta} - \boldsymbol{\mu}_1)'\mathbf{C}_1(\boldsymbol{\beta} - \boldsymbol{\mu}_1)$, which proves the result.

19. The group means $\bar{y}_j \sim N(\beta_j, (m\phi)^{-1})$ are sufficient and independent statistics for $\beta_1, \ldots, \beta_k$. Therefore, $E(\beta_j|\phi, \mathbf{y}) = E(\beta_j|\phi, \bar{y}_1, \ldots, \bar{y}_k) = E[E(\beta_j|\mu, \phi, \bar{y}_j)|\bar{y}_1, \ldots, \bar{y}_k]$. From Theorem 2.1, $E(\beta_j|\mu, \phi, \bar{y}_j) = (m\bar{y}_j + c_1\mu)/(m + c_1)$. So, $E(\beta_j|\phi, \mathbf{y}) = [m\bar{y}_j + c_1E(\mu|\phi, \bar{y}_1, \ldots, \bar{y}_k)]/(m + c_1)$. From Theorem 2.1 again, $E(\mu|\phi, \bar{y}_1, \ldots, \bar{y}_k) = \bar{\bar{y}}$. Therefore, $E(\beta_j|\phi, \mathbf{y}) = (m\bar{y}_j + c_1\bar{\bar{y}})/(m + c_1)$.

22. The required distribution has density $p(\theta_{t-1}|D_t) = \int p(\theta_{t-1}|\theta_t, D_t) p(\theta_t|D_t) \, d\theta_t$. By Bayes theorem and using the fact that y_t and θ_{t-1} are conditionally independent given θ_t, it follows that the first term in the integrand is $p(\theta_{t-1}|\theta_t, D_t) = p(\theta_{t-1}|\theta_t, D_{t-1}) \propto p(\theta_{t-1}|D_{t-1})p(\theta_t|\theta_{t-1}, D_{t-1})$. Application of Bayes theorem for the Normal *observation* θ_t and parameter θ_{t-1} gives $(\theta_{t-1}|\theta_t, D_t) \sim N[m_{t-1} + C_{t-1}(\theta_t - a_t)/R_t, C_{t-1} - C_{t-1}^2/R_t]$. Therefore, $(\theta_{t-1}|D_t)$ has mean $E(\theta_{t-1}|D_t) = E[E(\theta_{t-1}|\theta_t)|D_t] = m_{t-1} + C_{t-1}(m_t-a_t)/R_t$, and $V(\theta_{t-1}|D_t) = E[V(\theta_{t-1}|\theta_t)|D_t]+V[E(\theta_{t-1}|\theta_t)|D_t] = $

$C_{t-1} - C_{t-1}^2 (R_t - C_{t-1})/R_t^2$. The normality follows from the linear relation of θ_t on the conditional mean of θ_{t-1} and the marginal normality of θ_t.

List of distributions

This list includes the distributions that are most often used in Statistics and have appeared frequently in this book. They are listed in alphabetical order and are not divided into groups such as continuous versus discrete, univariate versus multivariate. This information will be clear from the context.

1. Bernoulli

   X is said to have a Bernoulli distribution with success probability θ, denoted by $X \sim Ber(\theta)$, if its probability function is given by

   $$p(x|\theta) = \theta^x (1 - \theta)^{1-x} , \; x = 0, 1,$$

   for $\theta \in [0, 1]$. This distribution has mean θ and variance $\theta(1 - \theta)$.

2. Beta

   X is said to have a Beta distribution with parameters α and β, denoted by $X \sim Beta(\alpha, \beta)$, if its density function is given by

   $$p(x|\alpha, \beta) = kx^{\alpha-1}(1 - x)^{\beta-1} , \; x \in [0, 1],$$

   for $\alpha, \beta > 0$. The constant k is given by

   $$k^{-1} = B(a, b) = \frac{\Gamma(a)\Gamma(b)}{\Gamma(a + b)} \text{ and } \Gamma(c) = \int_0^\infty x^{c-1}e^{-x}dx , \; c > 0.$$

   The functions $B(\cdot, \cdot)$ and $\Gamma(\cdot)$ are respectively known as the Beta and Gamma functions. This distribution has mean $\alpha/(\alpha + \beta)$ and variance $\alpha\beta/[(\alpha + \beta)^2(\alpha + \beta + 1)]$.

3. Beta-Binomial

   X is said to have a Beta-Binomial distribution with parameters n, α and β, denoted by $X \sim BB(n, \alpha, \beta)$, if its probability function is given by

   $$p(x|n, \alpha, \beta) = \binom{n}{x} \frac{B(\alpha + x, \beta + n - x)}{B(\alpha, \beta)} , \; x = 0, 1, \ldots, n,$$

   for $n \geq 1$, $\alpha, \beta > 0$. This distribution has mean $n\alpha/(\alpha + \beta)$ and variance $n\alpha\beta[1 + (n - 1)/(\alpha + \beta - 1)]/(\alpha + \beta)^2$. The expression for the variance can be rewritten in the form $n\bar{\theta}(1 - \bar{\theta})(1 + \epsilon)$, where $\bar{\theta} = \alpha/(\alpha + \beta)$ and $\epsilon = (n - 1)/(\alpha + \beta - 1)$. Since $\epsilon > 0$, $\forall n > 1$, the variance of the Beta-Binomial distribution is bigger than the variance of the Binomial, when one compares θ with $\bar{\theta}$, for $\alpha + \beta > 1$.

4. Binomial

X is said to have a Binomial distribution with parameter n and success probability θ, denoted by $X \sim Bin(n, \theta)$, if its probability function is given by

$$p(x|n, \theta) = \binom{n}{x} \theta^x (1 - \theta)^{n-x}, \ x = 0, 1, \ldots, n,$$

for $n \geq 1$, $\theta \in [0, 1]$. This distribution has mean $n\theta$ and variance $n\theta(1 - \theta)$. This family of distributions includes the Bernoulli as the special case $n = 1$.

5. Dirichlet

$\mathbf{X} = (X_1, \ldots, X_p)'$ is said to have Dirichlet distribution with parameter $\boldsymbol{\theta} = (\theta_1, \ldots, \theta_p)'$, denoted by $D(\boldsymbol{\theta})$, if its joint density function is given by

$$p(\mathbf{x}) = \frac{\Gamma(\theta_+)}{\prod_i \Gamma(\theta_i)} \prod_{i=1}^{p} x_i^{\theta_i - 1}, \ x_i \in [0, 1], \ i = 1, \ldots, p, \ \sum_{i=1}^{p} x_i = 1,$$

for $\theta_i \in [0, 1]$, $i = 1, \ldots, p$ and $\theta_+ = \sum_i \theta_i$. This distribution has mean $\boldsymbol{\theta}/\theta_+$ and variance-covariance matrix given by $\theta_+^{-2}(\theta_+ + 1)^{-1}[\theta_+\text{diag}(\boldsymbol{\theta}) - \boldsymbol{\theta}\boldsymbol{\theta}']$, where $\text{diag}(\boldsymbol{c})$ denotes a diagonal matrix having the elements of the vector $\boldsymbol{c}$ in the main diagonal. This family of distributions includes the Beta as the special case $p = 2$.

6. Exponential

X is said to have an Exponential distribution with parameter θ, denoted by $X \sim Exp(\theta)$, if its density function is given by

$$p(x|\theta) = \theta e^{-\theta x}, \ x > 0,$$

for $\theta > 0$. This distribution has mean $1/\theta$ and variance $1/\theta^2$.

7. Gamma

X is said to have a Gamma distribution with parameters α and β, denoted by $X \sim G(\alpha, \beta)$, if its density function is given by

$$p(x|\alpha, \beta) = kx^{\alpha-1} e^{-\beta x}, \ x > 0,$$

for $\alpha, \beta > 0$. The constant k is given by $k = b^\alpha/\Gamma(a)$. This distribution has mean α/β and variance α/β^2. This family of distributions includes the Exponential as the special case $\alpha = 1$.

The χ_p^2 distribution is equivalent to the $G(p/2, 1/2)$ distribution. Therefore, the $G(\alpha, \beta)$ distribution corresponds to the $2\beta\chi_{2\alpha}^2$ distribution.

8. Laplace or double exponential

X is said to have a Laplace (or Double exponential) distribution with parameter θ, denoted by $X \sim Lap(\theta)$, if its density function is given by

$$p(x|\theta) = \frac{\theta}{2} e^{-\theta|x|}, x \in R,$$

for $\theta > 0$. This distribution has mean 0 and variance $2/\theta^2$.

9. Multinomial

$\mathbf{X} = (X_1, \ldots, X_p)'$ is said to have Multinomial distribution with parameter n and probabilities $\boldsymbol{\theta} = (\theta_1, \ldots, \theta_p)'$, denoted by $M(n, \boldsymbol{\theta})$, if its joint probability function is given by

$$p(\mathbf{x} \mid \boldsymbol{\theta}) = \frac{n!}{\prod_{i=1}^{p} x_i!} \prod_{i=1}^{p} \theta_i^{x_i}, x_i = 0, 1, \ldots, n, \ i = 1, \ldots, p, \sum_{i=1}^{p} x_i = n,$$

for $\theta_i \in [0, 1]$, $i = 1, \ldots, p$, $\sum_{i=1}^{p} \theta_i = 1$. This distribution has mean $n\boldsymbol{\theta}$ and variance-covariance matrix given by $n[\text{diag}(\boldsymbol{\theta}) - \boldsymbol{\theta}\boldsymbol{\theta}']$, where $\text{diag}(\mathbf{c})$ denotes a diagonal matrix having the elemnts of the vector $\mathbf{c}$ in the main diagonal. This family of distributions includes the Binomial as the special case $p = 2$.

10. Negative Binomial

X is said to have a Negative Binomial distribution with parameters r, θ, denoted by $X \sim NB(r, \theta)$, if its probability function is given by

$$p(x|r, \theta) = \binom{r + x - 1}{x} \theta^r (1 - \theta)^x, \ x = 0, 1, \ldots,$$

for $r \geq 1$, $\theta \in [0, 1]$. This distribution has mean $r(1 - \theta)/\theta$ and variance $r(1 - \theta)/\theta$.

11. Normal

X is said to have a Normal distribution with mean μ and variance σ^2, denoted by $X \sim N(\mu, \sigma^2)$, if its density function is given by

$$p(x|\mu, \sigma^2) = (2\pi\sigma^2)^{-1/2} \exp\left\{ -\frac{1}{2\sigma^2}(x - \mu)^2 \right\}, x \in R,$$

for $\mu \in R$ and $\sigma^2 > 0$. When $\mu = 0$ and σ^2, the distribution is referred to as standard Normal.

$\mathbf{X} = (X_1, \ldots, X_p)'$ is said to have a Multivariate Normal distribution with mean vector $\boldsymbol{\mu}$ and variance-covariance matrix $\boldsymbol{\Sigma}$, denoted by $N(\boldsymbol{\mu}, \boldsymbol{\Sigma})$, if its density function is given by

$$(2\pi)^{-p/2} |\boldsymbol{\Sigma}|^{-1/2} \exp\left\{ -\frac{1}{2}(\mathbf{x} - \boldsymbol{\mu})'\boldsymbol{\Sigma}^{-1}(\mathbf{x} - \boldsymbol{\mu}) \right\}, \mathbf{x} \in R^p,$$

for $\boldsymbol{\mu} \in R^p$ and $\boldsymbol{\Sigma} > 0$, where $|\mathbf{A}|$ denotes the determinant of $\mathbf{A}$.

12. Pareto

X is said to have a Pareto distribution with parameters a and θ, denoted by $Pa(a,\theta)$, if its density function is given by

$$p(x|\theta, a) = a\theta^a / x^{1+a} , \ x > \theta,$$

for $a, \theta > 0$. This distribution has mean $a\theta/(a-1)$, when $a > 1$, and variance $a\theta^2/[(a-1)^2(a-2)]$, when $a > 2$.

13. Poisson

X is said to have a Poisson distribution with parameter θ, denoted by $X \sim Pois(\theta)$, if its probability function is given by

$$p(x|\theta) = e^{-\theta}\frac{\theta^x}{x!} , \ x = 0, 1, 2, \ldots,$$

for $\theta > 0$. This distribution has mean and variance given by λ.

14. Snedecor F

X is said to have Snedecor F (or simply F) distribution with ν_1 and ν_2 degrees of freedom, denoted $F(\nu_1, \nu_2)$, if its density function is given by

$$p(x|\nu_1, \nu_2) = \frac{\Gamma[(\nu_1 + \nu_2)/2]}{\Gamma(\nu_1/2)\Gamma(\nu_2/2)}\nu_1^{\nu_1/2}\nu_2^{\nu_2/2}x^{(\nu_1/2)-1}\left(\nu_2 + \nu_1 x\right)^{-(\nu_1+\nu_2)/2},$$

for $x > 0$ and $\nu_1, \nu_2 > 0$. This distribution has mean $\nu_2/(\nu_2 - 2)$, when $\nu_2 > 2$ and variance $[2\nu_2^2(\nu_1 + \nu_2 - 2)]/[\nu_1(\nu_2 - 4)(\nu_2 - 2)^2]$, when $\nu_2 > 4$. If $X_1 \sim \chi_{\nu_1}^2$ and $X_2 \sim \chi_{\nu_2}^2$ are independent then $(X_1/\nu_1)/(X_2/\nu_2) \sim F(\nu_1, \nu_2)$. This family of distributions includes the square of the Student-t as the special case $\nu_1 = 1$. If $\nu_2 \to \infty$, then $\nu_1 F(\nu_1, \nu_2) \xrightarrow{D} \chi_{\nu_1}^2$.

15. Student-t

X is said to have a Student-t (or simply t) distribution with mean μ and scale parameter σ^2 and ν degrees of freedom, denoted by $X \sim t_\nu(\mu, \sigma^2)$, if its density function is given by

$$p(x|\nu, \mu, \sigma^2) = \frac{\Gamma[(\nu + 1)/2]}{\Gamma(\nu/2)\sqrt{\pi}}\frac{\nu^{\nu/2}}{\sigma}\left[\nu + \frac{(x - \mu)^2}{\sigma^2}\right]^{-(\nu+1)/2} , \ x \in R,$$

for $\nu > 0$, $\mu \in R$ and $\sigma^2 > 0$. This distribution has mean μ, when $\nu > 1$, and variance $\nu/(\nu - 2)$, when $\nu > 2$. This family includes the Cauchy distribution as the special case $\nu = 1$.

$\mathbf{X} = (X_1, \ldots, X_p)'$ is said to have a multivariate Student-t distribution with mean vector $\boldsymbol{\mu}$ and scale matrix $\boldsymbol{\Sigma}$ and ν degrees of freedom, denoted by $t_\nu(\boldsymbol{\mu}, \boldsymbol{\Sigma})$, if its density function is given by

$$f(\mathbf{x}|\nu, \boldsymbol{\mu}, \boldsymbol{\Sigma}) = \frac{\Gamma[(\nu+p)/2]}{\Gamma(\nu/2)\pi^{p/2}}\nu^{\nu/2}|\boldsymbol{\Sigma}|^{-1/2}[\nu + (\mathbf{x} - \boldsymbol{\mu})'\boldsymbol{\Sigma}^{-1}(\mathbf{x} - \boldsymbol{\mu})]^{-(\nu+p)/2},$$

for $\mathbf{x} \in R^p$, $\nu > 0$, $\boldsymbol{\mu} \in R^p$ and $\boldsymbol{\Sigma} > 0$. This distribution has mean $\boldsymbol{\mu}$, when $\nu > 1$, and variance $\nu\boldsymbol{\Sigma}/(\nu - 2)$, when $\nu > 2$.

16. Uniform

X is said to have a Uniform distribution with parameters θ_1 and θ_2, denoted by $U[\theta_1, \theta_2]$, if its density function is given by

$$p(x|\theta_1, \theta_2) = \frac{1}{\theta_2 - \theta_1}, \; x \in [\theta_1, \theta_2],$$

for $\theta_1 < \theta_2$. When $\theta_1 = 0$ and $\theta_2 = 1$, the distribution is referred to as unit uniform. This distribution has mean $(\theta_1 + \theta_2)/2$ and variance $(\theta_2 - \theta_1)^2/12$.

References

Abramowitz, M. and Stegun, I. A., editors (1965). *Handbook of Mathematical Functions.* National Bureau of Standards Applied Mathematics Series, Number 55. Washington, DC: U.S. Government Printing Office.

Aitchison, J. and Dunsmore, I. R. (1975). *Statistical Prediction Analysis.* Cambridge: Cambridge University Press.

Aitkin, M. (1991). Posterior Bayes factors. *Journal of the Royal Statistical Society, Series B*, 53, 111–142.

Barndorff-Nielsen, O. E. (1983). On a formula for the conditional distribution of the maximum likelihood estimator. *Biometrika*, 70, 343–365.

Barnett, M. S. (1973). *Comparative Statistical Inference.* New York: Wiley.

Bartlett, M. S. (1947). Multivariate analysis. *Journal of the Royal Statistical Society, Series B*, 9, 76–97.

Bayes, T. (1763). An essay towards solving in the doctrine of chances. *Philosophy Transaction Royal Society London*, 53, 370–418.

Berger, J. (1985). *Statistical Decision Theory and Bayesian Analysis.* New York: Springer-Verlag.

Berger, J. and Delampady, M. (1987). Testing precise hypothesis. *Statistical Science*, 2, 317–352.

Berger, J. O., Liseo, B. and Wolpert, R. L. (1999). Integrated likelihood methods for eliminating nuisance parameters. *Statistical Science*, 14, 1–28.

Berger, J. O. and Pericchi, L. R. (1996). The intrinsic Bayes factor for model selection and prediction. *Journal of the American Statistical Association*, 91, 109–122.

Berger, J. and Sellke, T. (1987). Testing a point null hypothesis: The irreconcilability of significance levels and evidence. *Journal of the American Statistical Association*, 82, 112–122.

Bernardo, J. M. (1979). Reference posterior distributions for Bayesian inference (with discussion). *Journal of the Royal Statistical Society, Series B*, 41, 113–147.

Bernardo, J. M. and Smith, A. F. M. (1994). *Bayesian Theory*. Chichester: Wiley.

Bickel, P. J. and Doksum, K. A. (1977). *Mathematical Statistics*. Holden-Day.

Box, G. E. P. and Tiao, G. C. (1992). *Bayesian Inference in Statistical Analysis*. New York: Addison-Wesley.

Brazzale, A. R., Davison, A. C and Reid, N. (2007). *Applied Asymptotics: Case Studies in Small-Sample Statistics*. Cambridge: Cambridge University Press.

Broemeling, L. D. (1985). *Bayesian Analysis of Linear Models*. New York: Marcel Dekker.

Buse A. (1982). The likelihood ratio, Wald and Lagrange multiplier tests: An expository note. *The American Statistician*, 36, 153–157.

Cancho, V. G., Louzada-Neto, F. and Barriga, G. D. C. (2011). The Poisson-Exponential lifetime distribution. *Computational Statistics & Data Analysis*, 55, 677–686.

Casella, G. and George, E. I. (1992). Explaining the Gibbs sampler. *The American Statistician*, 46, 167–74.

Cordeiro, G. M (1987). On the correction to the likelihood ratio statistics. *Biometrika*, 74, 265–274.

Cordeiro, G. M. and Ferrari, S. L. P. (1991). A modified score test having chi-squared distribution to order n^{-1}. *Biometrika*, 78, 573–582.

Cox, D. R. and Hinkley, D. V. (1974). *Theoretical Statistics*. London: Chapman & Hall.

Cox, D. R. and Reid, N. (1987). Parameter orthogonality and approximate conditional inference (with discussion). *Journal of the Royal Statistical Society, Series B*, 49, 1-39.

Davison, A. C. and Hinkley, D. V. (1997). *Bootstrap Methods and Their Application*. Cambridge: Cambridge University Press.

Dawid, A. P., Stone, M. and Zidek, J. V. (1973). Marginalization paradoxes in Bayesian and structural inference (with discussion). *Journal of the Royal Statistical Society, Series B*, 35, 189–233.

de Finetti, B. (1937). La prévision: ses lois logiques, ses sources subjectives. *Annals of the Institute Poincaré*, 7, 1, 1–68.

de Finetti, B. (1974). *Theory of Probability* (vols 1 and 2). New York: Wiley.

DeGroot, M. H. (1970). *Optimal Statistical Decisions*. New York: Wiley.

Dempster, A. P., Laird, N. M. and Rubin, D. B. (1977). Maximum likehood estimation from incomplete data via the EM algorithm (with discussion). *Journal of the Royal Statistical Society, Series B*, 39, 1–38.

Draper, N. R. and Smith, H. (1966). *Applied Regression Analysis*. New York: Wiley.

Durbin, J. and Koopman, S. J. (2001). *Time Series Analysis by State Space Methods*. London: Oxford University Press.

Efron, B. (1979). Bootstrap methods: Another look at the jackknife. *Annals of Statistics*, 7, 1–26.

Efron, B. (1982). *The Jackknife, the Bootstrap and Other Resampling Plans*. Monograph 38. Philadelphia: SIAM.

Efron, B. and Morris, C. N. (1975). Data analysis using Stein's estimator and its generalizations. *Journal of the American Statistical Association*, 70, 311–319.

Fan, J. and Li, R. (2001). Variable selection via nonconcave penalized likelihood and its oracle properties. *Journal of the American Statistical Association*, 96, 1348–1360.

Ferguson, T. S. (1967). *Mathematical Statistics: A Decision-Theoretic Approach*. New York: Academic Press.

Ferrari, S. L. P., Botter, D. A., Cordeiro, G. M. and CribariNeto, F. (1996). Second and third order bias reduction for one-parameter family models. *Statistics and Probability Letters*, 30, 339–345.

Firth, D. (1993). Bias reduction of maximum likelihood estimates. *Biometrika*, 80, 27–38.

Fisher, R. A. (1936). The use of multiple measurements in taxonomic problems. *Annals of Eugenics*, 7, 179–188.

Gamerman, D. and Lopes, H. F. (2006). *Markov Chain Monte Carlo: Stochastic Simulation for Bayesian Inference* (2nd edition). London: Chapman Hall.

Gamerman, D. and Migon, H. S. (1993). Dynamic hierarchical models. *Journal of the Royal Statistical Society, Series B*, 55, 629–642.

Garthwaite, P. H., Jollife, I. T. and Jones, B. (2002). *Statistical Inference*, (2nd edition). Oxford: Oxford University Press.

Garthwaite, P. H., Kadane, J. B. and O'Hagan, A. (2005). Statistical methods for eliciting probability distributions. *Journal of the American Statistical Association*, 100, 680–701.

Geisser, S. (1993). *Predictive Inference: An Introduction*. London: Chapman & Hall.

Gelfand, A. E. and Smith, A. M. F. (1990). Sampling-based approaches to calculating marginal densities. *Journal of the American Statistical Association*, 85, 398–409.

Geman, S. and Geman, D. (1984). Stochastic relaxation, Gibbs distribution and the Bayesian restoration of images. *IEEE Transaction on Pattern Analysis and Machine Intelligence*, 6, 721–741.

Gilks, W. R., Richardson, S. and Spiegelhalter, D. J. (editors). (1995). *Markov Chain Monte Carlo in Practice*. London: Chapman & Hall.

Greene, W. H. (2008). *Econometric Analysis* (7th edition). Prentice Hall.

Harvey, A. C. (1989). *Forecasting, Structural Time Series Models and the Kalman Filter*. Cambridge: Cambridge University Press.

Heath, D. and Sudderth, W. (1976). De Finetti's Theorem on Exchangeable Variables. *The American Statistician*, 30, 188–189.

Hoerl, A. E. and Kennard, R. W. (1970). Ridge regression: Biased estimation for nonorthogonal problems. *Technometrics*, 12, 55–67.

James, G., Witten, D., Hastie, T. and Tibshirani, R. (2013). *An Introduction to Statistical Learning, with Applications in R*. New York: Springer.

James, W. and Stein, C. (1961). Estimation with quadratic loss. *In Proceedings of the Fourth Berkeley Symposium and Mathematical, Statistics and Probability, Volume 1*, Berkeley: University of California Press.

Jeffreys, H. (1961). *Theory of Probability*. Oxford: Clarendon Press.

Kalbfleisch, J. G. (1975). *Probability and Statistical Inference: Volume 2* (2nd edition). New York: Springer-Verlag.

Kyung, M., Gill, J., Ghosh, M. and Casella, G. (2010). Penalized regression, standard errors, and Bayesian lasso. *Bayesian Analysis*, 5, 369–412.

Laird, N. M. and Ware, J. H. (1982). Random-effects models for longitudinal data. *Biometrics*, 38, 963–974.

Lawley, D. N. (1956). A general method for approximating to the distribution of likelihood ratio criteria. *Biometrika*, 43, 295–303.

Lehmann, E. (1986). *Testing Statistical Hypothesis* (2nd edition). New York: Wiley.

Lehmann, E. E. L. and Romano, J. P. (2005). *Testing Statistical Hypotheses.* New York: Springer Science and Business Media.

Lindley, D. V. (1957). A statistical paradox. *Biometrika*, 44, 187–192.

Lindley, D. V. (1965). *Introduction to Probability and Statistics (Parts 1 and 2).* Cambridge: Cambridge University Press.

Lindley, D. V. (1980). Approximate Bayesian methods (with discussion). In *Bayesian Statistics* (Eds. J. M. Bernardo et al.), 223–245, Oxford: Oxford University Press.

Lindley, D. V. and Phillips, L. D. (1976). Inference for a Bernoulli process (a Bayesian view). *The American Statistician*, 30, 112–119.

Lindley, D. V. and Smith, A. F. M. (1972). Bayes estimates for the linear model (with discussion). *Journal of the Royal Statistical Society, Series B*, 34, 1–41.

Little, R. J. A. and Rubin, D. B. (1988). *Statistical Analysis with Missing Data.* New York: Wiley.

McCullagh, P. and Nelder, J. A. (1989). *Generalized Linear Models* (2nd edition). London: Chapman & Hall.

Metropolis, N., Rosenbluth, A. W., Rosenbluth, M. N., Teller, A. H. and Teller, E. (1953). Equation of state calculations by fast computing machine. *Journal of Chemical Physics*, 21, 1087–91.

Migon, H. S. and Tachibana, V. M. (1997). Bayesian approximations in randomized response model. *Computational Statistics and Data Analysis*, 24, 401–409.

Naylor, J. C. and Smith, A. M. F. (1982). Applications of a method for efficient computation of posterior distributions. *Applied Statistics*, 31, 214–235.

Neyman, J. and Pearson, E. S. (1928). On the use and interpretation of certain test criterion for purposes of statistical inference. *Biometrika*, 20, 175–240 and 263–294.

O'Hagan, A. (1995). Fractional Bayes factors for model comparison. *Journal of the Royal Statistical Society, Series B*, 143, 99–38.

Park, T. and Casella, G. (2008). The Bayesian lasso. *Journal of the American Statistical Association*, 103, 681–686.

Quenouille, M. H. (1949). Approximate test of correlation in time series. *Journal of the Royal Statistical Society, Series B*, 34, 1–41.

Quenouille, M. H. (1956). Notes on the bias estimation. *Biometrika*, 43, 353–360.

Raiffa, H. and Schlaifer, R. (1961). *Applied Statistical Decision Theory.* Boston: Harvard Press.

Rao, C. R. (1973). *Linear Statistical Inference* (2nd edition). New York: Wiley.

Ripley, B. D. (1987). *Stochastic Simulation.* New York: Wiley.

Severini, T. A. (1998). An approximation to the modified profile likelihood function. *Biometrika*, 85, 403–11.

Silvey, S. D. (1970). *Statistical Inference.* London: Chapman & Hall.

Tanner, M. A. (1996). *Tools for Statistical Inference - Methods for Exploration of Posterior Distributions and Likelihood Functions* (3rd edition). New York: Springer Verlag.

Terrell, G. R. (2002). The gradient statistic. *Computing Science and Statistics*, 34, 206–215.

Thisted, R. A. (1976). *Elements of Statistical Computing.* New York: Chapman & Hall.

Tibshirani, R. (1996). Regression shrinkage and selection via the Lasso. *Journal of the Royal Statistical Society, Series B*, 58, 267–288.

Tierney, L. and Kadane, J. B. (1986). Accurate approximations for posterior moments and marginal densities. *Journal of the American Statistical Association*, 81, 82–86.

Valliant, R., Dorfman, A.H. and Royall, R.M. (2000). *Finite Population Sampling and Inference: A Prediction Approach.* New York: Wiley.

Wald, A. (1943). Tests of statistical hypotheses concerning several parameters when the number of observations is large. *Transactions of the American Mathematical Society*, 54, 426–482.

West, M. and Harrison, P. J. (1997). *Bayesian Forecasting and Dynamic Models* (2nd edition). New York: Springer Verlag.

Wilks, S. S. (1938). The large-sample distribution of the likelihood ratio for testing composite hypotheses. *The Annals of Mathematical Statistics*, 9, 60–62.

Zellner, A. (1971). *An Introduction to Bayesian Inference in Econometrics.* New York: Wiley.

Zou, H. and Hastie, T. (2005). Regularization and variable selection via the elastic net. *Journal of the Royal Statistical Society, Series B*, 67, 301–320.

Index